ACS SYMPOSIUM SERIES **595**

# Biorational Pest Control Agents

## Formulation and Delivery


**Franklin R. Hall,** EDITOR
*Ohio State University*

**John W. Barry,** EDITOR
*Forest Service, U.S. Department of Agriculture*


Developed from a symposium sponsored
by the Division of Agrochemicals
at the 207th National Meeting
of the American Chemical Society,
San Diego, California,
March 13–17, 1994

American Chemical Society, Washington, DC 1995

**Library of Congress Cataloging-in-Publication Data**

Biorational pest control agents: formulation and delivery / Franklin R. Hall, John W. Barry, editor[s].

p. cm.—(ACS symposium series; 595)

"Developed from a symposium sponsored by the Division of Agrochemicals at the 207th National Meeting of the American Chemical Society, San Diego, California, March 13–17, 1994."

Includes bibliographical references and indexes.

ISBN 0–8412–3226–1

1. Biological agents for pest control—Congresses. 2. Pests—Biological control—Congresses. I. Hall, Franklin R. II. Barry, John Williard, 1934- . III. American Chemical Society. Division of Agrochemicals. IV. American Chemical Society. Meeting (207th: 1994: San Diego, Calif.) V. Series.

SB975.B59 1995
668′.65—dc20 95–13986
CIP

This book is printed on acid-free, recycled paper.

PRINTED IN THE UNITED STATES OF AMERICA

# Foreword

THE ACS SYMPOSIUM SERIES was first published in 1974 to provide a mechanism for publishing symposia quickly in book form. The purpose of this series is to publish comprehensive books developed from symposia, which are usually "snapshots in time" of the current research being done on a topic, plus some review material on the topic. For this reason, it is necessary that the papers be published as quickly as possible.

Before a symposium-based book is put under contract, the proposed table of contents is reviewed for appropriateness to the topic and for comprehensiveness of the collection. Some papers are excluded at this point, and others are added to round out the scope of the volume. In addition, a draft of each paper is peer-reviewed prior to final acceptance or rejection. This anonymous review process is supervised by the organizer(s) of the symposium, who become the editor(s) of the book. The authors then revise their papers according to the recommendations of both the reviewers and the editors, prepare camera-ready copy, and submit the final papers to the editors, who check that all necessary revisions have been made.

As a rule, only original research papers and original review papers are included in the volumes. Verbatim reproductions of previously published papers are not accepted.

*M. Joan Comstock*
Series Editor

# Contents

# Preface

AS THE PUBLIC'S AND RESOURCE MANAGERS' DEMAND for safer pesticides increases and legislative constraints continue to reduce the amounts of pesticide used in North American agriculture and forestry, there is increased opportunity for developing and improving environmentally friendly control agents and strategies. Improvements in field efficacy are a key feature of those technologies, but new advancements in formulation and delivery are *critical* to achieve successful commercialization of such agents. This book was designed to identify those critical needs of formulation and novel delivery systems for the baculoviruses, bacteria, fungi, pheromones, hormones, and nematodes being advocated for agricultural and forest pest-management strategies of the 1990s. Biorational pest control materials engender a very wide group of agents with an equally wide range of specific formulation and delivery needs. We sought to include this information in the symposium upon which this book is based, and it is presented herein.

The book is divided into six sections in addition to an overview of biorationals. In the first section, registration needs and data requirements are discussed. Data requirements for experimental use and full commercial registration of these biorational agents differ somewhat from those required for conventional chemical pesticide products. Formulation choice has a large impact on the data necessary to support a particular biological pesticide product for development as it relates to U.S. Environmental Protection Agency (EPA) regulatory requirements. EPA assessments for the biorational agents include an analysis of effects on nontarget species. As for chemical pesticides, these analyses are based on single-species testing, but allowances are made for the unique characteristics of biological pesticides. Canadian counterparts envision a buffer-zone guide for use patterns of biologicals that highlights the need for product specificity information when developing drift mitigation options in the registration process.

As discussed in Chapters 4 and 5, fundamental information needs mean that the use of "biorational" agents, such as entomopathogens for pest control, must be firmly based in ecological principles. Aspects of environmental release, such as timing and placement, must be based on ecological considerations for entomopathogens to be efficacious. Timing and placement depend on entomopathogen species characteristics, its population parameters, and the interacting ecosystem parameters and delivery requirements. Modeling these very dynamic sequences has significant appeal in that the process can aid the identification of fundamental gaps and requires scientists to revisit our knowledge about specific interactions and conventional paradigms associated with toxin placement.

The third section discusses delivery of biorationals, which is principally accomplished with existing technology, although specific needs, tactics, and inadequacies are noteworthy. Forest Service efforts that have used an aerial application model have merit in demonstrating potential solutions to the questions of environmental fate issues, and hence, risks and hazards of biorationals. An overview of factors affecting *Bacillus thuringiensis* (Bt) foliar deposition, rainfastness coverage, and persistence is followed by two specific efforts in active-ingredient rainwashing (as affected by formulation) and some recent research on baculoviruses and UV radiation losses modified by a viral enhancement factor and fluorescent brighteners.

Soil biorationals are discussed in Chapters 11, 12, and 13. Soil biorationals include several examples of industry efforts on *Metarhizium* to explore the reason for field inconsistencies. The myriad of interacting factors clearly demonstrates the need for very close determination of the specific formulation and delivery criteria needed to maintain a viable biological with a predictable field performance. U.S. Department of Agriculture (USDA) research scientists have delineated the lack of sufficient effort on formulations and delivery of microbials for soil pathogens, although cooperative research with industry has significantly advanced the knowledge in the newly emerging technology. Nematode research has led to a series of successful products using specific formulation–delivery conditions for each nematode species and is leading to requirements for less product in large-area applications.

As discussed in the fifth section, foliar uses of biorationals remain predominant, encompassing semiochemicals, which although not a large commercialization area, have demonstrated successful introductions using sex pheromones. Sprayable formulations offer quick application advantages but have shorter durations. Efficacy is the primary goal in the design of all semiochemical products, but optimizing formulations to minimize cost is the key to increasing their use in integrated pest-management programs, that is, making their use economically compelling rather than just economically competitive with conventional insecticides. Specific examples are given about formulation and delivery parameters of

*Metarhizium* and *Beauveria*, which require both a systematic examination of how fungi can be manipulated and an integration into the knowledge base of insect behavior and location. Viral biopesticides show promise, especially genetically engineered isolates of several viruses. Integration of formulations and delivery parameters to demonstrate reliable field performance for co-occlusion and preoccluded virus releases are critical for successful commercialization. Microbials can also be formulated as granules (with starch) to enable greater residual activity. These encapsulation technologies show promise for conditions of extreme moisture. The use of plant pathogenic microbes to control unwanted plants provides an environmentally friendly approach to weed management. Effective use of microbes as bioherbicides is dependent on developing formulation and delivery systems that support consistent performance.

Finally, in the last section, forest researchers report on upcoming strategies to use semiochemicals and microbials for management of forest pests. Pheromones of lepidopteran pests, such as gypsy moth and tip moths, saturate the air space and thereby inhibit mate location. This is commonly referred to as the "confusion" strategy. Pheromones of bark beetles, the other major group of forest insect pests, use the aggregation and antiaggregation pheromones in "trap-out" and "dispersion" strategies, respectively. Forest researchers also must blend the development of aerial application technology for microbial insecticides with the reduction of impacts to nontarget species. A five-year strategic plan will be prepared for a prototype effort with Bt var. kurstaki. A literature search of documented Bt impacts on nontarget arthropod species is completed and will be published by the Forest Service to aid the effective evaluation of potential hazards to nontarget species in the forest ecosystem.

A review of the chapters shows that we have need, resources, and opportunities to achieve the goal of producing safer pesticides to meet the demands of the public, the resource managers, and the USDA. Use of formulation and delivery knowledge for the biorationals still remains entrenched for the immediate future in the conventional wisdom and experiences of chemical pesticides. Genetic engineering of Bt in plants might be an effective delivery, but it remains unclear whether it is sustainable. We support the efforts of industry to fill the requests for more environmentally acceptable technologies and pest management. We urge that the fast-track registration goals from government be fulfilled in a harmonized global approach so that these more specific biorational pest control agents can be rapidly integrated into successful use patterns in place of more hazardous materials. We applaud these efforts, and we urge greater attention to detailed studies of formulation, delivery, and ecologically based parameters so that the potential goal of reducing pesticide use may become a reality. The questions and data gaps addressed in this book should be of value to those who wish to take on the opportunities presented by this vast assay of biorationals for crop and forest protection.

FRANKLIN R. HALL
Laboratory for Pest Control Application Technology
Ohio Agricultural Research and Development Center
Ohio State University
1680 Madison Avenue
Wooster, OH 44691

JOHN W. BARRY
Forest Health Protection
Forest Service
U.S. Department of Agriculture
2121–C Second Street
Davis, CA 95616

February 2, 1995

# Chapter 1

# An Overview of Biorational Pest Control Agents

**Franklin R. Hall[1] and John W. Barry[2]**

[1]Laboratory for Pest Control Application Technology, Ohio Agricultural Research and Development Center, Ohio State University, 1680 Madison Avenue, Wooster, OH 44691
[2]Forest Health Protection, Forest Service, U.S. Department of Agriculture, 2121–C Second Street, Davis, CA 95616

The last American Chemical Society (ACS) sponsored Symposium on Formulation Technology was held in New Orleans in 1987. Since then, we have had numerous developments in more sophisticated active ingredients (AI's) and formulations, including controlled release [T° and pH mechanisms (1), as well as U.S. Environmental Protection Agency (EPA) mandates on eliminating many of the conventional solvents. Pesticides are under increased review by EPA and increased scrutiny by the press and the public it serves. Concerns focus on pesticides in food, air and water and the resulting impact on man and other organisms. The United States has established a goal of implementing integrated pest management (IPM) on 75% of all agricultural lands by the year 2000, and has a goal to reduce pesticide use and to use safer pesticides. In addition, there are mandated legislative activities in Europe, e.g., requiring a 50% reduction of AI's in 10 years. In Sweden, for example, the reduction in insurance spraying has been achieved by using more active herbicides (less AI/ha), additional education of the farming community, improved risk-benefit analyses, mandatory and voluntary testing of sprayers, pest forecasting research and general increases in advisory services which, in general, have been paid for by redirecting government resources, and assessing some fees. However, the general net impact on the Swedish farmer economy, for example, has been minimal (A. Emmerman, personal communication).

These pesticide policy shifts, occurring across the globe suggest that there is a need to revisit how formulations and delivery can aid the efficiency and reduction aspect of pesticide use. More active compounds mean less AI/A, but an increased attention to delivery parameters. As the shift is made to "safer" products, the use of biorational materials, which encompass an implication of added safety to either man or the environment than the current arsenal of registered products seems to be an increased opportunity. Discussions persist over the semantics of whether a biorational is also a

0097–6156/95/0595–0001$12.00/0

biopesticide, is natural, or contains only living organisms, etc.,(2). The purpose of this book was to explore the knowledge status of the formulation and delivery questions associated with this broad range of crop control agents, identified as being quite distinct from conventional synthetic diversity. For purposes of clarity in this chapter, we thus have elected to call these control agents biorationals. The following chapters illustrate the wide range of these crop protection agent (CPA) opportunities.

The events leading the industry towards the biorationals in general are summarized as follows. EPA's Safer Pesticide Policy and goals for new pesticides for the period of 1993-1996 include:

- Safer materials
- Less persistence
- Less toxic to non-target species
- Less likely to contaminate groundwater
- Lower exposure (man and environment)
- More practical disposal technology

This set of goals follows the accepted paradigm that prevention of environment contamination is more effective and efficient than any cleanup protocol, as indicated by the general lack of successes in the broad ranging EPA Super-Fund Programs.
If one reviews the history of agricultural use of pesticides, we are faced with the question of why are pesticides utilized so heavily; Reichelderfer (3) and Barry summarize this (Table 1,a,b) which are appropriate for both agriculture and forestry.

Trends in both pesticides and adjuvants (Table 2 and 3) suggest that there are opportunities for the crop protection industry (4). However, it is clear that technology complexity will be higher and perhaps not as inexpensive as with those of the 1960's and 1970's. The economic, environmental and general pesticide policy influences on the development of biorationals continue to indicate that they will be of increasing interest to pesticide manufacturers (2). The dependency upon specific types of pesticides or CPA's is now modified by EPA's goals for "new" pesticides. However, paradigm for the development of pesticides in the crop protection industry (Table 4) shows the current economic rationale and opportunities for crop protection agents.

Nevertheless, biorationals in general are viewed favorably by registration authorities, have less stringent registration requirements and may cost considerably less to bring to market. Recent data show the rather slow registration pace of biorationals into the conventional markets. On the other hand, biorationals are projected to be significant components of the crop protection arsenal by the year 2000. For example, these projections are conservatively placed at a rate of growth of ca. 10%/yr to reach $130 million by 2000 (5). The biorational agent product mix is currently at $75 million with ca. 80% in Bt products, nematodes at 13% and all others at 7%.

**Table 1a. Socio-economic Reasons for Pesticide Use in Agriculture**

- Pesticides are generally cheaper, faster acting, and more convenient than other technologies, e.g., mechanical, cultural, etc.

- The riskiness of farming, especially in horticultural crops, is high, thus
  - > "Chemical Technology" is used to reduce the variability
  - > Pesticides save management time because of (a) high % off-farm income requirements, and (b) IPM, Sustainable Agriculture or Ecosystem Management, etc., requires more management

- The influence of current farm policy
  - > Depending upon Farm Support Programs
  - > And the <u>disincentives</u> for <u>diversification</u>

- Finally, farmers respond to market signals provided by price. Heretofore, environmental costs were not included in production costs at the grower level.

SOURCE: Reproduced with permission from reference 3. Copyright 1989.

**Table 1b. The management rationale for pesticide use in Forestry[a]**

- Establish, restore and maintain native ecosystems while controlling exotic plants, disease and insect pests.
- Maintain productivity of multiple uses of the forest consistent with environmental quality and economics.
- Support forest regeneration programs.
- Increase productivity in plantations and agroforestry.

[a] J. Barry, 1994.

While the advantages seem to be in favor of a rapid shift to the biorationals, one should note that these materials have been notoriously low in efficacy and erratic in performance with poor lab to field correlations. Chemists tended to view them as conventional chemicals but many biorationals did not behave as such (5). However, the various trends in policy direction, public general dissatisfaction with chemistry as we know it, suggests that if we are to shift an emphasis to biorationals, then more scientific information, different experimentation, and hence a different level of scientific inquiry has to be made if the industry is to efficiently seize the opportunities and increase the market penetration.

**Table 2. Pesticide Trends**

- Stable or Declining Market
- Application Developments
- Reduction from Grams vs. Lbs/Acre
- Lower Spray Volumes
- More Complex/Fragile AI's
- Different Carriers
- More Pre-Mixes, Tank Mixes
- Environmental/Worker/User Exposure Concerns
- More Complex Formulations
- Biotechnology
- Post-Emergence (Herbicide) Uses
- Product/Distribution Changes
- Regulations/Registration Requirements

SOURCE: Reproduced with permission from reference 4.


**Table 3. Trends in Adjuvant Technology**

- An Increasing Market Size
- For Improvement of AI Economics
- Changing Formulations
- Improve AI Performance
- Increasing Complexity of AI Systems
- Reduction of AI Risks
- Reduction of AI Utilized
- Are Function Specific
- From Herbicides to Other Products
- New Products
- Regulation/Registration

SOURCE: Reproduced with permission from reference 4.

**Table 4. Pesticide Development Paradigm[a]**

- Effective on Wide Range of Pests (Broad Spectrum)
- Patentable and Definable
- Provide an Effective Advantage
- Acceptable Toxicological and Environmental Impacts

[a] After Stowell, (6).

If one reviews technology advances in general, we can see the fundamental elements of change in society which will influence agricultural use of crop and tree protection agents (CPA's). For example, these trends in society include the following:

- Control → Management
- Conventional → Alternatives
- Single → Mixed
- Insurance → As Needed

The simple change from control to management, for example, was explored by Winder and Shamoun (7) as they attempted to resolve the issues associated with herbicide terminology, e.g., "cide" inferring mortality and urging the use of vegetation management as an option.

These changes suggest that:

- Information
- Analysis
- Decision-Making

will be even more important for not only agriculture and forestry industries, in general, but also the ultimate user of this new technology. Thus, we see "prescription farming," increased crop rotations, a wide array of tillage modifications for conservation goals, and, in general, variable application rate technologies emerging on the scene, all of which will and are influencing why and how we will utilize these new biorational agents in agriculture and forestry.

## BIORATIONAL AGENTS AND FORESTRY

The USDA Forest Service (Forest Service), in addition to managing 191 million acres of national forest and range lands, also operates greenhouses, nurseries, and seed orchards to support tree improvement, habitat restoration, and reforestation programs. The Forest Service is a minor user of pesticides but the agency does use a variety of pesticides that are listed in their pesticide use report for fiscal year 1993 (USDA Forest Service, 1993 (24)) as fungicides, fumigants, herbicides, algicides, plant regulators, insecticides, acaricides, and pheromones. Of these, the biorationals that the Forest Service uses are included under insecticides (bacterial and viral organisms) and pheromones. The Forest Service used pesticides on only 0.16 percent of the national forest system lands in fiscal year 1993. Of the acreage treated, only 7.2 percent was with biorational agents and most of this was *Bacillus thuringiensis* for control of *Lymanantria dispar* L. (gypsy moth). Although similar data are not available from other pesticide uses, we suspect that the percentages of biorational agent use is significantly lower than the 7.2% used on national forest system lands. Some of the same problem that limit the development and deployment of biorational agents in agriculture, limit their use in forestry. These include appropriate formulations and delivery systems, development and production costs, limited use thus limited profitability, registration difficulties, and lack of governmental use incentives. The latter two being the most serious impediments. Unlike agricultural growers, federal forestry managers, given the option of using biorationals, will likely be more willing to try this approach to pest management, due to the severe public scrutiny of federal land management. Private forest managers are responding to similar pressures. It is likely that the Forest Service will increase its need and use of biorational agents to manage forest and range lands, including eradication of exotic species that threaten management and reestablishment of native ecosystems. For the use of biorationals to become more commonly used and to replace conventional pesticides, the problems listed above will need to be effectively addressed. This may only be realized through partnerships among industry, academic, and government - especially the regulators.

## INFORMATIONAL NEEDS

If biorationals are to take on an increasing importance as part of the arsenal of CPA's available for agriculture and forestry, certain fundamental informational needs appear to be critical in optimizing the formulation and delivery parameters. For example, Bt strain efficacy continues to increase because of advances in biotechnology. However, the problem still remains - more efficacy is just increased wastage of technology UNLESS fundamental information about formulation and delivery parameters, (e.g., micrometeorology, target characteristics, drop size distribution and density by location and the interactions, etc.), and pest behavior to the presentation of the AI are understood and related to these advances. As suggested in the following "questions" for biologists (Table 5), these are fundamental inquiries that we must attempt to answer in order to utilize advances in the biologicals to their fullest advantage.

**Table 5. Biological Requirements for an Effective Dose-Transfer of Biorationals**

- What are we aiming at - Target identification: temporal and spatial.
- Where do we need to deposit this AI > why, how much?
- What is the optimum deposit form: quality and quantity, for effective bioavailability?
- Can we achieve this goal? <u>OR</u> How close can we get?

Thus, we have the need to interact at many levels of the science as well as among the disciplines. With the biorationals, it is becoming abundantly clear that interdisciplinary discussions and scientific inquiries must be planned and executed with this higher level of interaction in place in order to achieve these greater expectations of "doing more with less." Technology is only as good as the efficiency of the delivery process, otherwise it still represents a significant waste to the economy.

As Normand DuBois succinctly concluded at the 1994 workshop on Application Technology of Microbials in France,

> "It is no longer sufficient to view application technology as one influenced essentially only by the physical parameters of drop size, drop density coverage and formulation physical characteristics. Rather, successful and efficient application and use of microbial insecticides is also significantly influenced by (1) the mechanism by which the deposited dose is actually acquired and processed by the target pest, (2) formulation compatibility with the pest, the insecticidal agent and the delivery system used, and (3) by the strategies associated with vulnerability of the pest relative to its behavior and timing of application. Indeed the spray application window is significantly altered."

## OPPORTUNITIES

### Regulation

A new EPA Division - Biopesticides and Pollution Prevention, founded in October 1994, has 3 basic functions: biopesticides, integrated pest management (IPM), and pollution prevention or pesticide use reduction (8). It remains to be seen whether this emphasis will indeed aid the development and utilization of biorationals at the farm and forest level. It is clear that use reduction aimed at pollution mitigation, as well as the administration's IPM directives, all indicate the increased pressure to do it differently. The enclosed chapters suggest the development and aforementioned needs of formulation and delivery are still not at a highly effective level. Unless the innovator can forecast a "profit," the technology is not likely to see much progress as a useful product on the farm or in the forest. Easing (fast-track) or streamlining the registration process is espoused but, to date, only a few examples can be shown to have a truly fast track towards registration. We view the slow-pace of registrating biorationals as a serious impediment to advancing our goal of reducing pesticide use.

As pointed out by Weatherston and Minks (9) on the global regulation of semiochemicals, pheromones and other semiochemicals are recognized to be different from chemical insecticides. Not withstanding the contributions towards a reduction in data requirements, much more can be done to expedite and harmonize pheromone regulations. The authors further proposed a structure/activity and partnership approach (with a sharing of the existing database within EPA) for the evaluation of health and environmental risk assessments in order to reduce registrant costs and time delays.

The concerns are that the reregistration processes occurring throughout the world are forecast to remove ca. 20% of the older pesticides without suitable replacements. Thus, while the emphasis is on safer, less toxic "natural" materials, the regulatory structures have not provided enough of an agreed upon road map to encourage the industry to engage these opportunities at a faster pace as encouraged by Edwards and Ford (10). Placed in perspective, while the U.S. pesticide market was ca. $5.3 billion in 1990 (11), only 27% of that was the insecticide component and only 0.6% of that represents current sales for all semiochemical products. Consequently, the opportunities for control of insect (and mite) pests is certainly exceeded by herbicidal products (ca. 60% of total sales).

## Formulation and Delivery

Advances in formulation specifics are, of course, not revealed in this book for obvious business reasons. What is still missing is an in-depth discussion of the kinds of inquiry needed for each of the biological areas covered in order to overcome the aforementioned field inconsistencies and poor performance (efficacy). There are a few examples of the more basic in-depth array of questions being addressed within this publication, e.g., see Hall, et al., Rees et al., A. Sundarum, Lumsden, Fuxa, and Wood. Other readings should include: Baker and Drum (12), Green et al. (13), Carlton (14), Carlton et al. (15) and Kim (16). But the needed cooperative dialogue and effective research agendas among formulation/biologist/engineers, including industry/university/federal laboratory partnerships remains a serious weakness.

The microbial market currently accounts for <17% of the total insecticide market, but, encouragingly, is growing at ca. 10-25%/yr, compared to conventionals (17). These authors emphasize that further growth of the baculoviruses, for example, depends upon reducing production costs, developing practical, effective formulations, optimizing field performance, overcoming regulatory obstacles and educating the user and the public. A less studied arena, according to Starnes et al. (17) which was emphasized by the work herein of Hall et al., Fuxa and Sundram, is that greater attention must be given to the application needs of the biological, e.g., the presentation of the biological in an appropriate format in tune with biological information on population densities, spatial, and temporal needs in order to fully develop the area of knowledge-based strategies.

Wood and Hughes (18) point out the need for careful evaluations of the effects of biorationals on non-target organisms. The potential interactions between microbials and non-targets are particularly critical with recombinant baculoviruses where the infections and symptom assessments are vital in order to correctly evaluate ecological consequences.

Carlton (14) points out the fact that traditional biorationals have not gained wide acceptance by the agricultural community. The Federal forest ecosystem managers, on the other hand, have moved from conventional pesticide use to biological pesticides. On a much more limited scale, they have cooperated with researchers in classical biological control of exotic plants. As emphasized in this book, biorationals in general do not persist well, are difficult to deliver, and are limited in ranges of pests controlled. While these factors may be disadvantages in some agricultural situations, they are a non-problem or an advantage in forest use, particularly selectivity. Limited specific examples are given to illustrate some gains in efficacy, e.g., UV protection, etc., but clearly, there is no easy, general answer for these materials. As to engineered biorationals, Stowell (6) concludes that there are 2 major barriers to their development, field performance and regulation. Some indications of barrier mitigation are presented herein, but these remain significantly difficult problems to solve if the prognostications for safer, nonpersistent, more specific CPA products are to become reality.

Finally, as noted by Starnes et al. (17), considerations which would speed up the utilization of biologicals in agriculture and forestry include:

- Clear, consistent regulatory structures.
- State legislation consistent with federal regulations to reduce redundancies.
- Public understanding of biologicals.
- Greater attention to pest behavior, biology and population dynamics.
- Greater delivery precision.
- Education of the user for the need for greater monitoring, timing and attention to application.

To this list we might also include cooperation demonstration projects on the farm and in the forest.

## Concerns Not Addressed

Biorationals are not immune from pest resistance phenomena, which has been called "a problem reaching crisis proportions" (19), due in part to a lack of regulatory policy protecting against such risks, although it would also seem prudent and "politically correct" if we could work this out without more regulation. As an example, Higley and colleagues (19) cite the WHO malaria eradication program which has threatened human health by limiting management options for medical pests. It is exactly this lack of a national policy which not only acts as a disincentive to new technologies (i.e., engineered biorationals), but also challenges the basic questions of human health and environmental risks evaluation of economic and environmental impacts, and sustaining pest management. Potential pesticide policies and the criteria met by those policies are posed by Higley et al. (19) [Table 6]. In summary, criteria for an effective national pesticide policy should:

- Allow rescue
- Reduce true risk
- Reduce risk perceptions
- Encourage non-chemical alternatives
- Reduce costs to user and the public

Table 6. Proposed potential pesticide policies and criteria met by these policies

| | Criteria met by policy | | | | | |
|---|---|---|---|---|---|---|
| | Rationalizes pesticide use | | | | Minimizes economic impact | |
| Policy | Allows curative pesticide use | Reduces true risks of use use | Reduces perceived risks of tactics | Encourages use of non-chemical users | Reduces cost to pesticide | Reduces cost to government |
| | | | Policies concerning pesticides | | | |
| Maintain status quo | X | - | - | - | X | X |
| Cancel all pesticides | - | X | X | X | - | X |
| Cancel Class I poisons | X | X | ? | ? | ? | X |
| Improve existing regulation | X | X | ? | - | X | - |
| Provide financial support to reduce use | X | X | ? | X | X | - |
| Establish mandatory prescriptions | X | X | X | - | - | - |

| | | | | | | |
|---|---|---|---|---|---|---|
| Policies concerning nonchemical tactics | | | | | | |
| Create pest damage insurance | X | - | - | X | ? | ? |
| Revise commodity support programs | X | - | - | X | ? | X |
| Improve regulation of biologicals | X | - | - | X | X | - |
| Levy fees on pesticides | X | - | - | X | - | X |

SOURCE: Reproduced with permission from reference 19. Copyright 1992.

While the recent (1994) EPA/USDA memoranda of understanding (MOU) advances cooperation and exchange of information on pesticides, it is unclear who is leading the advances to be made in IPM, a program advocated by U.S. administration aimed at a goal of 75% of U.S. farmers on IPM by the year 2000. At the same time, we have the USDA moving towards "service centers" and USDA Natural Resources Conservation Service (formerly the Soil Conservation Service) promoting conservation and IPM practices. Current government strategies are being directed towards interdisciplinary, long-term, large area programs involving both Research (Basic and Applied) and Extension (Technology Transfer) with active grower participation. From the standpoint of enabling better understanding of biorational technologies, their benefits and risks, this approach should prove helpful.

Why a discussion about pesticide policy? Policy incurs pressure on economic incentives and scientific innovation for new products. Thus in today's global business world the registration parameters are vitally important for biorationals. Although registration harmony for biorationals (Table 7) appears similar, even slight interpretation differences can exacerbate the harmonization problem (20). The resulting higher costs <u>and delays</u>, in fact, will deter development of biorationals and extend our dependence upon conventional agrochemicals. Houghton (21) predicts severe danger of increasing politics of registration and that NAFTA and GATT also appear to be headed in the same direction. Houghton (1994) postulates that the EU is in severe danger on this matter and further suggests the need for a global forum to deliberate registration strategies in order to effectively monitor CPA use in a global context.

Not covered adequately in the ACS Symposium on Biorational Pest Control Agents were a number of issues, including: (a) variable application rate technology, (b) basic informational needs to address crops on a "treat as needed," or by other population density sensing systems which link to economic or ecological threshold baselines, (c) strategies which address the rapid dissemination of "delivery" expertise, which, on a world-wide basis, is critically low for the numbers of problems and opportunities to effectively exploit the advantages of safer biorationals, improved linkage between lab and field performance (22) and (e) the lack of a demonstrated support by government for sustained interdisciplinary programs aimed at the more complex problems posed by biorationals.

From Table 8, and other prognostications, one could easily assume that biotechchnology (via genetic engineering) was on the verge of replacing the need for external sprays with genetically engineered plant resistance/tolerance. However, several recent reports (23) suggest utmost caution on the basis of observations that "RNA recombination (possibilities) should be considered when analyzing the risks posed by virus resistant (and often gene transfer targets) transgenic plants". Hence, the planned large scale tests with extensive environmental monitoring will be <u>critical</u> determinants for the future of external CPA's. Meanwhile there are disturbing reports of a weakening of apple scab resistance in cultivars planted in Europe. Further, these genetic engineered

Table 7. Sections required in applications for registration of biopesticides in various countries

| | USA | UK | EU | CANADA | AUSTRALIA | NEW ZEALAND |
|---|---|---|---|---|---|---|
| 1. | Product analysis | Introduction | Identity of organism | Index | Introduction | Identification |
| 2. | Toxicology[1/] | Information on formulated product | Biological properties | Label | Chemistry, manufacture, biological properties | Toxicology |
| 3. | Residues | Identity of AI | Further information | Characteristics and specification | Toxicology | Residues |
| 4. | NTO hazards | Biological properties | Analytical methods | Human health and safety testing | Toxicokin-etics and metabolism | Environmental and wildlife data |
| 5. | Environ-mental fate | Manufacture formulation | Toxicology | Metabolism | Residues | Biological properties |
| 6. | Efficacy | Application | Residues | Residues | Occupational health and safety | Label |
| 7. | EUP data | Efficacy | Fate and behaviour in environ-ment | Environmental fate | Environmental | - |
| 8. | Label | Toxicology | Ecotoxi-cology | Environmental toxicology | Efficacy and NTO safety | - |
| 9. | - | Effects on humans | - | Efficacy and benefits | - | - |
| 10. | - | Residues | - | - | - | - |
| 11. | - | Environmental and wildlife hazards | - | - | - | - |

SOURCE: Reproduced with permission from reference 20. Copyright 1994 BCPC (UK).

[1/] Toxicology includes toxicological, pathogenicity and infectivity studies.

plants also need to have close examination of changes that will take place in the arthropod and disease organism populations that evolve under these new plant systems.

Table 8. Transgenic field releases being tested (cumulative since 1987)[a]

| | | |
|---|---|---|
| ■ | Herbicide Tolerance | 30% |
| ■ | Product quality | 24% |
| ■ | Insect resistance | 21% |
| ■ | Viral resistance | 14% |
| ■ | Other | 8% |
| ■ | Fungal resistance | 3% |

[a] after (23)

While biorationals offer some advantages of being environmentally acceptable and supported in principle by government policy, it remains unclear how governmental and industry will bridge the pest control gaps left by product elimination by reregistration demands and/or bans and lagging biorational registration uses. Conventional chemistry has left the growers quite contented with swiftness of action, economy of scale with field sustainable performance across a wide range of pests. The demands for higher levels of "educating the user" about how to use these biorationals are abundantly clear. Less clear is how we are going to effectively achieve this goal. Currently, three events effectively make growers change their ways:

1. Economics: They can make more money by changing their strategy from conventional pesticides to biorationals.
2. Regulation: Removal of the conventionals leaves the user with alternative [bans or label restrictions] choices - switching to biorationals or other strategies, including crop rotation, mechanical, or the means of pest control.
3. Liability: A subtle trend creeping into agriculture and forestry built upon perception by the public and neighbors about current agricultural practices and resulting in changes in grower utilization of AI's. On the other hand, users need liability protection as they take the risks and shift from traditional pesticides to biorationals. Here is where national policy issues need to be reinforced and clarified.

## DISCUSSION AND CONCLUSIONS

Several needs and issues surfaced during the two day symposium on Biorational Pest Control Agents which focused on formulation and delivery of biorational agents.

1. Clearly, biorational materials are not "magic bullets" and we as scientists have responsibility to communicate a balanced picture to the public. We have to do a better job of demonstrating the economy, safety, efficiency, and efficaciousness of biorational pest control agents.

2. Consistent with item 1, we need to **highlight biorational agent success stories**. There was discussion on this point and the general conclusion was that we need to publicize these successes to inform, and to develop public and user support for the biorational programs.

3. Technology transfer to potential users will need much more emphasis and innovation in order to support the replacement of conventional pesticides with biorationals. The challenges are mentioned in several of the papers included herein. Costs and benefits of the biologicals are quite different from conventional pesticides, but when addressing the formulation and delivery criteria, along with the reduced broad-spectrum (highly specific activity), and the need for much more monitoring and application specific timing and equipment, these environmental advantages quickly disappear.

4. "Biorational" may seem to be a misnomer in that it could infer that current technology is non-rational when, in fact, we have many examples of highly specific actives and some that are essentially environmentally benign. As noted by Winder and Shamoun (7), it will take time to build a common dictionary of terms so we can better communicate with the public and with other disciplines. However, in light of the current mandates by various legislatives around the world to reduce pesticide use by the year 2000 (25) and the current U.S. EPA Safer Program by U.S. EPA [less AI/A], it would seem logical that biologicals, in the broadest sense, have unique opportunities for development in the coming years. Certainly, there is a need for a clear definition of the more biorational approaches as espoused by biologicals to replace current conventional pesticides which may be more hazardous to man and the environment.

5. There is need for national leadership to encourage use of biorational materials over conventional pesticides. How do we go about organizing this - joint government, industry, and academic partnerships in wide-sweeping commodity programs? This is another communications issue. (See following item 6.)

6. The national association of biorational manufacturers (probably not its exact name) is an excellent media to inform and unite users, regulators, researchers, and manufacturers of biorational materials. (Note - they may be communicating but much more is needed.)

7. The Technology Transfer Act 99-502 of October 20, 1986, is a mechanism for Federal Cooperation with industry. It would appear that if there are successes with this Act, it certainly was not addressed by any of the authors. Clearly there are opportunities here and federal agencies should be taking advantage of the potential mechanisms for technology transfer of new technologies. This may relate to the

age-old problem of federal organizations not having to demonstrate practical, saleable technologies and of federal management lacking in providing incentives.

8. There was concern expressed about the low interest in biocontrol of exotic plants. Since ca. 60% of pesticides sold in the U.S. are herbicides, there would appear to be an opportunity to reduce herbicide use through the use of natural enemies. There will be additional efforts to induce other registration management strategies involving a complex series of cultivation, sprayed cropping, no-tillage technologies, crop tolerance (to completion) etc., all designed to reduce herbicide usage. However, it is clear that significantly higher proportions of private industry R & D dollars are focused on new (conventional) chemistry where past history has demonstrated financial success. Therefore, the risk to chemical companies to venture into the "uncharted waters" of biocontrol of weeds seems to most industries to be a bit severe and challenging R & D, marketing and profits. If countries continue to focus on the simple reductionist approaches of mandating 50% less, etc., then new herbicides with less AI will have a role. Clearly, the phenomenal success of ever lower rates of AI herbicides are outstanding, but nevertheless, also contribute to the challenge of efficient delivery. If frequency of application and suppression vs. eradication are also addressed, then perhaps alternative strategies using biocontrol techniques for weed control will also get an increased level of research activity. It is likely that success of the biocontrol of exotic plant programs will begin at the ground roots level with federal researchers initiating the technology transfer efforts through communities and private and public agents.

9. The current U.S. administration, the Canadian government, as well as others, are targeting pesticides and calling for reductions and IPM strategies. Biorationals have a major opportunity to fill the void but the economics and efficacy must be demonstrated to agriculturists and forest managers. Also, the regulatory requirements need to be adequately defined and "fast tracked" to encourage and support development, testing, and registration. There is a need to increase the coordination with Canada and other nations on regulatory harmonization of biorational management.

10. Formulation inadequacies are clearly a focus area and, indeed, a challenging one for many biologicals, especially in efficient delivery systems, UV protection, rainfastness, and repellency. Some companies formulate agents for application in existing spray hardware; others do not. Some materials can be efficiently applied by conventional systems; e.g., Bt, while others, e.g., sticky pheromone fibers, remain a challenge. Inerts, in some cases, can be more disruptive than the active ingredients of the biological formulation, which is an ongoing concern in formulation laboratories. The loss of conventional solvents adds to the dilemma presented to the plant protection industry. Finally, there is a clear lack of knowledge about the optimal presentation scenario for most biologicals; i.e., what droplet size, concentration, number/$cm^2$, target area characteristics, etc., are needed for these new agents. The lack of sufficient information about the target pest location and feeding characteristics limit the immediate potential successes of new strains, etc., developed as a result of biotechnology. For example, there is need

to understand why Bt does not provide consistent efficacy against the forest defoliating gypsy moth. If efficacy is not a big problem in agriculture, then the lack of efficacy in forestry may be formulation and delivery problems, given the higher release heights and forest biomass to be covered.

11. This same lack of knowledge also holds true for the identification of specific protocols for the design and conduct of field studies to evaluate impact of biorational materials on non-target organisms. It would seem that here is a potential opportunity for a "Spray Drift Task Force" approach to aid the development of generic principles of risk assessment towards non-target organisms. Such an approach could include both public and private agencies representing government, industry and academia.

12. The Forest Service is very concerned about the introduction and impact of exotic pests on native forest ecosystems. Exotic pests will likely receive a major forestry emphasis in the future. Native pests, and their damage, on the other hand, will more likely be viewed as important ecosystem functions, therefore, receiving less attention. Biorational materials will be needed to manage exotic pest and some native pests, e.g., management pests of seed orchards, and registration sites, nurseries, greenhouses, plantations in reforestation sites, and habitat restoration projects. The Forest Service strives for a healthy ecosystem where native pests are recognized as having a vital role in the ecosystem. The concern and need for direct intervention occurs when nature proceeds to eliminate a species, e.g., the Torrey pine near San Diego, or when man initiates activities, e.g., suppression of natural fire that might result in a serious bark beetle infestation. Some forest scientists are calling native pests "agents of change" and change is being viewed as a natural and beneficial ecosystem function. This, of course, is fine providing we don't need the forest commodity now and can wait, in some cases, many decades - therein is one of the developing sources of controversy, and the story will continue.

13. Support and partnerships among both the public and private sectors will be necessary to establish biorationals as proven and acceptable agents to manage agricultural and forest pests. The primary suggested approach is to inform the consuming public - the consumer of food and fiber products - about safer alternatives to conventional pesticides as the first step in gaining support of biorational agents. Scientists, by their own admission, have poor skills in public communications and those who have some experience find it most difficult to communicate in non-technical terms; nevertheless, it will be up to the scientist to communicate and to "sell" the safety and other advantages of biorational agents to the consumer. With public support of biorationals, governments will be more inclined to provide incentives and reduce regulation, and industry will be at less risk and have more of an economic basis to pursue research, development, and marketing of biorationals. While idealistic, this approach is preferable to governments regulating us out of conventional pesticides without proven and acceptable alternatives.

In summary, the authors believe that this book is an important step in leading us toward a heightened awareness of biorational agents and their utility as acceptable tools to manage agricultural and forest pests in the future.

## LITERATURE CITED

1. Wilkins, R. *Controlled Delivery of Crop Protection Agents*. Taylor & Francis, UK, 1993.
2. Smith, A. Why Bother With Pesticides. In *Agrow Suppl. Autumn*, ed., pp. 19-25. 1994.
3. Reichelderfer, K. Ag Economics, pers. comm. January 1989.
4. Underwood, J. Helena Symposium Proceedings, pp. 1-12, 1994.
5. Lisansky, S., and Combs, T. In *BCPC Pests and Diseases*. Vol. 3, pp. 1049-1054. 1994.
6. Stowell, L. In *Advanced Engineered Pesticides*, M. Dekker.... pp. 249-260. 1993.
7. Winder, R. and Shamoun, S. *Biol. Control* 1:339. 1991.
8. Ag Aviation. Newsletter, Aug. 1994.
9. Weatherston, I. and A. K. Minks. (In press) *Publication of Global Aspects of Regulation of Semiochemicals*. 1995.
10. Edwards, C. R. and Ford, R. E. In *Food, Crop Pests and the Environment*. APS Press, St. Paul, MN., pp. 13-55. 1992.
11. Maxey, F. In *U.S. Industrial Outlook*. U.S. Dept. Commerce Doc. No. S/N003-009-00586-8. 1991.
12. Baker, R., and Dunn, ed. *New Directions in Biological Control: Alternatives for Suppressing Agricultural Pests and Diseases*. Liss, NY. pp. 419-434. 1990.
13. Green, M., Moberg, W., and LaBaron, H., eds. *Managing Resistance to Agrochemicals to Practical Strategies*. ACS Series #4221. 1990.
14. Carlton, B. C. In *New Directions in Biological Control*. Alternatives for Suppressing Agricultural Pests and Diseases. Liss, NY. 1990.
15. Carlton, B. C., Gawron-Burke, C., and Johnson, T., In *Proceedings, Fifth Int. Coll. Invert. Path and Microbial Control*, SIP, pp. 18-22. 1990.
16. Kim, L., ed., Advanced Engineered Pesticides. M. Dekker, N.Y. 1993.
17. Starnes, R. L., Liu Li, C., and Marrone, P. *Amer. Ento.*, p. 82-91. Summer 1993.
18. Wood, A. and Hughes, P. *Science* Vol. 261:227. 1993.
19. Higley, L., Zeiss, M., Wintersteen, J., and Pedigo, L. *Am. Ento.*, p. 139-146. Fall 1992.
20. Lisansky, S. G. *BCPC Pests & Diseases*, Vol. 3, pp. 1397-1402. 1994.
21. Houghton, A. M. *BCPC Pests & Diseases*, Vol. 3, pp. 1389-1396. 1994.
22. Hewitt, H., Caseley, J., Copping, L., Grayson, B., and Tyson, D., ed. *Comparing Greenhouse and Field Pesticide Performance*. SCI Mono. No. 59, 323 pp. 1994.
23. Science, Vol 266. 2 Dec 1994, pp 1472-73.
24. USDA Forest Service. 1993. Report of the Forest Service Fiscal Year 1993. Washington, D. C.
25. Thonke, K. In *Aspects of Applied Biology* 18, pp. 327-329. 1988.

RECEIVED April 3, 1995

# Registration Needs and Data Requirements

Chapter 2

# Registration of Biologicals

## How Product Formulations Affect Data Requirements

**Michael L. Mendelsohn, Thomas C. Ellwanger, Robert I. Rose, John L. Kough, and Phillip O. Hutton**

**Office of Pesticide Programs, U.S. Environmental Protection Agency, 401 M Street S.W., Washington, DC 20460**

Among those pesticide products currently registered with the U.S. Environmental Protection Agency (EPA), there exists a subgroup of products known as biologicals. Biologicals include 1) certain microorganisms and 2) compounds classified as biochemical pesticides that act to control pests as defined in § 2 of the Federal Insecticide, Fungicide and Rodenticide Act. Biochemical pesticides are distinguished from conventional chemical pesticides by their nontoxic mode of action toward target organisms and by their natural occurrence, e.g., insect pheromones. Data requirements for experimental use and full commercial registration of these pesticide products are somewhat different than those required for conventional chemical pesticide products. Formulation choice has a large impact on the data necessary to support a particular biological pesticide product. This chapter stresses the importance of formulation choice in product development as it relates to EPA regulatory requirements.

## What Are Biological Pesticides?

Pesticides are defined in § 2(u) of the Federal Insecticide, Fungicide and Rodenticide Act as "(1) any substance or mixture of substances intended for preventing, destroying, repelling, or mitigating any pest and (2) any substance or mixture of substances intended for use as a plant regulator, defoliant or desiccant" except those articles considered to be new animal drugs or animal feeds bearing or containing a new animal drug. Biological pesticides include microbial and biochemical pesticides.

Microbial pesticides include the following microorganisms when they act as pesticides per FIFRA § 2(u): 1) eucaryotic microorganisms including protozoa, algae and fungi; 2) procaryotic microorganisms including bacteria and 3) viruses.

Biochemical pesticides are distinguished from conventional chemical pesticides by their nontoxic or indirect mode of action toward target organisms and by their natural occurrence or structurally similarity and functional equivalence to naturally occurring compounds, e.g., insect pheromones and certain growth regulators.

## EPA's Role

The process through which a pesticide use is approved for large scale experimental or full commercial use is managed at the federal level by the Environmental Protection Agency. A major part of EPA's review process is the evaluation of data submitted by the applicant regarding mammalian and nontarget organism toxicology and product identity/analysis. In order to justify the approval of large scale experimental or full commercial use of a biological pesticide, the Agency must determine that such use will not result in unreasonable adverse effects to the environment. In other words, the pesticide cannot pose too high a risk to human health or the environment. Applicants can often affect the risk of their product by the way they formulate.

## What Is Risk?

### *Risk = (Hazard) X (Exposure)*

In order to affect risk in a product, one must either affect the hazard of the product or its exposure to man and/or the environment. Either end can be accomplished using various formulation strategies. When the way in which a product is formulated reduces the hazard and/or exposure to a particular pesticide product, often not only is risk reduced, but also the need for some of the nontarget organism and human health data normally required for large scale experimental and full commercial use may be obviated.

## Nontarget Organism Data

Nontarget organism Tier I testing requirements for biochemical and microbial pesticides generally include avian acute oral, avian dietary, freshwater fish $LC_{50}$, freshwater invertebrate $LC_{50}$, nontarget plant, nontarget insect and honeybee tests according to Title 40 of the Code of Federal Regulations (40 CFR) Parts 158.690 and 158.740. The test substance for these Tier I tests is normally the technical grade active ingredient (TGAI). However, both the actual nontarget organism studies required and the test substance of those studies can be affected by formulation.

40 CFR Part 158.75(b) provides that, on a case-by-case basis, testing may be required with the following:

1) an intentionally added inert ingredient in a pesticide product;

2) a contaminant or impurity of an active or inert ingredient;

3) a plant or animal metabolite or degradation product of an active or inert ingredient;

4) the end-use pesticide product;

5) the end-use product plus any recommended vehicles and adjuvants; and

6) any additional substance which could act as a synergist to the product for which the registration is sought.

In addition to the TGAI, inert ingredients and manufacturing or formulation process changes have the potential to adversely affect nontarget organisms. Examples of these ingredients and their possible effects could include, but are not limited to, the following:

1. Bacterial cultures have been stored under mineral oil to preserve them. Oils have been used in some formulations. Oils have been used as mosquito larvicides because they form surface films which also affect other nontarget organisms that depend on the integrity of the water surface.

2. Plastic devices are used for controlled release of volatile pheromones that are classified biochemical pesticides. Depending on shape, they could result in intestinal damage to animals that eat them or trap animals that become entangled in them.

3. A change in fermentation, recovery or manufacturing process may result in concentration of byproducts that, at low levels may cause no adverse effects, but as concentration increases, effects occur.

4. A preservative, diluent or other formulation ingredient may be used that is potentially more toxic than the active ingredient.

5. The biochemical or microbial pesticide is combined with a conventional chemical pesticide which exacerbates adverse effects.

6. A granular, pellet or other formulation is introduced that is attractive to birds or other wildlife resulting in collection by animals and increased oral exposure compared to other formulations and uses of the same active ingredient.

## Product Specific Acute Toxicity Data Requirements

The acute toxicity studies required for labeling purposes to support registration of pesticide products include: acute oral, dermal and inhalation toxicity, eye and dermal irritation, and dermal sensitization. This data is required to label these products for human safety, i.e., precautionary statements (how to avoid exposure) and statements of practical treatment (first aid if exposure does occur). This same data can trigger Restricted Use Classification and Child Resistant Packaging requirements depending on the resulting toxicity categories.

Applicants may satisfy these requirements in any one of several ways. Applicants may elect to cite someone else's data providing the referenced studies used a test material similar, from an acute toxicological perspective, to the proposed end-use product. Applicants may elect to "bridge" their acute toxicity data to another data set by supplying a reduced set demonstrating the extent of similarity or difference compared to the existing data set. Applicants may also "bridge" their data by establishing that each of its active and inert ingredients are within a range of ingredient combinations between two registered products having complete data sets showing identical acute toxicity profiles. A third way of "bridging" data involves the quantity of water used in a formulation. A proposed simple water dilution of an existing registered product can utilize data from the registered product for those studies in Toxicity Categories III and IV (least toxic categories) since the product labeling can not reflect any less acute hazard. Studies would have to be submitted to fulfill requirements not covered by Toxicity Category III and IV studies. Alternatively, a proposed product containing less water than a registered product may utilize data from any studies already in Toxicity Category I (most toxic category) since the product labeling can not reflect any higher level of hazard. Again, studies would have to be submitted to fulfill the remaining requirements. Beyond citing existing studies on similar products or "bridging" using a combination of existing data and new data, registrants may elect to generate all their own data or have a contract laboratory generate it.

For microbial/biochemical pesticide products, the components/constituents of the formulation, including both active and inert ingredients, can have a great influence on the acute toxicity data requirements. It must be understood that the term "inert ingredient" applies only to the fact that the ingredient is pesticidally inert, and does not mean that the ingredient is assumed to be biologically inert. It has been the Office of Pesticide Programs' experience that it is frequently the inert ingredients rather than the active ingredients which drive the acute toxicity profile for a product. This is the case for conventional chemical pesticides and especially for biologicals where the active ingredients are more frequently relatively innocuous.

Each of the six acute toxicity data requirements for labeling must be satisfied by citing existing studies, submitting new studies or using a combination of cited and submitted studies. Still another option is a request for one or more specific waiver(s) accompanied by a detailed scientific justification. Waivers are addressed in two documents, 40 CFR Part 158.45 and a recently completed Acute Toxicity Waiver Guidance Document. The latter is available by phone at (703) 308-8341. Biological end-use products whose inerts are naturally occurring materials such as water, cereals, vegetable oils, sugars, corn cobs, etc. or are formulated in such a way as to substantially reduce human exposure are more likely to be granted requested waivers than products using inerts such as petroleum distillates, alcohols or surfactants. The former are less likely to pose any significant health risks. It should be noted that the dermal sensitization study required for both chemical and biochemical end-use pesticides is not required for microbial end-use pesticides. Instead, the registrant must submit any reported incidents of hypersensitivity which occur.

## Examples of Formulation Based Reductions in Mammalian Toxicity Data Requirements

As stated earlier, certain risk scenarios can be mitigated by altering the exposure to the hazardous substance since risk is the product of hazard times exposure. Potential hazards from biological pesticides, like other pesticides, can often be addressed by lowering the exposure through pesticide formulation technology.

The efficacy for many pheromone biochemical pesticides is dependent on being able to deliver the volatile compound at certain rates over an extended period of time to confuse mating behavior. Therefore, the efficacy can be enhanced by dispensing formulations that slow the rate of delivery. These range from cigarette filters, hollow tubes and twist ties to flakes, chopped fibers and microscopic granules. Taking advantage of this development period in pheromone technology, the EPA is encouraging pheromone pesticide manufacturers to develop their pheromone products with dispensers that reduce the risk of food contact or incorporation by being large enough to be easily seen and retrievable. On December 8, 1993, the Agency proposed exempting arthropod pheromones from the requirement of a tolerance when used in solid matrix dispensers at a rate not to exceed 150 grams active ingredient/ acre /year in accordance with good agricultural practices (58 FR 64539). Further, the Agency established an exemption from the requirement of a tolerance for all inert ingredients of certain types of semiochemical dispensers under 40 CFR Part 180.1122 on December 8, 1993 (58 FR 64494). With use of these dispensers the manufacturers avoid having to address the risks associated with repeated, direct food exposure. By reducing pesticide exposure, pheromone applicants have often successfully obviated the need for teratology and subchronic feeding studies on their product due to the formulation and use rate chosen.

Microbial pesticides have also been formulated into products that may alter the toxicological risks through reduced exposure. Some fungal pathogens of insects are known to produce toxins of unknown mammalian toxicity. Broadcast spray formulations of such a fungal agent would require the manufacturer to directly test the product for the presence of these toxins considering the extensive exposure to the fungal agent especially in raw food. Use of these same agents in enclosed traps greatly reduces the likely exposure to any toxin potentially produced by the fungus. The lessened exposure can mitigate against the need to test the product directly for toxins, especially if previous acute toxicology testing showed no untoward effects.

## Product Performance

Product formulation can affect performance in a variety of ways. Microbial pesticides are uniquely formulated in comparison to their traditional chemical counterparts for a variety of reasons.

Most bacteria and viruses are activated in the gut of the host, and thus must be eaten to be effective. Obviously any changes in the product formulation which would stimulate ingestion may also improve product performance. Likewise, the active ingredients of a product may be quite capable of providing control, but will never have the opportunity if the formulation inhibits ingestion. Arthropod organoleptics; studying the gustatory stimulation and inhibition of various formulations, has become a critical element in the production of microbial pesticides intended for insect control.

A second area of formulation technology affecting performance involves protection of the microbial active ingredient from degradation. For viruses and bacteria, this most commonly involves protection from the ultraviolet spectrum which can shorten the active life of the product on the leaf surface. The addition of U.V. inhibitors to viral pesticides is now becoming a commonplace practice to increase the duration of effectiveness.

One area where formulation technologies are similar between microbial and traditional chemical pest control agents are the elements of sprayability, spreadability, and stickability. For any product to be successful, it must be able to be applied with the equipment typically available by growers. This is no mean feat considering the coarse raw intermediate product that comes out of the fermentor. If a product doesn't stay in suspension, it may never make it out of the nozzle. No microbial agent can control pests from the inside of a 500 gallon tank. Once out of the nozzle (hopefully undamaged), the microbial formulation must be able to provide the required coverage with reasonable volumes of diluent. Even coverage is essential, for uneven coverage may result in uneven control. Finally, the formulation must have staying power, to stick to the plant. After delivery, the microbial will be subjected to rain, wind and other factors. Again, even the best product cannot perform its' intended function if the product has been washed to the soil while the pests are still up chewing on the plant.

Acknowledgment

The authors thank William Schneider for kindly presenting this chapter at the 1994 American Chemical Society National Meeting.

References

Betz, F., Forsyth, S., and Stewart, W. "Registration Requirements and Safety Considerations for Microbial Pest Control Agents in North America" In L. Marshall, L. Lacey, and E. Davidson, ed. Safety of Microbial Insecticides (CRC Press, Boco Raton, FL, 1990).

Mendelsohn, M., Rispin, A., and Hutton, P. "Environmental Protection Agency Oversight of Microbial Pesticides" In A. Rosenfield and R. Mann, ed. Dispersal of Living Organisms into Aquatic Ecosystems (Maryland Sea Grant College, College Park, MD, 1992).

Matten, S., Schneider, W., Slutsky, B., and Milewski, E. "Biological Pesticides and the U.S. Environmental Protection Agency In L. Kim, ed. Advanced Engineered Pesticides (Dekker, NY. 1993.)

U.S. Congress, Federal Insecticide, Fungicide, and Rodenticide Act of 1972, 7 U.S.C. 136 et seq., as amended October 24, 1988.

U.S. Congress, Federal Food, Drug, and Cosmetic Act, and as amended. 21 U.S.C. 201 et. seq., 1991.

Title 40, Code of Federal Regulations, Parts 158, 172, and 180, 1991.

Mendelsohn, M., Delfosse, E., Grable, C., Kough, J., Bays, D., and Hutton, P. "Commercialization, Facilitation and Implementation of Biological Control Agents: A Government Perspective" in C. Wilson and M. Wisniewski ed. Biological Control of Postharvest Diseases - Theory and Practice (CRC Press, Boco Raton, FL, 1994.)

RECEIVED January 31, 1995

# Chapter 3

# A Generic Approach to Minimizing Impact on Nontarget Species in Canada

**R. E. Mickle**

**Atmospheric Environment Service, 4905 Dufferin Street, Downsview, Ontario M3H 5T4, Canada**

In 1989, the Canadian Interdepartmental Task Force on Pesticide Drift was established to develop a generic approach for the registration and regulation of pesticides in Canada. The generic approach utilizes the vast data base existing in the open literature to characterize the downwind deposit profiles for varying application and meteorological conditions. Presently, models are being evaluated against the data base and will be used to interpolate for conditions not found in the data base. The generic approach recognizes the similarities in spray dynamics for different formulations and hence only incorporates product specific toxic data in linking fate to environmental impact. In this paper, the approach is used to compare the case for the spraying of biologicals and chemicals for forest insect control.

As of 1992, it was estimated that 3 million tons of pesticides were in use worldwide and that this number was increasing steadily (*1*). Published results (*2*) indicate that off-target losses on the order of 50 - 70 % can occur during the aerial application of insecticides to forests owing to evaporation and drift. The environmental risk of these toxins continues to be a concern. However, woven within the fabric of sustainable forests and increased yields is the continued use of pesticides. In recent years, the use of broad-spectrum chemical insecticides to control forest pests has been sharply curtailed in large part due to their perceived negative impact on the environment (*3*). In most jurisdictions in Canada, these traditional chemicals are being replaced by biologicals. As new products are developed, there is an increasing need to evaluate them in a timely manner in a regulatory sense and ensure their use in an environmentally acceptable fashion. Buffer zones (sometimes referred to as 'no-spray' zones) have become an important regulatory tool to protect the environment during the application of pesticides. Each province in Canada has developed, over a period of time, an approach to the setting of buffer zones (*4*) which has recognized environmental concerns as they arose. This has led to a non-uniformity in the buffer zones utilized on a provincial basis and also buffer zones which do not necessarily address the uniqueness

0097–6156/95/0595–0027$12.00/0
Published 1995 American Chemical Society

of the product toxicity and the ecosystem sensitivities. Recently, the product specificity of buffer zones for glyphosate and permethrin have been addressed through extensive field trials (*5,6*). Although ultimately the field approach may be the most accurate estimate of deposit for the meteorological conditions during which the trials were conducted, years of research have afforded an extensive data base encompassing a variety of combinations of controlling input parameters. Based upon these data, models have been developed (*7-9*) and verified and now offer the opportunity to be used to develop meaningful buffer zones on a product and use-strategy basis. From the data base and/or models a generic approach can be utilized to screen new pesticides for potential environmental impact and regulate the use of existing pesticides through the implementation of buffers zones which reflect the relationship between the toxicity of individual products and identified areas requiring protection.

**Generic Approach**

In 1989, the Canadian Interdepartmental Task Force on Pesticide Drift was formed to develop new regulatory guidelines for assessing the potential for drift associated with a given use strategy in the registration of pesticides (*10*). In essence, the new guidelines would include a tool to be used for drift prediction centred on the use of models and/or a data base. The outcome of the prediction exercise would be an assessment of the expected environmental concentrations which, in association with the environmental toxicology data already provided by the applicant, could be used to determine the environmental significance of the drift and if necessary to estimate buffer zones as a mitigative measure. Essentially (Figure 1), rather than requiring supporting field drift data, submissions would be assessed for drift potential utilizing a generic approach. Initially, the product submission would be assessed for drift potential taking into consideration the product volatility and application method. If there was a potential for off-target movement, then the submission would be reviewed for potential impacts from direct overspray otherwise it would be reviewed according to current procedures. For direct overspray, the proposed application rate would be utilized to estimate the expected environmental concentration or deposit and coupled with the toxicological data, the potential impact on the appropriate indicator species in the relevant ecosystem would be evaluated. If the direct overspray indicated a potential impact then the assessment would be expanded to include the potential for environmental impact due to off-target drift. In order to simplify the task of drift behaviour assessment, a generic approach would be followed (Figure 2) utilizing formulation properties, application vehicle and use pattern. Fundamental in the assessment of the potential for drift is an indication by the applicant of a recommended emission spectrum for the product under review. For that portion of the emission spectrum with droplet diameters greater than 150 μm, research (*11-13*) has shown that the bulk of the spray will impact the target area with maximum deposits occurring at a given flight line offset due to cross winds. Droplets with diameters less than 150 μm (*14*) can potentially drift significant distances from the flight line and hence impact sensitive areas down wind of the spray block. The volatile fraction of the proposed end formulation coupled with the relative humidity at the time of application enhance the potential for off-target movement due to evapor-ation. Emission height and meteorology have also been found (*2, 15*) to strongly

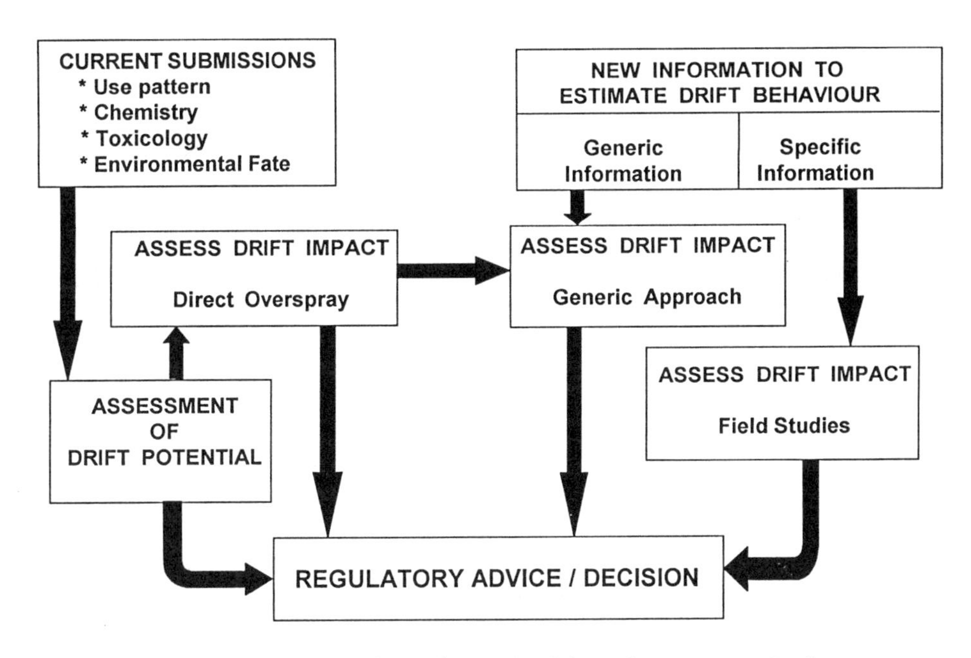


*Figure 1* *Proposed Regulatory Guidelines for Assessing Drift*

PESTICIDE IMPACT REVIEW

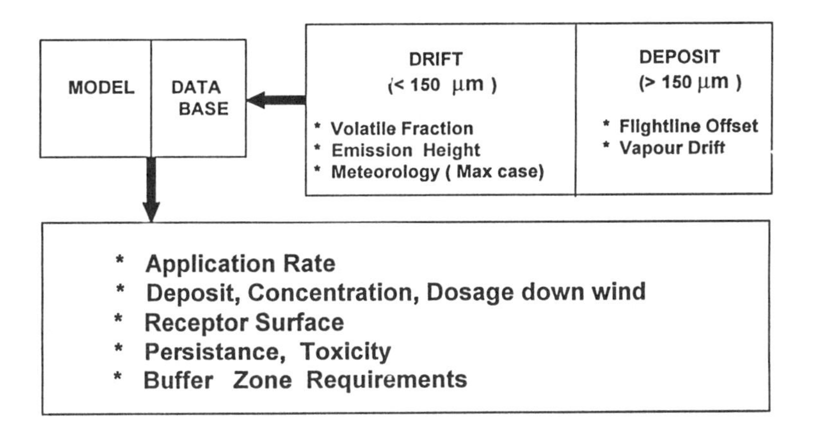


*Figure 2* *Generic Approach for Evaluating Pesticide Drift*

influence the ultimate deposit profile. Utilizing either models or a data base, the extent of drift for the proposed use strategy can be determined for an operational worse case scenario leading to a maximum off-target drift.

It is expected that the majority of cases will fit into the generic approach channel. If the data submitted indicate that the product under review will not lend itself to the generic approach (because the characteristics of the product do not match the categories established or the proposed use strategy is not applicable to the generic approach), field studies will have to be carried out and the results submitted by the applicant for review. After the extent of drift has been estimated by either the generic approach or field studies, mitigation options will be considered. The available options include buffer zones, restrictions on the type of application equipment, restrictions on the meteorological conditions during which the product can be applied or modification of the use pattern.

## Spray Drift Database

The initial step in developing the generic approach was the establishment of a data base (Riley, C.M., Research and Productivity Council, Fredericton, unpublished contract report) against which models could be assessed and which could be used to carry out sensitivity studies for those parameters most likely to influence off-target drift. Presently, the data base contains in excess of 200 data sets from nearly 50 references. The majority of the studies (nearly 200) include deposit and drift profiles from simulated aerial insecticide and herbicide spray applications relevant to a forest scenario. Spray deposit profile data are for natural and artificial targets representing the vegetation canopy and for ground deposit. Drift measurements include mass transport profiles to heights of 200-300 m above ground level at distances of 200 m and 600 m downwind of the spray line.

Figures 3a, b give examples from the data base (Data Ref: 1047/32_1992 and 1049/05_1987) for aerial applications with aircraft heights around 36 m ((a) = 35 m, (b) = 36.7 m) above ground (characteristic of aircraft heights over a mature forest) and spraying with equivalent application rates in cross-winds ( (a) = 2.6 $ms^{-1}$ at 53° to the flight line and (b) = 3.5 $ms^{-1}$ at 77° to the flight line). The two different nozzle types produce emission spectra with volume median diameters ($D_{v0.5}$) of 150 (a) and 59 (b) µm. The deposit has been plotted as a function of downwind distance from the spray line for foliar, ground and total deposit. The total deposit has been fitted with a least squares power law fit (*16*) to predict deposit beyond 800 m. A comparison of the two deposit profiles demonstrates that for the larger droplet emission spectrum (D8-46), the peak deposit occurs close to the spray line with deposit dropping an order of magnitude over the next 200 m. For the other case study (b), the maximum deposit, at a greater down wind distance from the spray line, is an order of magnitude lower and the decline in deposit with distance is much slower having decreased an order of magnitude in 800-1000 m. Beyond 100 m, the deposit from the AU4000 emission is considerably greater (up to 6 X) than the deposit from the spray using the D8-46 nozzles. Clearly with the smaller emission spectrum, the spray is spread over greater distances with less material depositing close to the spray line. For the larger droplets (D8-46), the deposit close to the spray line mainly reaches the ground with only about 20% being deposited on

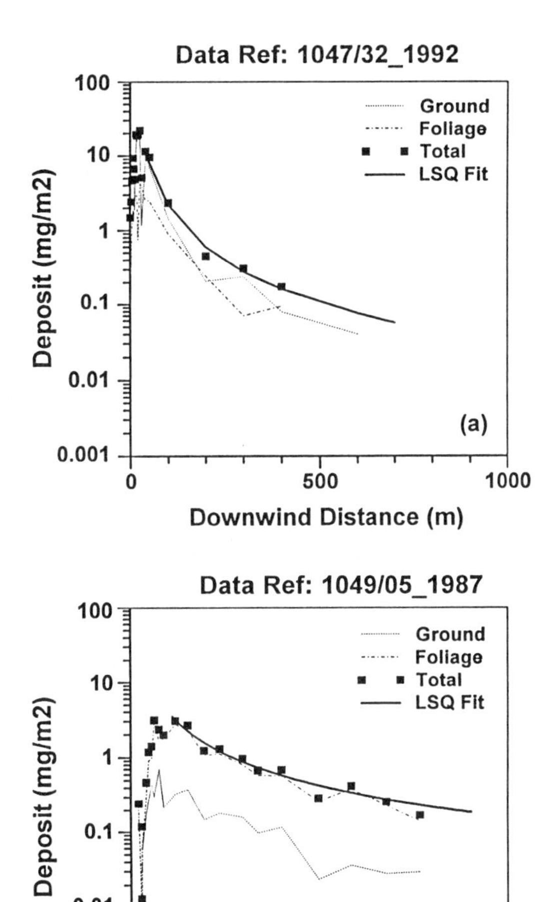


***Figure 3*** *Downwind Deposit for Aerial Applications with Different Emission Spectra (a) D8-46 (b) AU4000*

foliage. The distribution of the deposit reflects the density of the canopy at the spray site (in this case, the leaf area index (LAI), the ratio of foliage area to ground area, was 0.3). Beyond 200 m, the deposit is distributed evenly between foliage and ground deposit. For the smaller drop emission spectrum (AU4000), the bulk of the deposit is into the forest canopy (LAI = 3.2) with about 10% passing through the canopy to impact the forest floor. The partitioning of the deposit between canopy and ground is relatively constant over the full downwind distance. The partitioning of the deposit has been found to be a function of wind speed (Figure 4) at the canopy top. These data sets are taken from spray trials over a mature forest ($D_{v0.5}$=59, 68 µm, maximum tree height of 20 m, (LAI) of 3.2 $m^2$ of foliage / $m^2$ of ground (*2*)) and over a reforested block ($D_{v0.5}$=150 µm, maximum foliage height of 2.5 m, LAI of 0.3 $m^2$ of foliage/ $m^2$ of ground (Riley, C.M., Wiesner, C.J., Research and Productivity Council, Fredericton, unpublished data)). With increasing wind speed, the ratio of foliage to ground deposit increases significantly so that a greater fraction of the total deposit is deposited into the canopy. The effects of increased impaction efficiency are less significant with increased volume median diameter. However, with increasing distance from the spray line, a higher fraction of the remaining spray cloud will deposit on foliage due to the reduced $D_{v0.5}$ of the cloud.

Utilizing the data from a single swath, the effects of track spacing on deposit uniformity (Figure 5) within the spray block and the impact on off-target drift can be ascertained. While maintaining a constant line source strength and increasing track spacing, the average deposit (upper curve in Figure 5) across the block increases to approach the target application rate. However, with increased track spacing, the variation in the deposit (vertical bars) increases thereby leading to a less uniform deposit within the block. The ratio of the deposit variation to the mean deposit is the coefficient of variation (COV) which minimizes at an optimal track spacing and continues to increase beyond due to the increase in the variation of deposit. Coefficients of between 0.2 and 0.3 (*16*) have been suggested as appropriate for deposit variation. Using the optimum track spacing (as given by the track spacing that produces the highest minimum deposit), a composite deposit profile (Figure 6) can be developed from the single swath data set. In this case, the ratio of deposit to application rate has been plotted as a function of downwind distance from the first spray line. The boundary of the field commences on the upwind side of the field at the location of the peak deposit from the first spray line and ends at the position of the peak deposit from the last spray line. The deposit increases from a value of 40% of the target application rate at a distance of 40 m from the initial spray line and increases across the block to a maximum near 90% at a distance 230 m into the spray block. As the width of the block is increased, the average deposit would continue to approach the target application rate. However, operationally a typical track spacing is 21 m leading to a less uniform deposit (Figure 7) within the spray block (COV = .142). The minimum deposit approaches 65% of the target application rate, significantly lower than the 85% associated with the lower track spacing. Also with the increase in deposit variation, areas within the target receive deposits which are greater than the target application rate. Increased track spacings leading to COVs of 0.2 and 0.3 would increase the extent of both the over deposit and under deposit within the spray block. Beyond the downwind edge of the block, the off-target deposit falls off in accordance with the contributions from the spray

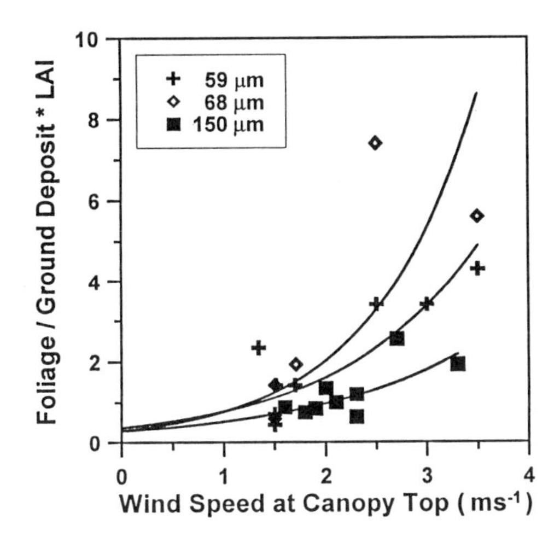


***Figure 4*** *Partitioning of Deposit within a Canopy*

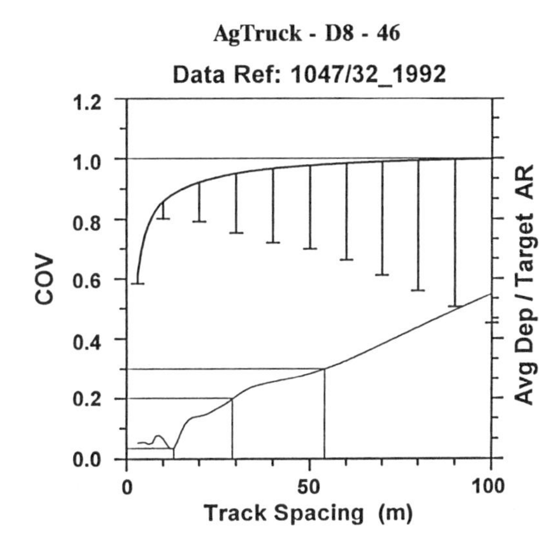


***Figure 5*** *Average Deposit and Variation across a Block*

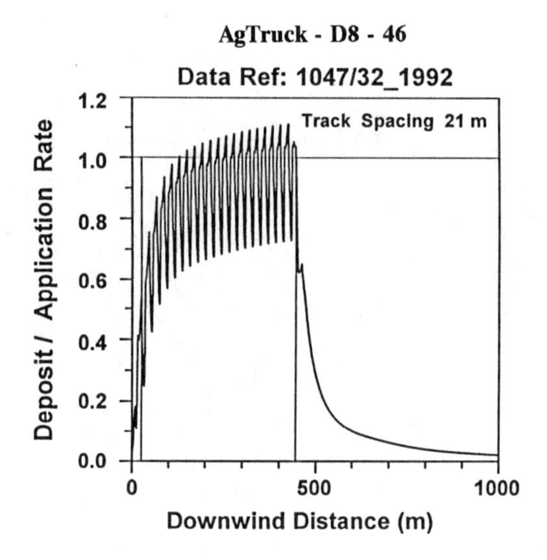


***Figure 6*** *Normalized Deposit for an Optimal Track Spacing*

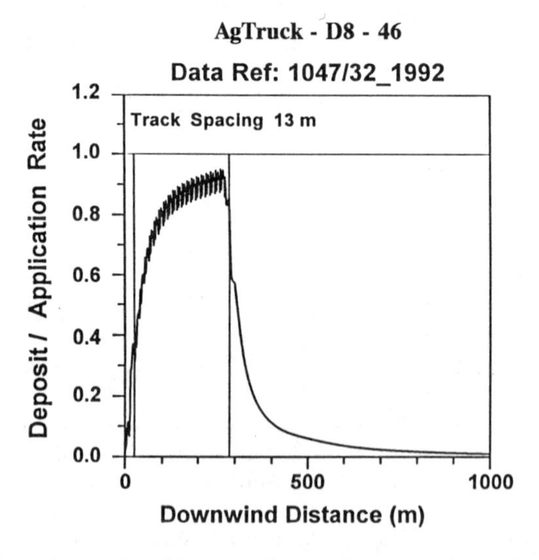


***Figure 7*** *Normalized Deposit for an Operational Track Spacing*

lines within the block. Varying the track spacing has little effect on the off-target deposit profile and hence does not impact the buffer zone requirements.

Figure 8 presents the estimated buffer zones as a function of spray block width for an aerial application scenario with D8-46 atomizers. The isopleths are for acceptable off target deposits as a percentage of the target application rate. For example, if an off-target deposit of 10% of the application rate is acceptable then the buffer zone requirements increase rapidly from 60 m to 220 m as the spray block width is increased to 1 km. For blocks greater than 1 km in width the buffer zone would remain relatively constant. The isopleths also indicate the strong relationship between buffer zone requirements and product toxicity. A product with a higher toxicity, reflected as a lower acceptable deposit, would result in a substantially increased buffer zone which is more strongly influenced by block width. Hence, the use of these products could possibly be limited to small blocks. Figure 9 presents a similar graph but for a use strategy involving an emission spectrum producing significantly smaller droplets. In comparison to the previous example, the buffer zones are considerably larger reflecting the effects of increased drift due to the reduced volume median diameter of the emitted spray. Now, tolerable off-target deposits that were 10% of the application rate would require buffers greater than 200 m and increasing rapidly with block width. With a block width of 1000 m, the buffer is over 900 m. Even at an acceptable off-target deposit of 50% of the application rate, buffer zones are 160 m. The results clearly demonstrate that the buffer zone approach strongly favours the use of environmentally safe pesticides in order to be able to effectively protect the crop of interest and minimize the potential impact on the environment. Operationally by decreasing the volume of fines in the emitted spray (increasing the volume median diameter), the requirements for buffer zone width can be reduced.

Ultimately, models will be used to interpolate for those cases not covered in the data base. Presently in Canada, the Interdepartmental Task Force is completing a verification study on FSCBG (*9*), AgDISP (*8*) and PKBW2 (*7*). Sensitivity studies (*17*) have highlighted the importance of aircraft height, emission spectrum and meteorology on deposit profiles. Accurate models can be used to develop a more thorough understanding of the application and hence can be used to optimize spray strategies for the product of choice given the environmental constraints. Ultimately, they will be used to assess the proposed use strategy for pesticide registration and development of buffer zones as a means of environmental protection. In reality, once the generic approach has been established, an assessment of the proposed use strategy could be encompassed in a limited number of case studies combining spray emission size categories with those operational parameters that would produce the maximum drift.

## Application of the Technique to Insecticide Spraying of Forests

Historically, buffers have developed in each jurisdiction based on data on pesticide toxicity, spray deposit and drift coupled with public opinion and political pressure. This has lead to a range of buffer zones (Figure 10) for the use of biologicals in Canada which in certain instances span orders of magnitude. Buffer zones around aquatic areas vary from no buffer requirements to buffers in excess of 3 km. There is, from a scientific perspective, little rationalization behind such differences unless the areas

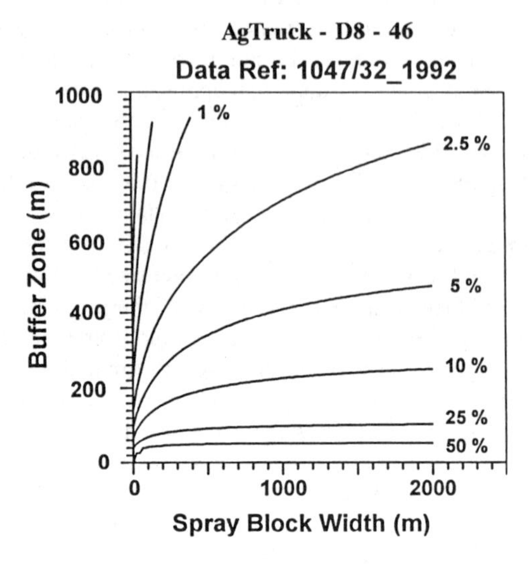


*Figure 8* *Proposed Buffers for Aerial Applications using Medium Drops*

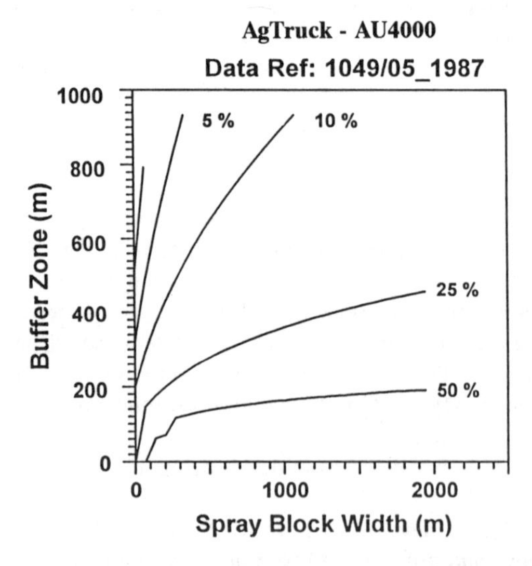


*Figure 9* *Proposed Buffers for Aerial Applications using Small Droplets*

BUFFER ZONES
BIOLOGICALS

| AREA OF CONCERN | BUFFER ZONE RANGE |
|---|---|
| HABITATION | |
| Residential | 600 m |
| WATER | |
| Potable | 50 - 300 m |
| Lakes | 0 - 50 m |
| Municipal Intake | $10^+$ - 3200 m |
| Fish Hatcheries | 0 - 50 m |
| Fish Bearing Waters | 0 - 50 m |
| SENSITIVE CROPS | |
| Berry Patches | 0 m |
| Organic Farms | Specific Buffers |
| ENDANGERED SPECIES | 0 - 500* m |

* Eagle Nesting

***Figure 10*** *Range of Buffers in Canada for the Aerial Application of Biologicals*

requiring protection, the pesticides being used or the off-target deposit from the application technique being employed differ greatly. Laboratory and outdoor stream studies (*18,19*) on the effect of Bacillus thuringensis var. kurstaki (Btk) on representative aquatic insects ( Ephemeroptera, Plecoptera, Trichoptera) showed a 24-h LC50 greater than 600 IU/ml. The forestry application for the control of spruce budworm in Canada is 30 BIU/ ha. Complete deposit of a direct overspray represents a target deposit of 300 IU/ $cm^2$ which when mixed through a water column of 10 - 50 cm leads to a maximum concentration of 30 IU/ ml some twenty times lower than the LC50 for these aquatic insects. The generic approach has indicated a potential over-application up to 25% under operational spray conditions due to track space variations. This can be further increased by flight lines that may overlap causing multiple applications into the same receptor area. However, even this type of variability is still not sufficient to produce a potential concentration close to the LC50 for these aquatic insects. An environmental review of Btk for use in forest pest management concluded that "laboratory and field studies indicate that Btk is specific to Lepidoptera larvae and does not pose a threat to humans and the environment" (Surgeoner, G.A., Farkas, M.J., University of Guelph, unpublished report). A second insecticide used within Canada for budworm control is fenitrothion which is applied at an application rate of 210 g a.i./ha. The median static 24-h LC50 value for aquatic insects (Trichoptera, Plecoptera, Neuroptera, Diptera, Odonata) was found to be 68 µg/L (*20*) or approximately 30% of the target application rate mixed through a water column of 10 cm. In some jurisdictions within Canada, the trend in forestry insecticide spraying is towards the utilization of nozzles producing a fine spray ($D_{v0.5}$ < 100 µm) in order to maximize deposit into the target foliage (*21*). Figure 9 can be used to characterize the buffer zone requirements for such a scenario. Tolerable off-target deposits that were 30% of the target application rate would require buffer zones of 160 m for small blocks with buffer zones increasing to 400 m for spray blocks up to 2000 m wide. By comparison, utilizing a somewhat larger spectrum (Figure 8) would result in buffer zones between 40 and 80m for varying block sizes. The appropriate receptor representing the specific environmental sensitivity within an area requiring protection and an acceptable off-target deposit are critical in developing actual buffer zones. The level of acceptable deposit will be strongly influenced by the level of protection required, whether it be half the population (LC50) or the whole population (No Observable Effect Level - NOEL). However, the generic approach does highlight the need to address product specific toxicity information in developing mitigative options in the registration and regulation process.

**Summary**

The Canadian Interdepartmental Task Force on Pesticide Drift is completing the assessment of a generic approach for the registration and regulation of pesticides in Canada. Drawing on extensive field trials, the approach has been used to highlight the influence of emission spectra on the resultant buffer zone requirements to protect sensitive environmental areas. For spray strategies employing small droplet emission spectra, the cross-wind width of the spray block strongly influences the size of the associated buffer zone. In practice, with the categorization of the proposed emission spectrum, the generic approach collapses to a select number of cases relating the proposed strategy with the operational conditions which promote maximum drift. The case example between the use of a biological and chemical in forestry insect control highlights the need to regulate the use of pesticides on a product specific basis. Given the specificity of existing field trials, the use of models will be required to interpolate for conditions relevant to assessing the maximum drift potential. Ultimately, the precise buffer requirements will be reflected in the choice of representative receptors and the level of protection to be afforded.

## Literature Cited


1. Tardiff,R.G. Methods to Address Adverse Effects of Pesticides on Non-Target Organisms; John Wiley and Sons Ltd.: Chichester, West Sussex, 1992.
2. Picot, J.J.C.; Kristmanson, D.D.; Mickle, R.E.; Dickison, R.B.B.; Riley, C.M.;Wiesner, C.J. Trans of the ASAE, 1994, 36, No 4: 1013-1024.
3. Ennis, T.; Caldwell, E.T.N. Tortricid Pests, Their Biology, Natural Enemies and Control; van der Geest, L.P.S. ; Evenhuis, H.H., Ed.; Elsevier Science Publishers, Amsterdam, pp. 621-641.
4. Buffer Zones: Their Application to Forest Insect Control Operations; Trial, JG ISBN: 0-662-15732-X; Minister of Supply and Services Canada, Ottawa, 1987.
5. Payne, N.J. Pestic. Sci., 1992, 34: 1-8.
6. Payne, N.J.; Helson, B.V.; Sundaram, K.M.S.; Fleming, R.A. Pestic. Sci., 1988, 24: 147-161.
7. Picot, J.J.C.; Kristmanson, D.D.; Basak-Brown, N. Proc. of ACAFA Symposium on Aerial Application in Forestry, 1987: 161-170.
8. Bilanin, A.J.; Teske, M.E.; Barry, J.W.; Ekblad, R.B. Trans of the A.S.A.E., 1989, 32: 327-334.
9. Teske, M.E.; Bowers, J.F.; Rafferty, J.E.; Barry, J.W. Environmental Toxicology and Chemistry, 1993, 12: 453-464.
10. Fortin, C.; Baril, A.; Chang, F.Y.; Constable, M.; Crabbe, R.; Ernst W., Maloney, P.; Mickle, R.E.; Payne N.; Samis, S.; Wan M. Proc ACAFA Symposium and Trade Show, 1990: 46-51.
11. Anderson, D.E.; Miller, D.R.; Wang, Y. J Applied Meteorology, 1992, 31: 1457-1466.
12. Picot, J.J.C.; Kristmanson, D.D.; Basak-Brown, N. Trans of the A.S.A.E., 1986, 29: 90-96.
13. Mickle, R.E.; Hoff, R. Proc ACAFA Symposium and Trade Show, 1990: 169-171.
14. Akesson, N.B.; Hitch, R.K.; Jones, A.W. NAAA/ASAE Joint Tech. Session, 1992, AA92-001, 30 pp.
15. Teske, M.E.; Barry, J.W.; Thistle Jr, H.W. In Environmental Modelling; Zannetti, P., Ed.; Computational Mechanics Publications, Boston, 1994, Vol II; pp 11-42.
16. Parkin, C.S.; Wyatt, J.C. Crop Protection, 1982, 1: 309-321.
17. Teske, M.E.; Barry, J.W. ASAE Summer Meeting, 1992, 921086: 8 pp.
18. Eidt, D.C. Can Entomology, 1985, 117: 829-837.

19. Kreutzweiser, D.P.; Holmes, S.B.; Capell, S.S.; Eichenberg, D.C. Bull Environ Contam Toxicol, 1992, 49: 252-258.
20. Symons, P.E.K. Residue Rev, 1977, 38: 1-36.
21. Irving, H.J.; Kettela, E.G.; Picot, J.J.C. Proc ACAFA Symposium and Trade Show, 1990: 133-139.

RECEIVED January 14, 1995

# Basic Information Needs

## Chapter 4

# Ecological Factors Critical to the Exploitation of Entomopathogens in Pest Control

**James R. Fuxa**

**Department of Entomology, Louisiana Agricultural Experiment Station, Louisiana State University Agricultural Center, Baton Rouge, LA 70803**

The use of "biorational" agents such as entomopathogens for pest control is based in ecology. Aspects of environmental release such as timing and placement must be based on ecological considerations for entomopathogens to be efficacious. Timing and placement depend on entomopathogen species characteristics (life cycle, host specificity, portal of entry, site of attack, searching ability, virulence, speed of action, reproductive capacity, transmission) and population characteristics (population density, distribution, spread, persistence). The target pest also has species (pest category, r- or K-selection, number of generations, behavior) as well as population characteristics (population density, distribution, age structure, quality) critical to agent application. Important ecosystem characteristics include the habitat and its stability, the crop or resource and its value, agricultural practices, the pest complex, abiotic and biotic environmental variables, and environmental risks. Ecological considerations for application of biorational agents can be complex, but not to a degree that should prevent their implementation.

Certain "biorational" agents might be useful in pest management. These biorational agents are microorganisms or chemicals found in nature, or chemicals synthesized by man to mimic natural chemicals. Microorganisms include viruses, bacteria, protozoa, fungi, nematodes, and some of their by-products (toxins). Biochemicals include

0097–6156/95/0595–0042$13.50/0

semiochemicals, hormones, natural plant regulators, natural insect growth regulators, and enzymes. The major rationale for development of such agents for pest control is their environmental safety. For example, the insect pathogens are host-specific to the degree that they have caused virtually no environmental harm upon release (1). On the other hand, these "biorational" agents, though having excellent potential for suppressing pest populations, can be inherently difficult to use.

One reason agents such as the entomopathogens can be difficult to use is due to problems in delivery of the agent in a timely manner and suitable location such that the agent is in a position to exert its pest-suppressive action. For example, it has been estimated that a virus preparation sprayed for insect control can lose 70% of its activity before it even impacts the target foliage (2). The requirements for such delivery are based heavily in ecology. Pathogens, like other living organisms, must fit into an ecological niche if they are to survive and function. Additionally, the concept of pest control, integrated pest management, is based in ecology.

The purpose of this paper is to outline the ecological considerations -- the pathogen population, pest population, and ecosystem or environmental factors -- that might affect the use of entomopathogens in pest management, with emphasis on delivery, or timing and placement of release.

**Basic Ecology**

In order to discuss ecology relative to pest management with entomopathogens, it is necessary to establish certain definitions and concepts. "Ecology" is difficult to define to everyone's satisfaction. A definition proposed by Andrewartha (see 3) is suitable: "ecology is the scientific study of the distribution and abundance of organisms." Regardless of the exact definition, an important concept in ecology is that every individual in a population of a given species is part of the environment of other individuals of that species (4). The individual organism is a basic unit for study (5). A group of individuals comprises a population, defined by Mayr as "the group of potentially interbreeding individuals at a given locality," which results in a situation whereby all members of a local population share in a single gene pool (5). Another concept that has been adopted by agricultural scientists in general and biological control specialists in particular is that of the "ecosystem" (6). The ecosystem is comprised of the community (a level of organization higher than the population, consisting of coexisting interdependent populations) and its physical environment (5). An organism's environment often is separated into biotic and abiotic, or physical, factors (e.g., 7), though environmental factors can be divided in other ways (4). The linkages among populations and other factors in an ecosystem are complex and lead to fluctuations in population densities.

Epizootiology of entomopathogens and pest management, including microbial control of insects, are heavily based in ecological principles. Epizootiology can be defined as the science of causes and forms of the mass phenomena of disease at all levels of intensity in a host population (8). In other words, it is the study of animal disease at the population level. Epizootiology encompasses the total environment including the host and pathogen populations and thus is heavily allied with ecology. If entomopathogens and other "biorational" agents are to be used for pest population suppression, they must be integrated into pest management. Pest management incorporates a wide variety of approaches aimed at maintaining pest populations below economic injury thresholds, or the numbers of insects that will cause damage equal to the cost of artificial control. Thus, a prime purpose of microbial control is to increase disease levels in insect pest populations. The time and place where these two populations (pest and pathogen) interact, and subsequent population dynamics, depend heavily on their ecology.

The approach by which an entomopathogen is used to suppress a pest population is important to the manner in which the pathogen is utilized. There have been three such approaches in which entomopathogens are artificially produced and released (9). In the microbial insecticide approach, relatively large numbers of pathogen units are released for quick suppression of the pest population. Residual effects are not significant, and subsequent increases in the pest population to damaging levels require additional releases of the pathogen. The seasonal colonization approach amounts to a "booster shot;" the release results in replication of the pathogen and suppression of more than one pest generation. This may or may not be aimed at immediate knockdown of the pest population, and subsequent releases are required, usually in each new growing season. Introduction-establishment results in permanent pest population suppression; the entomopathogen species or strain becomes a permanent part of the ecosystem in which it is released.

Ecological considerations for the application of entomopathogens in pest management include pathogen characteristics, pest characteristics, and the ecosystem.

**Pathogen Attributes**

Different entomopathogens have characteristics that affect every phase of developing and utilizing them for microbial control. Some characteristics, such as their capability to be produced inexpensively in large quantities for release, relate only indirectly to the application and efficacy of pathogens in the ecosystem and will not be discussed in this paper. The "ecological" factors that relate to application fall into two major groups: characteristics of the species or strain and characteristics of the population.

**Pathogen Species or Strain.** Certain characteristics of an entomopathogen that relate to ecology of application are inherent to the pathogen species or strain and thus function on the level of an individual pathogen unit's attack on the insect host. Most characteristics of pathogen species or strains (subspecies) affect timing of application, though some affect placement.

Perhaps the most obvious timing consideration for an entomopathogen is that its life cycle be temporally synchronized with that of its host insect in a way suitable for pest management. This is potentially a greater problem with respect to the persistent approaches to microbial control -- seasonal colonization and, especially, introduction-establishment. The pathogen might be unable to function, for example, if it is quiescent while the pest is active. There are examples of poor synchrony in attempted biological control with parasitoids (10). One example in natural pest population supression by an entomopathogen is that of the fungus *Nomuraea rileyi* in caterpillar pests of soybean in the southeastern United States. This fungus is consistently a major mortality agent in this soybean system, but the high mortality usually occurs too late in the season to prevent crop damage (11). This example represents a lack of synchrony of the pathogen with respect to pest management rather than with respect to the host life cycle, but it presents an opportunity for fungal application to correct for the asynchrony.

Host specificity has a subtle effect on application. All five major pathogen groups have at least a few species that are relatively host specific, though such specificity is perhaps most common among the viruses and protozoa. Host specificity limited to pest insects is, of course, a major advantage for pest control due to environmental safety. However, a high degree of specificity is a severe problem in commercial development, because narrow host ranges restrict market size as well as usefulness in pest complexes (12). If an entomopathogen has a wide host range, then "secondary" hosts can support viral replication and, therefore, its persistence in the environment. The availability of a greater number of hosts, due to the greater number of susceptible species, simply increases the inoculum that can result. There is circumstantial evidence that this can occur with certain viruses with relatively wide host ranges (3). In such a case, long-term control, for example through seasonal colonization, might require less frequent and less extensive applications of the pathogen.

Another inherent characteristic of entomopathogens is the "portal of entry" into the host insect. Bacteria, viruses, and protozoa virtually always must be ingested when they are artificially applied for insect control. Nematodes and fungi invade primarily through the external integument or through body openings other than the mouth, though many can invade after being ingested and a few must be ingested in order to infect the insect. A third portal has been hypothesized to be important in the longer term approaches to control (13-14); this is by vertical, or parent-to-offspring, transmission, which is thought to occur in many of the viruses and protozoa. A critical problem to any organism with an internally parasitic lifestyle is invasion of the host or, in other words, having the opportunity to contact a host and the capability to surmount that host's external barriers (e.g., the integument) to invasion.

Portal of entry is critical to timing and placement of application for microbial control, and it interacts with the lifestyle and feeding habits of the insect host as well as certain environmental factors (discussed below). For a pathogen that must be ingested, application must be synchronized to the insect's feeding so that the pathogen will be ingested before the insect is too old to be infected and before the pathogen can be inactivated by environmental factors such as sunlight. For example, timing of application is important to the efficacy of the bacterium *Bacillus thuringiensis* (15-16). Timing and placement are critical for an insect that feeds in protected locations, such as a fruit feeder like *Heliothis* spp., because the pathogen must be ingested while the insect feeds on the surface of the fruit. After it burrows, delivery of the pathogen so that it can be ingested becomes virtually impossible. Synchrony with the host's habits is not as critical for pathogens with an external portal of entry, though other factors can still make timing important. Entry through the integument also affects the host insects that can be targeted for control; plant-sucking insects are virtually immune to control attempts by pathogens that must be ingested. It has been hypothesized that a vertical transmission portal of entry lowers the threshold of host population density for successful introduction-establishment (17-18); this lower threshold in turn can allow greater flexibility in timing and placement of release.

In addition to portal of entry, the site of invasion by pathogens can be limited to certain stages, ages, and tissues of the host insect, which in turn can affect timing and placement of application. The vast majority of entomopathogens in nature attack only one stage, usually the feeding stage of the host (e.g., larvae), and not others (e.g., eggs or adults). In fact, most entomopathogens are even more limited; insect larvae usually become more difficult to infect as they age (i.e., "maturation immunity") (3). Greater rates of feeding compensate to some degree for this reduced susceptibility (3), but not enough to completely counteract it. This again affects timing of application. For best results, the pathogen must be applied to maximize contact between it and the most susceptible ages of the target insect. The importance of timing due to maturation immunity has been best demonstrated for *B. thuringiensis* (e.g., 15-16). Tissues of the host insect that are attacked by the pathogen can affect application. If midgut is a site of infection and large quantities of the pathogen are voided through the gut, this can contribute to widespread distribution of the pathogen. For example, certain viruses can be released at very limited locations and then spread throughout a pest insect's geographical range by this mechanism (3).

A few pathogens, particularly certain nematodes and fungi, have limited searching ability. Nematodes such as *Steinernema carpocapsae* and *Romanomermis culicivorax*, and aquatic fungi such as *Lagenidium giganteum* can move short distances and actively contact the host insect prior to the infection process. This capability, particularly with nematodes targeted against certain terrestrial insects, can be an important factor in placement of a field application. Most entomopathogens must be timed and placed accurately in order to infect an insect preparing to burrow into a plant part or soil. Once an insect has burrowed, it can be virtually impossible to deliver most pathogens to a suitable contact point with the host, particularly if the pathogen must be ingested. If a pathogen such as a nematode can search even a distance of 2-3 cm., it only has to be delivered into the general vicinity of the host insect for infection to take place. For example, heterorhabditid nematodes have a tendency to move downward in soil, giving them good potential for control of such "cryptic" insects as Japanese beetles, billbugs, and root weevils (19).

Virulence is the disease producing power of a pathogen, a characteristic often associated with strains within a pathogenic species. Virulence is sometimes measured in terms of time required to kill a host but is more often measured as number of pathogen units required to kill a certain portion of a group of host insects. Virulence can affect application in two ways. First, a more virulent pathogen requires fewer pathogen units at the point of contact with a host insect in the field than a less virulent pathogen. This in turn affects placement of application. Thus, increased virulence is often an objective of recombinant-DNA research (9). Second, virulence requirements can change with the approach to microbial control; it has been hypothesized that virulent strains may not be ideal for the longer-term approaches, particularly introduction-establishment (20). For example, highly virulent strains of the bacterium *Bacillus popilliae* do not establish and control Japanese beetles as well as strains with moderate virulence (16). Thus, virulence can affect placement indirectly, since application requirements are less critical for relatively permanent control by introduction-establishment.

One of the simplest way to categorize entomopathogens, and one that can affect timing of application, is by speed of action. In this respect, entomopathogens fall into two categories: "quick damage" and "slow" pathogens. The quick pathogens stop insect feeding within 24 h, though death may take several days. These mostly include those that produce toxins (e.g., *B. thuringiensis*) or those that initiate a bacterial septicemia (e.g., *S. carpocapsae*). The slow pathogens debilitate their hosts after 3-4 d or more by more typical "parasitic" action. These comprise the great majority of natural strains of entomopathogens, including virtually all the viruses, fungi, and protozoa, and many of the nematodes and bacteria. A major emphasis of recombinant-DNA research of entomopathogens is to improve this slow action, because, until the insect is debilitated by a slow pathogen, it continues to feed (cause damage) and be observed by users who expect quick action. For example, slow action has been cited as a reason for failure of certain pathogens in the pesticide market (21-23). In certain cases, timing of application can alleviate this problem. The nuclear polyhedrosis virus (NPV) of *Anticarsia gemmatalis*, a soybean pest, was significantly more effective in reducing pest numbers and damage when it was sprayed several days before pest numbers reached thresholds developed for application of chemical pesticides (24).

One of the basic characteristics of a species from an ecological viewpoint is its innate capacity for increase, which relates partly to its reproductive rate. The "parasitic" lifestyle of many entomopathogens, due to its risky nature, particularly in host-to-host transfer, often results in a high reproductive rate. For example, as many as three generations of an entomopathogenic virus may be produced in the field within one generation of the host insect (25), with numbers of viral polyhedral inclusion bodies commonly exceeding $10^9$ per host insect (26). High rates of pathogen replication can affect application in the long-term approaches to control; the extensive environmental contamination that can result from high pathogen reproduction or replication rates can reduce frequency and amount of pathogen application.

Transmission has been called "one of the key ecological factors that must be understood before entomopathogens can be manipulated" (25). In nature, host-to-host transfer can be a "weak link" in the life cycle of these pathogens, providing a natural point to enhance their activity through timely and accurate delivery. In the microbial insecticide approach to control, transmission is replaced by artificial application. Thus, it is critical that pathogen timing and placement account for the ecology of natural transmission.

Most natural strains of entomopathogens are used for seasonal colonization or introduction-establishment. For these approaches, natural transmission is critical (9, 25, 27-28). Most insect pathogens have some stage specialized for survival outside the host so that they can transfer to a new host. These pathogens are transported by a variety of abiotic and biotic factors (29).

Transmission can affect placement of a pathogen for microbial control. The best examples come from a technique called "autodispersal," whereby the pathogen is applied over a limited area and is then dispersed by infected or contaminated insects. An example is the dispersal of a baculovirus for long-term control of palm rhinoceros beetle, *Oryctes rhinoceros* (30-31). Vertical, or parent-to-offspring, transmission, found in many protozoa and viruses, might provide a method for pathogen application and dispersal for microbial control (28). For these approaches to control, transmission might lower the threshold of host population density for successful introduction and lower the minimum introduction rate of the pathogen (the amount that must be delivered to the point of contact with the insect) (17-18, 32). Vertical transmission is perhaps the most efficient route of transfer of a pathogen between hosts (3).

**Pathogen Population Characteristics.** The population is a basic unit of ecology of a species, and the pathogen population is a unit basic to application for microbial control. Basic population characteristics include density, distribution, and spread. Persistence is basically a characteristic of the taxon but usually is measured as a population parameter.

Population density is one of the most basic characteristics in ecology, and pathogen population density is one of the most important factors in disease epizootics. Due to the interest in microbial control and epizootiology, there are numerous examples of dose-related response by insects to entomopathogens both in the field and laboratory (e.g., 33-37). However, there is virtually no information about actual dosages delivered to insects in the field (38), though there theoretically are thresholds for pathogen population density to develop epizootics (39). It is intuitively obvious that, all other factors being equal, a greater pathogen population density simply increases the chance of contact between a pathogen and an uninfected host. Thus, the pathogen population density delivered at the appropriate time and place through microbial control technology will impact heavily on success of the control effort.

The distribution of an organism is another basic ecological parameter. The effect of pathogen distribution in microbial control is not clear. Pathogens most likely have clumped distributions in nature (40); this may explain how insects encounter high doses in nature and why poor infection rates often result when a pathogen is sprayed evenly into an ecosystem. Harper (25) pointed out that much of a spray application is wasted. Of course, this is also the case in nature; parasitism is a risky life style, and most pathogen transmissive units never encounter a host, which is a major reason that they replicate in such large numbers. This implies that pathogen clumping might benefit microbial control (3). Degree of clumping, as determined by spray droplet size, has affected control experiments (37). On the other hand, "thorough" coverage, presumably meaning even distribution, has been called an important goal for application technology (16). Thus, there is a good possibility that distribution is a critical parameter for placement of microbials, yet it also is clear that much research is necessary in this area.

Spread of a pathogen population is closely related to the intrinsic capability of that pathogen to be transmitted. Fuxa (29) reviewed the abiotic and biotic environmental agents that transport pathogens, as well as the effect of biotechnology on transport. However, dispersal and transport of pathogens after environmental release is only poorly understood (29, 41). Yet capability for spread clearly can affect placement of application. For example, the NPV of *A. gemmatalis* spreads at a rate of approximately 1 m per day after its release into soybean. It has been estimated that this virus can be sprayed at intervals of 22 m and 100 m to provide satisfactory insect control through seasonal colonization and introduction-establishment, respectively (42).

Environmental persistence is a characteristic intrinsic to pathogen species and strains, but it usually is measured as a population parameter. Persistence is one of the few factors critical to all the approaches to microbial control (20). The pathogen must persist at the point of contact with the insect long enough to be encountered; for the long-term approaches to control, persistence somewhere in the habitat is an obvious prerequisite. Entomopathogens generally persist only short time periods, measured in terms of one or a few days, on exposed surfaces (41); sunlight quickly kills all pathogens, and moisture is critical to nematodes and fungi. It is intuitively obvious, therefore, that timing and persistence are critically interrelated in entomopathogen application; the release must be timed so that the insect encounters the pathogen before it is inactivated. For the longer-term approaches, persistence through recycling reduces the number of times that an entomopathogen must be released.

**Pest Attributes**

The pest insect is a component of the environment, one as critical as any to the timing and application of entomopathogens. A pest can be defined simply as "an organism detrimental to man" (43). No organism is intrinsically a pest, but it becomes one when its lifestyle somehow conflicts with man. Pest attributes, like those of a pathogen, can be divided into species (or strain) and population characteristics.

**Pest Species or Strain.** Pests can be divided into categories which can impact timing for microbial control. Key pests are those organisms that appear yearly at such high levels that control measures are necessary if economic losses are to be avoided; populations of occasional pests grow to damaging levels on an occasional basis, when natural regulating factors do not keep the population restrained (44). These categories influence application of microbial control agents in much the same way as other control agents, such as chemicals. Application for key pests can almost be timed on a scheduled basis, as, for example, with *Heliothis* spp. in cotton, though pest scouting is still required. Sampling for pests and timing control measures in response to population increases is usually the best procedure for occasional pests.

Another way to conceptualize pests populations is according to the theory of r-K selection. This theory holds that there is a continuum of species based on their life histories. At one end are the r-selected species, those that take advantage of temporary habitats and usually characterized by swift development, early breeding, high reproductive rate, relatively small size, and polyphagy. K-selected species specialize in fully exploiting special niches in stable habitats, and they are characterized by low reproductive rates, good competitive abilities, low numbers of high-quality progeny, large size, effective defenses against natural enemies, and long life-spans (45-46). Several authors have suggested that the r-K continuum has a bearing on microbial control (25, 46-48). Anderson (32) suggested that applications of a microbial once every few years could suffice for control of certain forest pests (K-selected), whereas frequent applications generally are necessary in row crops (r-selected pests). Entwistle (46) proposed that the r-K continuum should be considered in application strategies for viruses in microbial control. For r-selected pests, autodissemination and early introductions (i.e., seasonal colonization) are possible approaches; lattice introductions (spot application at intervals to lower costs, depending on viral spread to evenly distribute itself) and single sprays are not likely to be useful; and multiple applications are often necessary. For K-selected pests, early introductions are unlikely to work; autodissemination, lattice introductions, and single sprays often are valuable; and multiple sprays are sometimes essential.

Another pest species characteristic is its number of generations. Univoltine insects produce one generation per year, and multivoltine produce more than one. Multivoltine pests may require either persistent agents or multiple applications. For example, briquets dispensing *B. thuringiensis* over a prolonged time period were developed for mosquitoes with a continuous succession of generations (15). Multiple applications of pathogens, or pathogens that can recycle and control pests for the remainder of a growing season, would be necessary for other multivoltine pests, whereas insects with feeding stages present only a few days, such as western corn rootworm (*Diabrotica virgifera virgifera*), present only a narrow window of opportunity for timing an application (19).

Behavior is, arguably, the most important host species characteristic with respect to timing and placement of entomopathogen application. Behavior can be age-related. For example, young larvae of many insects, such as *Pieris rapae* and *Trichoplusia ni*, have a very slow feeding rate (and, therefore, feed over a smaller area) compared with the older larvae (3). Thus, placement will be critical for these insects. Some insects are gregarious; there are several examples of disease spreading through a colony of gregarious caterpillars or sawfly larvae if a relatively few can be infected (3). Insects that feed on plant surfaces usually are comparatively easy to target for pathogen application, whereas timing as well as placement is much more difficult for an insect that burrows into a fruit or some other structure (49). For example, *B. thuringiensis* products have had relatively little success for control of fruit feeders in cotton, corn, and apple, but have provided good control of leaf or surface feeders in cabbage, avocado, stored products, and forest situations (50). Plant surface feeding can be broken down another step; certain insects, such as *Agrotis ipsilon*, feed on hypogeal parts of the plant and on leaves close to the ground, where spray penetration can be poor (50). Sucking insects limit the pathogens available for possible control to those that can invade through the external integument; however, such "contact" action can somewhat simplify placement of application (49). Insects that inhabit soil obviously can cause delivery problems if they are not at or near the surface, though this disadvantage is perhaps counteracted to some degree by the fact that many entomopathogens survive well and cause epizootics in soil (49). Aquatic insects such as mosquito larvae pose special problems in placement of a pathogen because these larvae usually frequent and feed in certain specific depths of water, ranging from the surface to the bottom substrate (15, 49). On the other hand, water is a medium conducive to the activity of certain pathogens, certain fungi and nematodes, that can search for a host (pest) insect, lessening the importance of delivering the pathogen to the exact location of the insect. Social insects have a variety of behaviors that can make delivery difficult, particularly if one or more queen insects must be killed. Such behaviors were recently reviewed for ants (51). Pest behavior is potentially more important to placement and timing of microbials than conventional pesticides, many of which work by contact and possibly fumigant action in addition to ingestion (52).

**Pest Population Characteristics.** Like the pathogen, the pest has population as well as taxon characteristics that affect microbial control. Insect control with chemical pesticides concentrates on pest population characteristics, particularly with respect to timing of application, so it is not surprising that this also is a consideration for microbials.

Population density is a key factor in defining a pest; many species of insects compete with man for various resources, but, in most cases, only the ones that become sufficiently numerous to cause economically significant damage are considered pests. It follows that density also is a key factor in determining the timing of application of a control agent. Insect pest management is based heavily on such timing in the framework of economic injury levels. An economic injury level is the pest population density at which economic losses begin to surpass the costs of control. The economic threshold, also known as the action threshold, is the pest population density at which a certain control action must be taken to prevent the pest population from reaching the economic injury level. If action thresholds are adhered to by growers or resource managers, then pesticides are applied only when needed, thus reducing their harmful side-effects.

The problem for microbials is that action thresholds are dynamic and depend on a variety of factors, including the activity and speed of the control agent. The decision-making process for application of microbials, including their timing of release, is heavily dependent on a large body of research of the relatively quick-acting chemical pesticides. The action thresholds developed for chemical pesticides will not always work for microbials. This is particularly true of the slow pathogens. For example, *A. gemmatalis* NPV sprayed a few days before its host population reaches the action threshold for chemical insecticides was significantly more effective in reducing pest numbers and damage to soybean than virus sprayed at the action threshold (24). For direct pests (those that directly damage the final product, such as a fruit), the timing required to apply a slow pathogen in relation to an action threshold may be so precise as to be impractical, because the economic injury level is very low. Another problem with action thresholds, even for chemical insecticides, is that they are difficult to determine for insects that vector diseases or are simply annoying to man (53).

Another aspect of host density is the concept of host density dependence. Populations of biological control agents generally increase and decrease in response to similar increases or decreases in the density of their host or prey populations. Such density dependence is almost certainly true for entomopathogens in general (47). Density dependence has an indirect bearing on timing of application of a microbial pesticide. Pathogens released in the longer-term approaches to control cannot be released when host population density is too low, because there will likely be a threshold of host population density below which the pathogen would not be able to replicate to a sufficient degree to control subsequent generations of the pest. There is evidence for such thresholds in releases of NPV for suppression of *A. gemmatalis* populations in soybean (42).

Pest population distribution is not as well studied as pest density, though distribution can be an important consideration in application. Pest distribution or dispersion can affect pathogen application in two ways. The first is intuitively obvious but has not been researched: if a host population has a clumped distribution, then as much of the pathogen population as possible must be delivered to those clumps to contact the target insects and avoid wastage. The second has only an indirect effect on targeting; clumped host populations can be conducive to pathogen transmission, at least within the clumps. For seasonal colonization and introduction-establishment, this can reduce the pathogen inoculum that must be delivered to infect a number of insects. For example, host clumping was conducive to prevalence of disease caused by an entomopathogenic fungus (54), protozoan (55) and viruses (3, 56). Pest distribution also can affect pathogen delivery due to the insects' location in the habitat. For example, vertical distribution of a target insect in soil would affect the selection of the type of pathogen to use (e.g., 19) and/or the way in which a pathogen is delivered (e.g., soil surface spray, or use of a seeder or light tilling for inoculation into soil).

Age structure of a pest population is another factor in timing of application. As discussed above (see "Pathogen Species and Strain Characteristics"), the great majority of pathogens rapidly decrease in effectiveness as the host insect becomes older. Thus, it follows that the age structure of a host population will affect timing; the pathogen must be applied when a sufficient proportion of the host population is young enough to be infected and at a sufficient dosage to infect the hosts as they age. For example, single applications of *B. thuringiensis* and certain viruses have controlled univoltine pests with relatively uniform age structure (57). Repeated applications of *B. thuringiensis* and viruses were necessary for insects with overlapping generations and more complex age structure.

Pest population quality is a major consideration in application, not only in the sense of population quality at the time of application quality, but also in maintaining a certain quality. "Quality" in this sense primarily refers to susceptibility of the insects to the pathogen. Insects can develop resistance to pathogens just as they do to chemical insecticides. Also, the state of nutrition as well as physical and biotic stressors influence the susceptibility of insects to various pathogens (58-59). Stress usually increases susceptibility and thus decreases the number of pathogen units that must be delivered to the insect in order to initiate disease.

Resistance can be defined as the development of an ability in a strain of insects to tolerate doses of pathogens that would prove lethal or cause disease in the majority of individuals in a normal population of the same species. Thus, it is a "strain" characteristic. However, in terms of application of an entomopathogen for insect control, it is better discussed as a population characteristic. Resistance has been demonstrated for all five pathogen groups (60-62). Resistance has developed to *B. thuringiensis* in the field (63-64), and resistance might be developing in the field to a viral insecticide (65). The proportion of resistant versus susceptible insects in the pest population will impact the amount that must be delivered to the target insect; resistant insects will require a heavier dosage. Furthermore, resistance impacts application in the sense that a certain quality (susceptibility) of the pest population be maintained. It has been well accepted that, in the case of chemicals, application can increase the degree of resistance in pest populations in the field (66-68). Pesticides that are applied in a manner that will kill (or render incapable of reproducing) a high proportion of a pest population over a wide area and long time period exert a great selective pressure on that population, in many cases leading to rapid development of resistance and reduced efficacy of future applications of that chemical and perhaps others. There is concern that certain application strategies for microbials, such as engineering the gene for *B. thuringiensis* δ-endotoxin into crop plants or other bacterial species (47), amounts to indiscriminate application of a persistent insecticide that will rapidly lead to resistance.

### The Ecosystem

The pathogen population in microbial control is a biological entity interacting as part of its own environment with numerous environmental components which can greatly impact when and where a pathogen is applied for insect control. One way such factors can be analyzed is according to their intrinsic nature: abiotic, biotic, and interacting (ecosystem) factors. An additional consideration for application is that of environmental risk assessment. There can be problems in the distinction between abiotic and biotic factors and consideration of their individual effects on pathogen populations (4, 69); for example, food and shelter can be difficult to classify in such a manner, particularly for operational ecology (4). Nevertheless, classification into abiotic and biotic factors has been useful in discussions of entomopathogen ecology. Other environmental factors are comprised of both living and non-living components, such as habitat stability or the type of ecosystem. It must be emphasized that all these factors, including the host and pathogen populations, interact in a complex manner. In terms of pathogen application, non-host, non-pathogen environmental factors have the greatest effect on pathogen survival and transport between host insects, a lesser effect on host susceptibility, and an interacting effect on many of the pathogen and host parameters, particularly population-level factors, already discussed.

**Abiotic Factors.** Sunlight is one of the most critical environmental factors. Virtually every entomopathogen is killed or inactivated quickly when fully exposed to sunlight (7, 69-70). Pathogens lose their activity within days or even hours in many situations, most notably on aerial plant surfaces exposed to sunlight (28, 71). On the other hand, sunlight can stimulate the growth of a few pathogens, such as certain fungi (7). The effect of sunlight on application is obvious; the pathogen must be applied in a manner such that it contacts the host insect before it is inactivated. Pathogens often have been formulated to increase their persistence in sunlight (28), but the benefits of such formulations generally have not been worth the additional cost (9). Timing of application to the dusk-to-dawn time period has often been recommended as an inexpensive method to delay exposure of the pathogen to sunlight (28), though this might not be useful with an ingested pathogen applied against a pest that feeds only during daylight hours.

Temperature can affect entomopathogens in two ways: it can affect pathogen survival before it invades the insect, and it can then affect the relationship between pathogen and host. High temperatures can inactivate pathogens before their contact with the pest insect (70) as well as decrease the susceptibilty of the pest (7). Low temperatures can decrease feeding rates of mosquitoes, which in turn reduces consumption of *B. thuringiensis israelensis* toxin (15). On the other hand, once the insect is infected, high temperature also can intensify or accelerate disease development or simply affect the life cycle of a pathogen (7, 54, 72). Thus, it is sometimes helpful to apply a pathogen in the early morning or evening in order to avoid high temperatures (e.g., 73).

Humidity or a surface film of moisture affects the survival and activity of certain entomopathogens, particularly fungi and nematodes (7, 69-70). Although humidity level can be a constraint for such pathogens, this often is at the "microhabitat" level rather than at the overall atmospheric level. For example, fungal infections can be initiated in insects at relatively low macrohumidities if microhumidity at the surface of the host integument or foliage is sufficiently high (70). The nematodes in particular are known for their failures when pest control is attempted in situations where the larvae are subject to desiccation; a major reason for the recent success of certain nematodes is simply that they are targeted against insects that live in moist microhabitats, such as soil or burrows in plant structures. Thus, certain pathogens must be applied at a time and place when humidity is high enough for them to survive and infect the host insect.

Precipitation can be detrimental to persistence or advantageous to dispersal of entomopathogens and thus is a factor in timing of application. Rainfall washes some pathogens from plant surfaces before they can infect insects, though other pathogens are not washed away by precipitation (69, 74). On the other hand, rainfall can disperse pathogens, for example throughout a tree canopy or into soil (7, 28-29). Thus, theoretically, application over a relatively limited portion of a canopy might result in more widespread targeting of a pest population though timing might be difficult due to unpredictability of precipitation. It is clear that it is important to know which pathogens are washed off plant surfaces by precipitation and which are not.

Effects of air or water currents on pathogen application are similar to those of precipitation in the sense that they are largely negative but have some potential to contribute to pathogen dispersal (29). Drift of inoculum during application is a problem with any type of control agent; wind can cause a control agent to miss the target or be applied in an uneven manner. Thus, spray applications often are released at a time of day when wind velocity is diminished. A slight wind can be beneficial by increasing thouroughness of pathogen dispersal throughout a plant canopy. Though practical application might be difficult, it has been proposed that prevailing wind currents or even some sort of fan might be used to widely distribute pathogens after actual application limited to a relatively small area (28). Water currents play a role similar to air currents. Water transports bacterial pathogens (29), which can affect application either by diluting the pathogen at the intended point of contact with the host or by dispersing the pathogen from a relatively limited point of release.

Soil usually protects entomopathogens and often is a reservoir for long-term control (49, 74). Due to relatively long persistence in soil (some pathogens live for years), timing of soil application often is less critical than in more exposed situations. Physical soil or substrate structure can affect placement of application. For example, baculoviruses adsorb to clay particles, which can affect their movement in soil (3, 74). Similarly, *B. thuringiensis israelensis* crystal toxin adsorbs to particles of mud and organic materials, lessening its chance of ingestion by mosquito larvae (75).

**Biotic Factors.** The crop or resource being protected from the pest can affect application in two ways. The first is its economic value. If the plant part subject to damage has a low economic injury level, such as a fruit sold for produce, then little damage can be tolerated. Pathogen application must be timed so that sufficient pathogen units are placed at the point of contact with the pest insect before even "cosmetic" damage can occur. This, of course, also relates to pest behavior, habitat stability, and other factors discussed previously. If the leaf of an apple tree is damaged, or if the crop is fed to domestic animals rather than humans, timing and placement are not as critical, and repeat applications (for multivoltine pests) may not be required. Application of a relatively expensive microbial insecticide to low-value crops, such as pasture grasses, usually is not justifiable economically; however, such crops can be very amenable to long-term approaches to control, particularly if the pathogen can be applied in a cost-effective manner, such as lattice introduction. Franz (7) and Burges (49) extensively discussed type of crop in relation to microbial control.

The second manner in which the crop or resource being protected affects application is biological. Several aspects of a host plant affect pathogen application, including its physical structure, growth characteristics, and chemistry. Physical structure can affect pathogens in several ways. Closed canopies as opposed to more open growth patterns, including different shapes or sizes of leaves, affect penetration and deposit of a sprayed pathogen on various surfaces (e.g., 16). Rapid growth of a plant can result in rapid dilution of an applied pathogen, necessitating more frequent application (74).

Plant allelochemicals can enhance or inhibit activity of pathogens against host insects, which might affect the dosage that must be delivered to the target (16, 27, 50, 69). Also, different host plants can affect pathogen persistence (74), though it is not always known whether this is due to allelochemicals or to microscopic or macroscopic plant structure, perhaps by shading the pathogen from sunlight. The pH of dew on certain types of plants decreases persistence of viruses (74), which in turn can affect timing and dosage applied to the plant.

The pest complex is an important but largely unknown factor in the application of both chemical and microbial pesticides. Very often, more than one pest species damage a crop or resource simultaneously. In such cases, timing of application becomes confused, because economic damage might occur before the population of any one pest species reaches its threshold for insecticide application. This is not as serious a factor in timing applications of entomopathogens, primarily because many such pathogens are so host-specific that they will not suppress populations of more than one of the pest species. However, problems in timing and dosage delivered to target pests can arise with those entomopathogens with relatively broad host ranges, such as *B. thuringiensis* or the fungus *Beauveria bassiana*, particularly when two pests in a complex have different levels of susceptibility (76).

Various biotic agents in the environment can affect application by transporting a pathogen. Predatory and parasitic arthropods, various scavengers, birds, and mammals have transported various entomopathogens after their release (3, 29, 77-78). The entomopathogenic viruses in particular are known for dispersal in this manner. Generally, the animals eat infected or dead host insects; the pathogen survives passage through the gut and is deposited in a new location in the animal's feces. Though this type of transport is not in practical use in pathogen application, it has the potential to reduce the amount of inoculum and the area treated with a pathogen for microbial control. For example, the NPV of *A. gemmatalis* can be released at intervals of >20 m and still provide short- or long-term insect control due to transport by predatory arthropods (42).

Effects of other biotic agents are not as well studied as transport agents. The $\delta$-endotoxin of *B. thuringiensis* is degraded by certain soil microbes and, potentially, leaf colonizing bacteria (79). Nematode-trapping fungi might reduce the efficacy of nematodes (70). Another unknown is the outcome of releasing new pathogen strains (whether natural or genetically engineered) for insect control in locations with indigenous strains of the same pathogen species (9); there is virtually no information about whether such new strains could compete for a niche in such a situation. All of these instances might require an application strategy to avoid antagonism with such biotic agents. On the positive side, there is evidence that, in certain situations, a pathogen can be applied in a limited manner to infect a non-pest insect and build a greater, more widespread inoculum before infestation of the crop by the pest insect (80).

**Ecosystem Factors.** Several authors have proposed the idea that habitat stability affects the success of classical biocontrol with parasitoids, predators (45), and pathogens (e.g., 20, 25). Stable habitats are those that do not undergo various kinds of upheaval. For example, a permanent body of water is more stable than one that dries periodically; temperate climates, with their different seasons, are less stable than the tropics; and, particularly relevant to biological control, forests, grasslands, orchards, and various perennial crops offer a more stable habitat than annual crops, particularly row-crop agriculture in which even the soil is disturbed. Stability generally is favorable for pathogen persistence, whether in an abiotic environmental component, such as soil, or in a biotic component, such as a host population that is present for much of the year. Two successful, long-term examples of control with viruses have been attributed partly to habitat stability (25). Stable habitats are thought to offer advantages for the seasonal colonization approach as well as introductions (7). In addition, unstable habitats tend to foster r-selected pest species (48), which can indirectly affect pathogen application. Thus, stable habitats are thought to be conducive to the long-term approaches to control, which, in turn, often require release of fewer pathogen units due to the potential for pathogen persistence and spread.

Cultural practices, or, in other words, normal agricultural operations, can indirectly affect pathogen application. For example, modifying planting practices, such as row spacing or planting date, can lead to an early closure of the plant canopy in a crop such as soybean, which in turn can increase the effectiveness of a fungus for insect control by increasing intracanopy humidity and shading (81) or, possibly, increase difficulty in delivering a pathogen to an insect inside the canopy. Thus, canopy closure, partially dependent on farming practices, might be a factor in deciding when to apply the fungus. Use of chemical pesticides also can be a factor. The fungus *B. bassiana* was not harmed by a fungicide commonly used in potatoes provided that conidia were applied at least one day after the fungicide (82). Irrigation can affect efficacy of entomopathogens for insect control (28), for example, by raising humidity or moisture levels, and thus might dictate when a pathogen is applied.

**Environmental Risks.** Environmental risks are becoming a concern in pathogen application. Entomopathogens are safe to almost all non-target organisms (1), which is the major reason for their research and development for insect control. One concern is that the activity of genetically engineered entomopathogens might not be predictable in the environment; therefore, they carry a somewhat greater risk (83). Application techniques have been proposed for field-testing to greatly restrict any possible transport of the microorganism outside the release site (29). Concern over natural strains of entomopathogens is that they might reduce populations of arthropods closely related to the target pest species; most such non-target arthropods would not be severely affected, but a few might be endangered species or might play a crucial role in the ecosystem (84). Timing and placement of application are potential means to reduce risks to these non-target organisms without hampering pest population suppression. For example, application of a pathogen with a broad insect host range (e.g., *B. bassiana*) might be delayed during a blossom period when pollinating insects are present (7).

**Summary of Ecological Factors in Application**

It is clear that entomopathogens and other biorational agents whose activity is closely related to the pest insect's life system must be integrated into a wide variety of ecological factors as well as agricultural and resource-management practices. This fits into the concept of integrated pest management (IPM), whereby all suitable control techniques are integrated with one another and with other crop production practices to suppress (not eliminate) pest populations below economic injury levels while maintaining the integrity of the ecosystem. This control concept is heavily based in ecology.

The factors that affect timing and placement of entomopathogen application are summarized in Table 1. It should be emphasized that this categorization is somewhat arbitrary, depending not only on one's point of view but also on the specific pest-pathogen system under consideration. For example, soil is a somewhat minor factor for appplication in many entomopathogen-pest-crop systems, but it might be critical to placement for control of a soil pest. Similarly, environmental risks are a major consideration in microbial control for regulatory reasons, but they currently have only secondary importance in application for reasons of efficacy.

The list of factors in Table 1 is somewhat overwhelming at first glance, but this will not necessarily be the case in research and development of a particular entomopathogen. This summary was developed from a review of many pest-pathogen systems, and it covers factors related to all three major approaches to microbial control. Any one pest-pathogen system certainly will not

**Table 1. Categories and Specific Ecological Factors Affecting Entomopathogen Timing and Placement of Application for Microbial Control**

| Environmental Component | Factors Affecting: Timing | Factors Affecting: Placement |
|---|---|---|
| | PRIMARY IMPORTANCE | |
| Pathogen species or strain | life cycle<br>site of attack<br>speed of action | portal of entry<br>searching ability<br>virulence<br>transmission |
| Pathogen population | persistence | density<br>distribution<br>spread after release |
| Pest species or strain | pest category<br>behavior | behavior |
| Pest population | density<br>age structure | distribution<br>quality |
| Ecosystem | sunlight<br>humidity<br>resource (crop): economic<br>resource (crop): biological | sunlight<br>humidity<br>air currents<br>water depth, currents<br>resource (crop): economic<br>resource (crop): biological |
| | SECONDARY IMPORTANCE | |
| Pathogen species or strain | host range<br>portal of entry<br>reproduction rate<br>transmission | host range<br>site of attack<br>reproduction rate |
| Pest species or strain | r-K continuum<br>generations/yr. | r-K continuum |
| Pest population | quality | |
| Ecosystem | temperature<br>precipitation<br>soil<br>pest complex<br>biotic agents: antagonists or synergists<br>habitat stability<br>management practices<br>environmental risks | precipitation<br>soil<br>biotic agents: transport<br>management practices<br>environmental risks |

include all these complexities to a significant degree. Nevertheless, one lesson is clear. Application technology has been based primarily on a great deal of research of pathogen formulation; breakthroughs certainly are possible in formulation and research should be continued. However, timing and placement of application have been relatively ignored, and success in microbial control will depend to a large degree on ecological factors in relation to application technology.

**Acknowledgments**

[1]Approved for publication by the Director of the Louisiana Agricultural Experiment Station as manuscript number 94-17-8132.

**Literature Cited**

1. Laird, M; Lacey, L. A.; Davidson, E. W., Eds. *Safety of Microbial Insecticides*; CRC Press: Boca Raton, FL, 1990.
2. Lewis, F. B.; Etter, D. O., Jr. In *Microbial Control of Insect Pests: Future Strategies in Pest Management Systems*; Allen, G. E., Ignoffo, C. M., Jaques, R. P., Eds.; NSF-USDA-Univ. Florida, Gainesville, 1978, pp. 261-272.
3. Evans, H. F. In *The Biology of Baculoviruses*; Granados, R. R., Federici, B. A., Eds.; CRC Press, Boca Raton, FL, 1986, Vol. 2; pp. 89-132.
4. Andrewartha, H. G.; Birch, L. C. *The Distribution and Abundance of Animals*; Univ. Chicago Press: Chicago, IL, 1954.
5. Price, P. W. *Insect Ecology*; Wiley & Sons: New York, 1975.
6. Price, P. W. In *Biological Control in Crop Production*; Papavizas, G. C., Ed.; BARC Symposium V; Allanheld, Osmun Publ.: London, 1981; pp. 3-19.
7. Franz, J. M. In *Microbial Control of Insects and Mites*; Burges, H. D., Hussey, N. W., Eds.; Academic Press: London, 1971; pp. 407-444.
8. Fuxa, J. R.; Tanada, Y. In *Epizootiology of Insect Diseases*; Fuxa, J. R., Tanada, Y., Eds.; Wiley & Sons: New York, 1987; pp. 3-21.
9. Fuxa, J. R.; In *IPM Systems in Agriculture*; Rajak, R. L., Upadhyay, R. K., Mukerji, K. G., Eds.; Aditya Book Private Ltd.: India, In Press.
10. Hågvar, B. *Acta Entomol. Bohemoslov.* **1991**, *88*, 1-11.
11. Fuxa, J. R. In *ESA Handbook of Soybean Insects*; Higley, L. G., Boethel, D. J., Eds.; Entomol. Soc. Amer.: Lanham, MD, In Press.
12. Fuxa, J. R. *Biotech. Adv.* **1991**, *9*, 425-442.
13. Andreadis, T. G. In *Epizootiology of Insect Diseases*; Fuxa, J. R., Tanada, Y., Eds.; Wiley & Sons: New York, 1987; pp. 159-176.

14. Fuxa, J. R.; Richter, A. R. *Environ. Entomol.* **1991,** *20*, 603-609.
15. Becker, N; Margalit, J. In *Bacillus thuringiensis, an Environmental Pesticide: Theory and Practice*; Entwistle, P. F., Cory, J. S., Bailey, M. J., Higgs, S., Eds.; Wiley & Sons: New York, 1993; pp. 147-170.
16. Tanada, Y; Kaya, H. K. *Insect Pathology*; Academic Press: San Diego, 1993.
17. Anderson, R. M.; May, R. M. *Nature.* **1979,** *280*, 361-367.
18. Anderson, R. M.; May, R. M. *Science.* **1980,** *210*, 658-661.
19. Georgis, R. In *Entomopathogenic Nematodes in Biological Control*; Gaugler, R., Kaya, H. K., Eds.; CRC Press: Boca Raton, FL, 1990; pp. 173-191.
20. Fuxa, J. R. *Mem. Inst. Oswaldo Cruz.* **1989,** *84* (Supl. III), 81-88.
21. Huber, J. In *The Biology of Baculoviruses*; Granados, R. R., Federici, B. A., Eds.; CRC Press, Boca Raton, FL, 1986, Vol. 2; pp. 181-202.
22. Jaques, R. P.; Laing, J. E.; Laing, D. R.; Yu, D. S. K. *Can. Entomol.* **1987,** *119*, 1063-1067.
23. Jutsum, A. R. *Phil. Trans. R. Soc. Lond. B.* **1988,** *318*, 357-373.
24. Richter, A. R.; Fuxa, J. R. *J. Econ. Entomol.* **1984,** *77*, 1299-1306.
25. Harper, J. D. In *Epizootiology of Insect Diseases*; Fuxa, J. R., Tanada, Y., Eds.; Wiley & Sons: New York, 1987; pp. 473-496.
26. Shapiro, M. In *The Biology of Baculoviruses*; Granados, R. R., Federici, B. A., Eds.; CRC Press, Boca Raton, FL, 1986, Vol. 2; pp. 31-61.
27. Tanada, Y. In *Regulation of Insect Populations by Microorganisms*; Bulla, L. A., Jr., Ed.; Ann. N.Y. Acad. Sci.; N.Y. Acad. Sci.: 1973, Vol. 217; pp. 120-130.
28. Ignoffo, C. M. In *Biological Control in Agricultural IPM Systems*; Hoy, M. A., Herzog, D. C., Eds.; Academic Press, Orlando, FL, 1985; pp. 243-261.
29. Fuxa, J. R. In *Risk Assessment in Genetic Engineering*; Levin, M. A., Strauss, H. S., Eds.; McGraw-Hill: New York, 1991; pp. 83-113.
30. Bedford, G. O. *Agric. Ecosystems Environ.* **1986,** *15*, 141-147.
31. Zelazny, B.; Lolong, A.; Pattang, B. *J. Invertebr. Pathol.* **1992,** *59*, 61-68.
32. Anderson, R. M. *Parasitology.* **1982,** *84*, 3-33.
33. Hagen, K. S.; van den Bosch, R. *Annu. Rev. Entomol.* **1968,** *13*, 325-384.
34. Ferron, P. *Annu. Rev. Entomol.* **1978,** *23*, 409-442.
35. Kalmakoff, J.; Crawford, A. M. In *Microbial and Viral Pesticides*; Kurstak, E., Ed.; Dekker: New York, 1982; pp. 435-448.
36. Morris, O. N. In *Microbial and Viral Pesticides*; Kurstak, E., Ed.; Dekker: New York, 1982; pp. 239-287.

37. Tanada, Y.; Fuxa, J. R. In *Epizootiology of Insect Diseases*; Fuxa, J. R., Tanada, Y., Eds.; Wiley & Sons: New York, 1987; pp. 113-157.
38. Pinnock, D. E. In *Baculoviruses for Insect Pest Control: Safety Considerations*; Summers, M., Engler, R., Falcon, L. A., Vail, P. V., Eds.; Am. Soc. Microbiol.: Washington, DC, 1975; pp. 145-154.
39. Onstad, D. W. *Biological Control*. **1993**, *3*, 353-356.
40. Burges, H. D.; Hussey, N. W. In *Microbial Control of Insects and Mites*; Burges, H. D., Hussey, N. W., Eds.; Academic Press: London, 1971; pp. 687-709.
41. Fuxa, J. R. *Bull. Entomol. Soc. Am.* **1989**, *35*, 12-24.
42. Fuxa, J. R.; Richter, A. R. *Environ. Entomol.* **1994**, *23*, in press.
43. Rabb, R. L. In *Concepts of Pest Management*; Rabb, R. L., Guthrie, F. E., Eds.; North Carolina St. Univ.: Raleigh, NC, 1970; pp. 1-5.
44. Woods, A. *Pest Control: a Survey*; Wiley & Sons: New York, 1974.
45. Hokkanen, H. In *CRC Critical Reviews in Plant Sciences*; CRC Press: Boca Raton, FL, 1985, Vol. 3; pp. 35-72.
46. Entwistle, P. F. In *Biological Plant and Health Protection*; Franz, J. M., Lindaur, M., Eds.; G. Fischer Verlag: Stuttgart, 1986; pp. 257-278.
47. Fuxa, J. R. *Annu. Rev. Entomol.* **1987**, *32*, 225-251.
48. Evans, H. F.; Entwistle, P. F. In *Epizootiology of Insect Diseases*; Fuxa, J. R., Tanada, Y., Eds.; Wiley & Sons: New York, 1987; pp. 257-322.
49. Burges, H. D. In *Microbial Control of Pests and Plant Diseases 1970-1980*; Burges, H. D., Ed.; Academic Press: London, 1981; pp. 797-836.
50. Navon, A. In *Bacillus thuringiensis, an Environmental Pesticide: Theory and Practice*; Entwistle, P. F., Cory, J. S., Bailey, M. J., Higgs, S., Eds.; Wiley & Sons: New York, 1993; pp. 125-146.
51. Oi, D. H.; Pereira, R. M. *Flor. Entomol.* **1993**, *76*, 63-74.
52. Falcon, L. A. *Annu. Rev. Entomol.* **1976**, *21*, 305-324.
53. Lacey, L. A.; Harper, J. D. *J. Entomol. Sci.* **1986**, *21*, 206-213.
54. Carruthers, R. I.; Soper, R. S. In *Epizootiology of Insect Diseases*; Fuxa, J. R., Tanada, Y., Eds.; Wiley & Sons: New York, 1987; pp. 357-416.
55. Maddox, J. V. In *Epizootiology of Insect Diseases*; Fuxa, J. R., Tanada, Y., Eds.; Wiley & Sons: New York, 1987; pp. 417-452.
56. Bedford, G. O. In *Microbial Control of Pests and Plant Diseases 1970-1980*; Burges, H. D., Ed.; Academic Press: London, 1981; pp. 409-426.

57. Jaques, R. P. In *Regulation of Insect Populations by Microorganisms*; Bulla, L. A., Jr., Ed.; Ann. N.Y. Acad. Sci.; N.Y. Acad. Sci.: 1973, Vol. 217; pp. 109-119.
58. Briese, D. T. In *The Biology of Baculoviruses*; Granados, R. R., Federici, B. A., Eds.; CRC Press, Boca Raton, FL, 1986, Vol. 2; pp. 237-263.
59. Watanabe, H. In *Epizootiology of Insect Diseases*; Fuxa, J. R., Tanada, Y., Eds.; Wiley & Sons: New York, 1987; pp. 71-112.
60. Briese, D. T. In *Pathogenesis of Invertebrate Microbial Diseases*; Davidson, E. W., Ed.; Allanheld, Osmun Publ.: Totowa, NJ, 1981; pp. 511-545.
61. Entwistle, P. F.; Cory, J. S.; Bailey, M. J.; Higgs, S. *Bacillus thuringiensis, an Environmental Biopesticide: Theory and Practice*; Wiley & Sons: New York, 1993.
62. Fuxa, J. R. In *Parasites and Pathogens of Insects*; Beckage, N. E., Thompson, S. N., Federici, B. A., Eds.; Academic Press: San Diego, 1993, Vol. 2; pp. 197-209.
63. McGaughey, W. H. *Science*. **1985,** *229*, 193-195.
64. Tabashnik, B. E.; Cushing, N. L.; Finson, N.; Johnson, M. *J. Econ. Entomol.* **1990,** *83*, 1671-1676.
65. Fuxa, J. R.; Abot, A. R.; Moscardi, F.; Sosa-Gómez, D. R.; Richter, A. R. *Resistant Pest Management*. **1993,** *5*, 40-41.
66. Brown, A. W. A.; Pal, R. *Insecticide Resistance in Arthropods*; World Health Organization: Geneva, 1971.
67. Curtis, C. F.; Cook, L. M.; Wood, R. J. *Ecol. Entomol.* **1978,** *3*, 273-287.
68. Georghiou, G. P.; Taylor, C. E. *Proc. Int. Congr. Entomol., 15th, 1976*; 1977, pp. 759-785.
69. Benz, G. In *Epizootiology of Insect Diseases*; Fuxa, J. R., Tanada, Y., Eds.; Wiley & Sons: New York, 1987; pp. 177-214.
70. Roberts, D. W.; Fuxa, J. R.; Gaugler, R.; Goettel, M.; Jaques, R.; Maddox, J. In *Handbook of Pest Management in Agriculture*; Pimentel, D., Ed.; CRC Press: Boca Raton, FL, 1991, Vol. 2; pp. 243-278.
71. Podgwaite, J. D. In *Viral Insecticides for Biological Control*; Maramorosch, K., Sherman, K. E., Eds.; Academic Press: Orlando, FL, 1985; pp. 775-797.
72. Weiser, J. In *Microbial and Viral Pesticides*; Kurstak, E., Ed.; Dekker: New York, 1982; pp. 531-557.
73. Schmiege, D. C. *J. Econ. Entomol.* **1963,** *56*, 427-431.
74. Young, S. Y. III; Yearian, W. C. In *The Biology of Baculoviruses*; Granados, R. R., Federici, B. A., Eds.; CRC Press, Boca Raton, FL, 1986, Vol. 2; pp. 157-179.
75. Ohana, B.; Margalit, J.; Barak, Z. *Appl. Env. Microbiol.* **1987,** *53*, 828-831.

76. McEwen, F. L.; Glass, E. H.; Davis, A. C.; Splittstoesser, C. M. *J. Insect Pathol.* **1960,** *2,* 152-164.
77. Harper, J. D. In *The Biology of Baculoviruses;* Granados, R. R., Federici, B. A., Eds.; CRC Press, Boca Raton, FL, 1986, Vol. 2; pp. 133-155.
78. Leisy, D. J.; Fuxa, J. R. In *Opportunities from Natural Sources in Crop Protection;* Copping, L. G., Ed.; Critical Reviews in Applied chemistry; Elsevier: Oxford, In Press.
79. Gelernter, W.; Schwab, G. E. In *Bacillus thuringiensis, an Environmental Pesticide: Theory and Practice;* Entwistle, P. F., Cory, J. S., Bailey, M. J., Higgs, S., Eds.; Wiley & Sons: New York, 1993; pp. 89-104.
80. Ignoffo, C. M. *J. Invertebr. Pathol.* **1978,** *31,* 1-3.
81. Sprenkel, R. K.; Brooks, W. M.; Van Duyn, J. W.; Dietz, L. L. *Environ. Entomol.* **1979,** *8,* 334-339.
82. Loria, R.; Galaini, S.; Roberts, D. W. *Environ. Entomol.* **1983,** *12,* 1724-1726.
83. Fuxa, J. R. In *Safety of Microbial Insecticides;* Laird, M., Lacey, L. A., Davidson, E. W., Eds.; CRC Press: Boca Raton, FL, 1990; pp. 203-207.
84. Lockwood, J. A. *Environ. Entomol.* **1993,** *22,* 503-518.

RECEIVED January 31, 1995

# Chapter 5

# Modeling the Dose Acquisition Process of *Bacillus thuringiensis*

## Influence of Feeding Pattern on Survival

**Franklin R. Hall[1], A. C. Chapple[2], R. A. J. Taylor[3], and R. A. Downer[1]**

[1]Laboratory for Pest Control Application Technology, Ohio Agricultural Research and Development Center, Ohio State University, 1680 Madison Avenue, Wooster, OH 44691

[2]Ecogen Europe SRL, 3a Parco Technologico Agro-alimentare dell'Umbria, Frazione Pantalla, 06095 Todi (PG), Italy

[3]Department of Entomology, Ohio Agricultural Research and Development Center, Ohio State University, 1680 Madison Avenue, Wooster, OH 44691

Insecticide formulations and adjuvants are manipulated to optimize the pesticide deposit characteristics on the plant surface. The toxicity, deposit quality and quantity, together with the insect's pattern of feeding, determine the insecticide's efficacy. A model of the dose-transfer process, The Pesticide Drop Simulator, was used to investigate the effect of feeding and walking parameters of simulated insect foliar feeders on their survival when exposed to a leaf surface treated with a biological insecticide. Survival in the model was found to be most influenced by the speed of walking, the major determinant of the distance apart of feeding holes on the leaf. This result was obtained without explicitly simulating avoidance behavior, but is in agreement with findings where avoidance has been observed. These results serve to remind us how important the feeding, locomotory and searching behavior of defoliators is in the efficacy of pesticides. This conclusion is especially relevant to biological insecticides. The role of modeling in general, and the utility of PDS in particular, in the evaluation of pesticide formulations and additives is also discussed.

The efficiency with which pesticides are utilized in agriculture and horticulture is extremely poor (*1*). In part, this is because fields usually have to be treated as a whole, regardless of the distribution of the pest within the field (*2,3*). Even when an infested plant is sprayed with an insecticide, for example, little will be deposited where the pest will encounter it. Of the effective deposit, only a fraction will be acquired by the pest (*4*), and still less will reach the susceptible site within it (*1,5,6*). Estimates vary as to how much of the pesticide sprayed actually reaches its intended target and results in mortality (biological efficiency). Some broad spectrum post-emergent foliar-applied herbicides may achieve >1% biological efficiency (*1*). However, this is a best case; a worst case assessment suggests that <0.001% of the insecticide permethrin applied against diamondback moth larvae (a worldwide pest of cabbage and other Cruciferae)

0097–6156/95/0595–0068$12.00/0

reaches its intended target (*5*). Of the >99% excess pesticide, much of the spray falls from the plant to contaminate the soil or drift from the area. In addition to the environmental cost and increased production cost of excessive pesticide use, insecticide resistance can result from continuous low level exposure (*7*).

The pesticide deposit has two components: deposit quantity (mass per unit area) and deposit quality (deposit size distribution and the spatial distribution). Deposit quantity is a rough guide to the distribution of the AI through a canopy. However, one 800μm diameter droplet of a contact insecticide deposited on a leaf will not give the same biological result as the same volume deposited as 512 100μm randomly or uniformly distributed droplets. Hence, deposit quality is a key component of the application process. Laboratory studies suggest that biological efficiency of insecticides is inversely proportional to drop size: small drops work better for the same amount of AI (*8*). Thus, it should be possible to optimize the distribution of droplets on the plant to achieve a desired efficacy while reducing the total AI applied by manipulation of the nozzle dynamics and/or adjuvant characteristics. However, field data do not support a correlation between good deposition of insecticides (*1*) and biological result. In fact, the same amount of AI is usually required with small droplets as with large to get the same control in the field (*9-11*). One cause of reduced efficacy of small droplets in the field is drift which is also inversely proportional to droplet size.

The addition of adjuvants to pesticides remains the standard procedure for improving transfer efficiency by altering impaction characteristics. The effect of formulations and adjuvants on the atomization process has been addressed in some detail at the Laboratory for Pest Control Application Technology (LPCAT) over the last four years (*8,12-14*) as part of our objective to investigate the entire dose-transfer process. Dose-transfer can be defined as the process from atomization to biological effect, including but not limited to atomization, transport to the target, impaction, and retention (application), dose acquisition by the target (transfer), and degradation and non-target fate of AI (attrition). The dose-transfer efficiency is the product of the application and transfer efficiencies, and the biological efficiency is the dose-transfer efficiency adjusted by any loss in potency due to pesticide age or other factors.

A synthesis of our knowledge of the dose-transfer process has been compiled in the form of a simulation model of *Bt* transfer to the diamondback moth on cabbage (Pesticide Drop Simulator, PDS; *15*). Our model relates the deposition characteristics (quantity and quality) to the biological effect by simulating the feeding of diamondback larvae on cabbage leaves treated with specified size and spatial distributions of pesticide. Although the specific insecticide modelled is *Bt*, the model is sufficiently flexible to accommodate other foliar applied insecticides. Equally, the model is not restricted to diamondback larvae. The feeding module can be programmed to simulate a wide range of feeding behaviors.

To simulate the fate of a defoliating insect on a pesticide treated leaf, variables incorporated into the simulation include; a statistical description of the relevant behavior of the insect, including feeding and locomotory behavior; a statistical description of the spatial and size distributions of the pesticide droplets on the leaf; a deterministic description of the temporal evolution of the chemical potency, which in turn may depend on the time series of incident solar (usually ultra violet) radiation. Other desiderata include a mathematical (deterministic or stochastic) description of the digestive process as it influences absorption of the toxin and any subsequent moribund behavior; a feedback loop permitting changes in insect behavior in response to toxin intake; inheritance of behaviors from one experimental cycle to the next to simulate selection of pesticide resistance. Most of these latter features have been incorporated in the model, but were not used in these simulations.

The behavioral response of target insects to insecticide deposits can also be an important determinant of insecticide efficacy (*16,17*). Several responses to the insecticide have been identified: pre-contact response, including odor-mediated response, and post-contact hypersensitive response to the chemical. Head *et al* (*17*) found that changes to simple behavioral rules governing locomotory behavior made in response to the presence of insecticide droplets can lead to increased movement, increasing the probability of avoidance. Movement and the aggregative response are key components of the survival of all species (*18,19*). Thus, by extension, searching and avoidance behavior are important determinants of the survival of insects exposed to insecticides.

**Table I. Examples of feeding damage simulated by the interaction of the distributions of feeding and walking bout and rate**

| | *Short Feeding Bout* | | *Long Feeding Bout* | |
|---|---|---|---|---|
| *Walking Bout-Rate* | *Low Feeding Rate* | *High Feeding Rate* | *Low Feeding Rate* | *High Feeding Rate* |
| *Short-Low* | 1. many small regular holes aggregated | 2. many small irregular holes aggregated | 3. many large regular holes aggregated | 4. many large irregular holes aggregated |
| *Long-Low* | 5. few small regular holes aggregated | 6. few small irregular holes aggregated | 7. few large regular holes aggregated | 8. few large irregular holes aggregated |
| *Short-High* | 9. many small regular holes dispersed | 10. many small irregular holes dispersed | 11. many large regular holes dispersed | 12. many large irregular holes dispersed |
| *Long-High* | 13. few small regular holes dispersed | 14. few small irregular holes dispersed | 15. few large regular holes dispersed | 16. few large irregular holes dispersed |

Chapple *et al* (*20*) found in simulations of diamondback larvae that survival and the area eaten grows as the variance of the distribution of drop size increases. The reason is that as the size variance increases, the number mean must decrease because of the cubic relationship between volume and diameter. The smaller number of large droplets are less effective than close cover by many small drops, exactly in accordance with laboratory results. What the laboratory results do not suggest, but the simulation model apparently does, is an interactive effect of droplet distribution (governed by nozzle dynamics and adjuvant characteristics) and feeding/locomotory behavior on survival. Consequently, we ask the question in this paper, do different foraging strategies result in different mortalities for the same spatial and size distributions of droplets on the leaf?

## Simulation Experiments with PDS

The Pesticide Drop Simulator (*15*) was used to investigate the possible consequences of different feeding patterns on susceptibility of foliar feeding insects to insecticide. Four behavioral parameters govern the feeding pattern of simulated insects in the simulator: duration of alternating bouts of feeding and walking, the instantaneous rate of consumption of leaf surface, and the speed of locomotion, called feeding bout length, walking bout length, feeding rate, and walking rate, respectively. The balance of feeding bout and walking bout define how much time is spent on these two basic activities, while the rates respectively determine how much is eaten and how far apart are the feeding spots. These four parameters can be made deterministic, or be described by stochastic functions. In addition, probability density functions can be defined for the turning rate and turning bias. Given the large number of stochastic distributions that can be used to describe these parameters, the number of *quantitatively* different feeding patterns that can be defined is effectively infinite. However, 16 *qualitatively* different feeding/locomotory patterns can be selected using PDS to simulate different feeding strategies (Table I).

Table I summarizes the feeding damage resulting from qualitatively different combinations of simulated behavior. For example, the choice of short feeding bout and feeding rate relative to walking results in a large number of small pock marks on the leaf, simulating mite damage. A longer feeding bout combined with higher feeding rate result in feeding damage comprising large regular shaped holes such as is produced by many lepidopterous larvae. Simulating with high feeding rates results in feeding holes with irregular outlines. High rates of feeding combined with short bouts result in long narrow feeding holes, similar to the damage created by leafminers. The shape of the distribution also affects the pattern of damage on the leaf. High coefficient of variation results in irregularly-shaped feeding holes.

## The Influence of Insect Feeding Behavior on Efficacy

To investigate the possibility that the pattern of feeding might influence the efficacy of a pesticide application, we chose 16 patterns of feeding and walking corresponding to the qualitative patterns defined in Table I. To simplify matters, the uniform distribution was chosen as the distribution for the behavioral components. We recognize that this is an improbable distribution in nature, however we justify it by its mathematical tractability. The uniform distribution has mean and variance defined in terms of its parameters in a particularly simple way:

$$Mean = M(X) = a + \frac{1}{2} \cdot (b - a) \tag{1a}$$

$$Variance = V(X) = \frac{1}{12} \cdot (b - a)^2 \tag{1b}$$

where $a$ and $b$ are the lower and upper bounds respectively. The amount eaten per unit time and shape of feeding hole are the important variables of feeding pattern. The total amount eaten is derived from the sums, products and ratios of the feeding and locomotory parameters. Thus, the simplicity of the uniform distribution makes it comparatively easy to define the expected distribution of amount eaten in terms of the feeding and locomotory parameters. The sum, $Z$, of two random variables, $X$ and $Y$, have mean and variance:

$$M(Z) = M(X) + M(Y) \tag{2a}$$

$$V(Z) = V(X) + V(Y) \tag{2b}$$

The mean and variance of the product, $Z = XY$ are:

$$M(Z) = M(X) \cdot M(Y) \tag{3a}$$

$$V(Z) = M^2(Z) \cdot \left\{ \frac{V(X)}{M^2(X)} + \frac{V(Y)}{M^2(Y)} + 2 \cdot \frac{Cov(X,Y)}{M(Z)} \right\} \tag{3b}$$

The mean and variance of the ratio, $Z = X/Y$ are:

$$M(Z) = M(X)/M(Y) \tag{4a}$$

$$V(Z) = M^2(Z) \cdot \left\{ \frac{V(X)}{M^2(X)} + \frac{V(Y)}{M^2(Y)} - 2 \cdot \frac{Cov(X,Y)}{M(Z)} \right\} \tag{4b}$$

Equations 3b and 4b differ only in the sign of the covariance term, *Cov(X,Y)*, which in this case is assumed to be zero.

**Table II. Parameter values of uniformly distributed behavioral parameters used in simulations to examine the effect of insect feeding pattern on susceptibility to insecticides**

| *Behavior Pattern* | *Level* | *lower (a)* | *upper (b)* | *Mean* | *Stdev* |
|---|---|---|---|---|---|
| *Feeding bout (min)* | short | 0 | 5 | 2.5 | 1.443 |
| | long | 5 | 10 | 7.5 | 1.443 |
| *Feeding rate (pixels/min)* | low | 0 | 10 | 5 | 2.887 |
| | low | 5 | 15 | 10 | 2.887 |
| | high | 10 | 20 | 15 | 2.887 |
| | high | 15 | 25 | 20 | 2.887 |
| *Walking bout (min)* | short | 5 | 15 | 10 | 2.887 |
| | short | 17.5 | 27.5 | 22.5 | 2.887 |
| | short | 30 | 40 | 35 | 2.887 |
| | short | 42.5 | 52.5 | 47.5 | 2.887 |
| | long | 25 | 35 | 30 | 2.887 |
| | long | 62.5 | 72.5 | 67.5 | 2.887 |
| | long | 100 | 110 | 105 | 2.887 |
| | long | 137.5 | 147.5 | 142.5 | 2.887 |
| *Walking rate (pixels/min)* | low | 0 | 10 | 5 | 2.887 |
| | high | 5 | 15 | 10 | 2.887 |

Table II shows the parameter values used for feeding and locomotion. They were chosen such that the mean consumption rate was 60 pixels per hour. Distances and rates are measured in pixels for convenience: transforming to conventional units gives inconvenient numbers. The simulations were run for 48 hours, so the expected amount eaten in the absence of pesticide was 2880 pixels. Table IIIA shows the expected means and standard deviations of the distributions of feeding and walking, and of the derived parameters. Note that although the variances of the primary distributions are the same for all simulations, the resulting variances of amount eaten are not all the same, despite the constant expected amount eaten. It is the effect of this variation in resulting behavior we are principally concerned with, for it is this variation which is the raw material of selection and evolution (*21*).

The other model variables were fixed as follows:

1. 100 insects were simulated for 48 hr for each of the 16 feeding patterns;
2. feeding occurred throughout the 48 hr period without diurnal periodicity;
3. the pesticide simulated was *Bt*, with attrition defined by

$$P(t) = \exp\{-.0.0001 \cdot F(t)^{0.5}\} \tag{5}$$

where *P(t)* is the amount remaining at time *t* and *F(t)* is the total UV flux at *t*, starting at 6 am on 22 June (details are in *15*);

4. the number of droplets on the screen (leaf) was 100, distributed at random (Poisson spatial distribution);
5. the size spectrum was distributed as a 3-parameter lognormal with parameters, $\mu$ = 4.47, $\sigma$ = 1.02, and A = 3.64, which describes *Bt* in a water carrier very well (*15*);

**Results of PDS Simulations**

In 1600 simulations performed using PDS survival ranged from 71 to 91% (Table IIIB). The highest survival rate was obtained with Runs 5 and 9; few small regularly shaped aggregated feeding damage, and many small regularly shaped dispersed feeding damage. The highest mortality was with Runs 11 and 15; many large regularly shaped dispersed feeding damage, and few large regularly shaped dispersed feeding damage.

The net consumption rates for surviving and fatal simulations were not significantly different (surviving simulations, 43.63±0.38 (mean±standard error) pixels/hr; fatal, 46.78±3.35; $t$=0.93, $\alpha$>0.2). However, the total amount eaten by the two groups were significantly different (surviving, 2094±7.4; fatal, 93.4±4.1; $t$=27.9, $\alpha$<0.001). Within each group, analysis of variance of consumption rate with Run as the treatment resulted in significant differences between the treatments for both surviving simulations ($F_{15,1304}$=134.9, $\alpha$<0.001), and fatal simulations ($F_{15,264}$=1.81, $\alpha$<0.05). Because the total number of pixels eaten in fatal simulations was so highly variable within each treatment, until corrected for time alive (the consumption rate), there was no significant differences between treatments ($F_{15,264}$=0.96, $\alpha$<0.5). The total time alive of surviving simulations was the same for all treatments (2880 hr), thus the analysis of variance of treatment differences in total pixels consumed is the same as for consumption rate ($F_{15,1304}$=134.9, $\alpha$<0.001).

Because the parameters were constrained to produce the same *expected* net consumption rate, the distinction between many and few feeding holes is somewhat blurred. Not surprisingly, there was very little difference in survival between simulations with large and small feeding holes. However, the difference was greater than between regular and irregular feeding holes; shape of the feeding damage had absolutely no effect on survival. In line with Head *et al's* (*17*) result, we found a small difference in survival between aggregated and dispersed: simulations with aggregated feeding damage had generally higher survival rates than those with dispersed feeding damage. Thus higher walking rate conferred some protection, regardless of the walking bout length, and *in the absence of explicit avoidance behavior.*

Considering all simulations regardless of whether the insect survived or died, the observed consumption rate was consistently lower than the expected, regardless of the specifics of the simulated behavior. The rates ranged from 65% to 83% of the expected consumption rate. The simulations with faster locomotory movement (Runs 9-16) were consistently higher than the slower simulations (Runs 1-8). This is because smaller steplengths result in increased contacts with existing holes, thus increasing the time required to find an intact piece. Very large steplengths should result in observed consumption rates approaching the expected rate. It also explains why the mean consumption rates for fatal simulations are higher than for the survivors; less has been eaten, so less searching is necessary to find intact leaf.

Interestingly, the observed variances are substantially less than expected. The expected standard deviations are approximately equal to the expected consumption rate, but this clearly overestimates the actual standard deviations. The standard deviations for survivors is very low, while that of the fatal simulations is approximately half the expected standard deviation. The

Table III. The effect of input feeding and walking behavior variables (A) on expected and observed consumption rates (B) by simulated foliar-feeding insects with different feeding patterns

A

| Run | *Feeding Bout* | | *Feeding Rate* | | *Walking Bout* | | *Walking rate* | | *Pixels/Bout* | |
|---|---|---|---|---|---|---|---|---|---|---|
| | *Mean* | *Stdev* | *Mean* | *Stdev* | *Mean* | *Stdev* | *Mean* | *Stdev* | *Mean* | *Stdev* |
| 1 | 2.5 | 1.44 | 5.0 | 2.89 | 10.0 | 2.89 | 5.0 | 2.89 | 12.5 | 10.2 |
| 2 | 2.5 | 1.44 | 15.0 | 2.89 | 35.0 | 2.89 | 5.0 | 2.89 | 37.5 | 22.8 |
| 3 | 7.5 | 1.44 | 5.0 | 2.89 | 30.0 | 2.89 | 5.0 | 2.89 | 37.5 | 22.8 |
| 4 | 7.5 | 1.44 | 15.0 | 2.89 | 105.0 | 2.89 | 5.0 | 2.89 | 112.5 | 30.6 |
| 5 | 2.5 | 1.44 | 10.0 | 2.89 | 22.5 | 2.89 | 5.0 | 2.89 | 25.0 | 16.1 |
| 6 | 2.5 | 1.44 | 20.0 | 2.89 | 47.5 | 2.89 | 5.0 | 2.89 | 50.0 | 29.8 |
| 7 | 7.5 | 1.44 | 10.0 | 2.89 | 67.5 | 2.89 | 5.0 | 2.89 | 75.0 | 26.0 |
| 8 | 7.5 | 1.44 | 20.0 | 2.89 | 142.5 | 2.89 | 5.0 | 2.89 | 150.0 | 36.1 |
| 9 | 2.5 | 1.44 | 5.0 | 2.89 | 10.0 | 2.89 | 10.0 | 2.89 | 12.5 | 10.2 |
| 10 | 2.5 | 1.44 | 15.0 | 2.89 | 35.0 | 2.89 | 10.0 | 2.89 | 37.5 | 22.8 |
| 11 | 7.5 | 1.44 | 5.0 | 2.89 | 30.0 | 2.89 | 10.0 | 2.89 | 37.5 | 22.8 |
| 12 | 7.5 | 1.44 | 15.0 | 2.89 | 105.0 | 2.89 | 10.0 | 2.89 | 112.5 | 30.6 |
| 13 | 2.5 | 1.44 | 10.0 | 2.89 | 22.5 | 2.89 | 10.0 | 2.89 | 25.0 | 16.1 |
| 14 | 2.5 | 1.44 | 20.0 | 2.89 | 47.5 | 2.89 | 10.0 | 2.89 | 50.0 | 29.8 |
| 15 | 7.5 | 1.44 | 10.0 | 2.89 | 67.5 | 2.89 | 10.0 | 2.89 | 75.0 | 26.0 |
| 16 | 7.5 | 1.44 | 20.0 | 2.89 | 142.5 | 2.89 | 10.0 | 2.89 | 150.0 | 36.1 |

B

| Run | *Expected Consumption Rate (Pixels/hour)* | | *Observed Consumption Rate (Pixels/hour)* | | | | | | | |
|---|---|---|---|---|---|---|---|---|---|---|
| | | | *All simulations* | | *Surviving simulations* | | | *Fatal simulations* | | |
| | *Mean* | *Stdev* | *Mean* | *Stdev* | *N* | *Mean* | *Stdev* | *N* | *Mean* | *Stdev* |
| 1 | 60 | 51.8 | 38.88 | 9.22 | 88 | 38.72 | 3.05 | 12 | 41.89 | 28.59 |
| 2 | 60 | 56.4 | 37.14 | 9.98 | 88 | 36.97 | 3.78 | 12 | 41.71 | 28.98 |
| 3 | 60 | 49.7 | 44.53 | 11.17 | 86 | 44.28 | 3.14 | 14 | 48.74 | 30.62 |
| 4 | 60 | 56.2 | 42.14 | 13.27 | 81 | 41.52 | 4.14 | 19 | 49.21 | 39.91 |
| 5 | 60 | 54.9 | 38.34 | 8.77 | 91 | 38.14 | 3.91 | 9 | 45.47 | 27.57 |
| 6 | 60 | 57.2 | 38.64 | 10.41 | 83 | 38.13 | 3.89 | 17 | 44.43 | 29.10 |
| 7 | 60 | 54.4 | 42.69 | 12.23 | 79 | 42.02 | 3.86 | 21 | 49.08 | 24.75 |
| 8 | 60 | 57.1 | 42.71 | 12.84 | 83 | 42.36 | 4.44 | 17 | 47.81 | 30.27 |
| 9 | 60 | 51.8 | 44.41 | 9.12 | 91 | 44.48 | 2.24 | 9 | 43.24 | 33.28 |
| 10 | 60 | 56.4 | 45.04 | 13.95 | 82 | 44.86 | 4.05 | 18 | 47.52 | 31.45 |
| 11 | 60 | 49.7 | 49.19 | 17.60 | 72 | 49.15 | 2.04 | 28 | 49.51 | 33.82 |
| 12 | 60 | 56.2 | 49.83 | 14.95 | 80 | 50.27 | 2.42 | 20 | 48.12 | 27.94 |
| 13 | 60 | 54.9 | 45.19 | 12.85 | 80 | 45.18 | 3.12 | 20 | 45.40 | 26.25 |
| 14 | 60 | 57.2 | 44.50 | 13.61 | 84 | 44.58 | 4.41 | 16 | 43.55 | 34.18 |
| 15 | 60 | 54.4 | 50.47 | 16.76 | 71 | 50.27 | 2.21 | 29 | 51.61 | 29.10 |
| 16 | 60 | 57.1 | 50.27 | 13.62 | 81 | 50.25 | 3.38 | 19 | 50.64 | 28.03 |

overestimate is certainly due to the assumption of no correlation between feeding and walking bout lengths being false. In fact, because the two bout types must sum to 48 hr for survivors, there is a negative time correlation approaching unity. For the fatal simulations, however, the total time is not a constant, but a random variable (derived from interaction of all the other behavioral factors), so the correlation between bout lengths is less than unity and the observed consumption rate variance is higher.

**Discussion**

May (*22*) defined analytical or "strategic" models at one end and pragmatic or "tactical" models at the other end of a continuum. Tactical models describe or simulate systems in detail, and are intended to answer specific, usually applied questions, while strategic models sacrifice detail in favor of generalizations to provide a conceptual framework for investigating general principles. Despite its simulation model structure, PDS is not a tactical model. It was designed as a strategic model capable of elucidating general principles of the dose transfer process. Like most tactical models, PDS is quite detailed in its construction, and unlike most strategic models, it is not based on a small number of axioms from which an analytical result is to be deduced, but from a relatively large set of empirical relationships. However, like analytical models, the objective of PDS is to argue inductively from the specific to the general. Because its fundamental purpose is strategic, PDS's output constitutes predictions about the way an actual system might work. These predictions become the basis for experimental verification. Validation is thus a continuing process, rather than the *a priori* step in model acceptance as it would be for a tactical model.

The simulation model and approach described identifies critical parameters using *Bt* as a model system. It is not a substitute for experimental and observational science, indeed its construction would not have been possible without the hard laboratory and field work underpinning its logic. However, its synthetic approach to science substantially reduces the parameter space we must explore experimentally, in order to obtain answers to the near intractable questions posed by the need to improve the efficiency of agricultural chemical spray delivery. No basic (analytical or numerical) theory of dose-transfer has yet been derived, so that there is no general theory from which efficacy can be predicted from basic physico-chemical parameters of the spray. Without such a model, and the understanding which it implies, the successes of agricultural spray application technology have been obtained through atomization changes or through formulation/adjuvant changes arrived at by trial and error. PDS is, so far, the nearest thing there is to a theory of dose-transfer, but its acceptability as a tool suffers from the stigma of empiricism, and the criticism of validation omission.

The wasteful application of pesticides is the principal cause of most of the environmental problems associated with their use. Increasing biological efficiency by a factor of two would result in a halving of the environmental contamination: ground water contamination and food residues would be significantly reduced. Such an improvement has been mandated in several European countries by the end of the decade. Achieving this without loss in yields could be achieved with an improved understanding of the dose-transfer process, so that biological efficiency can be predicted from the set of physico-chemical parameters describing the spray and its toxicity. A number of pesticides now regarded as "high-risk" might possibly be reclassifiable with a 50% reduction in application rates. Reductions in pesticide applied could also help to slow the rate of resistance development in insects by reducing the frequency of exposure to sub-lethal doses (*7*).

Our model of the dose transfer of *Bt* to the diamondback moth on cabbage has identified some of the factors impeding attempts to improve spray application, and also areas that may lead to improvements (*23,24*) in dose-transfer efficiency. An important point to emerge is that the use of a surfactant to increase the number of droplets, without appreciably increasing the total amount of pesticide, can substantially reduce the life expectancy of simulated larvae, in keeping with experimental results using monodispersed drops (*8*). Polymers, on the other hand, tend to increase the lifespan of simulated larvae by a factor of ten over a water control. This appears to be a cost borne when attempting to reduce spray drift by increasing the drop size. It should be noted that despite its comparative failure, the polymer still killed about 1/3 of the test population, but because a significant proportion survived the damage was also significantly higher (*15*).

These results underscored a point already well-known to us, that droplet quality (spatial and size distributions) can have profound effects on the efficacy of a foliar-applied insecticide. This confirmation by the model, constitutes a partial validation, but its main contribution was to remind us how important the feeding, locomotory and searching behavior of defoliators could be in the efficacy equation. The simulations reported here make some specific predictions about the relationship between behavior and susceptibility to insecticides, predictions which need to be verified experimentally, and thereby validate the model a little further. But, the clearest message to emerge from these results is that the weak point in the entire dose-transfer process is the behavior of the target, which in general is very poorly documented. Particularly relevant is the fact that biological pesticides do not generally have the robustness and immediacy of effect of their chemical counterparts. Therefore the interaction of deposit quality and insect behavior is particularly important when considering biologicals, making PDS most appropriately used to investigate the possible consequences of biological applications.

We conclude with the reminder that the intersection of droplet quality, which is manipulated by adjuvant, formulation, and nozzle geometry, and insect behavior, which is species specific and frequently stage specific, is a high dimensional space which will only be understod by much more basic biology and ethology research.

## Literature Cited


1 Graham-Bryce I. J. In *Pesticide Chemistry: Human Welfare and the Environment. Volume 1. Synthesis and Structure-Activity Relationships.* Doyle, P.; Fujita, T., Eds.; Pergamon Press, Oxford, UK, 1983.
2 Hislop E. C. *Asp. Appl. Biol.* 1987, 14, 153-165.
3 Door G. J.; Pannell D. J. *Crop Prot.* 1992, 11, 384-391.
4 Adams A. J.; Hall F. R. *Pest. Sci.* 1990, 28, 337-343.
5 Adams A. J.; Hall F. R. *Crop Prot.* 1990, 9, 39-43.
6 Ratcliffe, S. L.; Yendol, W. G. *J. Environ. Sci. Health* 1993, B28, 91-104.
7 Rousch, T. T. *Pest. Sci.* 1989, 26, 423-441.
8 Adams A. J.; Chapple A. C.; Hall F. R. In *Pesticide Formulations and Application Systems.* Bode, L. E.; Hazen, J. L.; Chasin, D. G., Eds.; 10th Symposium. ASTM TP 1078, American Society for Testing and Materials, Philadelphia, PA, 1990.
9 Arnold A. J.; Cayley G. R.; Dunne Y.; Etheridge P.; Greenaway A. R.; Griffiths D. C.; Phillips F. T.; Pye B. J.; Rawlinson C. J.; Scott G. C. *Ann. Appl. Biol.* 1984, 105, 369-377.

10 Arnold A. J.; Cayley G. R.; Dunne Y.; Etheridge P.; Griffiths D. C.; Jenkyn J. F.; Phillips F. T.; Pye B. J.; Scott G. C.; Woodcock C. M. *Ann. Appl. Biol.* 1984, 105, 361-367.
11 Arnold A. J.; Cayley G. R.; Dunne Y.; Etheridge P.; Griffiths D. C.; Phillips F. T.; Pye B. J.; Scott G. C.; Vojvodic P. R. *Ann. Appl. Biol.* 1984, 105, 353-359.
12 Chapple A. C.; Hall F. R. *Atomiz. Sprays* 1994, in press.
13 Chapple A. C.; Downer, R. A; Hall F. R. *Crop Prot.* 1994, in press.
14 Chapple A. C.; Hall F. R.; Bishop B. L. *Crop Prot.* 1994, in press.
15 Taylor, R. A. J.; Chapple, A. C.; Hall F. R. In *Pesticide Formulations and Application Systems*. Berger, P. D.; Devisetty, B. N.; Hall, F. R., Eds.; 13th Symposium, ASTM STP 1183, American Society for Testing and Materials, Philadelphia, PA, 1993.
16 Royalty, R. N.; Hall, F. R.; Taylor, R. A. J. *J. Econ. Entomol.* 1990, 83, 792-798.
17 Head, G.; Hoy, C. W.; Hall, F. R. *Environ. Entomol.* 1994, in press.
18 Taylor, R. A. J. *Environ. Entomol.* 1987, 16, 1-8.
19 Turchin, P. *J. Anim. Ecol.* 1989, 58, 75-100.
20 Chapple A. C.; Taylor, R. A. J.; Hall, F. R.; Downer, R. A. In *Proceedings of the Conference Comparing Glasshouse and Field Pesticide Performance*, 1994, in press.
21 Taylor, R. A. J. *J. Anim. Ecol.* 1979, 48, 577-602.
22 May, R. M. *Stability and complexity in model ecosystems.* Princeton University Press, Princeton, NJ, 1974.
23 Chapple A. C. *Dose transfer of* Bacillus thuringiensis *to the diamondback moth (Lepidoptera: Plutellidae) via cabbage: a synthesis.* PhD Thesis, Department of Entomology, OARDC, Ohio State University, Ohio, USA, 1993.
24 Hall, F. R.; Chapple, A. C.; Taylor, R. A. J.; Downer, R. A. *J. Environ. Sci. Health* 1994, in press.

RECEIVED February 2, 1995

# Delivery and Environmental Fate

# Chapter 6

# Delivery Systems for Biorational Agents

**William E. Steinke and D. Ken Giles**

**Biological and Agricultural Engineering Department, University of California, Davis, CA 95616–5294**

Application of many biorational pesticides and pest control agents is currently accomplished with existing spray technology. Such equipment was developed for application of broad spectrum chemical pesticides and may not be suitable for biorational agents. Proper handling of the formulation and creation of an active and biologically optimized deposit or release is essential for the efficacy of biorational materials. The paper reviews existing practices for the application of biorational agents such as bacteria, viruses, fungi, pheromones, predators, and parasites. Inadequacies of the existing systems are analyzed. An assessment of future needs for handling, metering, dispersal, and collection of these materials is presented.

For purposes of this paper on delivery systems for biorational agents, we define biorationals to be biological products or organisms which must be produced outside the target field, or if naturally occurring at the target, are augmented from sources outside the target field. This is intended to include production and dispersion of agents such as viruses, bacteria, fungi, predators, parasites, and other organisms which are biological in origin and are capable of controlling pests in an agricultural, forest, urban, or controlled (greenhouses, modified atmosphere storage) environment. Intentionally excluded from this discussion are strategies such as in-field nurseries, application of pesticides which are not biological in origin, cultural practices, mechanical or physical control, and use of predators such as raptors. Synthetic pheromones, artificially multiplied and processed viruses and bacteria, products of insectaries, and similar biologically based pest control agents are all included in this discussion.

Most application systems currently in use for biorationals are based upon technology developed for application of broad spectrum pesticides. The limited exceptions are new equipment and techniques developed specifically for one biorational product. The development of application techniques and equipment for broad spectrum pesticides occurred simultaneously with development of the pesticides, accelerating rapidly in the 1950's and 1960's, with fewer innovations reaching the market in the 1980's and 1990's. Such liquid application systems generally have been designed with the principle of delivering a wide range of droplet sizes over the entire surface of the target field, plant, or animal. Dry material application systems also follow the same principle, that is, dispersing materials over the entire area to be treated with minimal change in particle size from that as produced by the formulator. As in liquid application systems,

0097–6156/95/0595–0080$12.00/0

the same equipment is used to deliver a broad range of products, with minor adjustments used for calibration.

Recent developments have predominately focused on reducing or eliminating some of the undesirable aspects of pesticide application. Drift, re-entrainment, off-target movement, and surface or groundwater contamination have been the driving forces in pursuing many innovations such as low-pressure nozzles, boom shields, orchard sprayer sensors and controls, tower sprayers, low volatility products, and equipment for reduced volume per unit area applications.

Requirements for product efficacy have often taken a secondary or even tertiary role when maximizing safe use procedure. As one example, regulations often specify a minimum nozzle size based on drift considerations. If the product is no longer efficacious with fewer, larger drops applied per unit area, the product will simply lose favor with growers and applicators. Potential solutions such as application equipment or technique development have often come only after the search for a new pesticide, if at all.

However, the increasing use of biorational materials provides incentive and opportunity to examine both the equipment used and the equipment needed for delivery of these products. For efficacy, many have special handling requirements, such as limited shearing action while in the delivery system, pH or temperature restrictions, or desired deposit characteristics. New delivery systems are being developed and must continue to be developed for use of biorationals to increase, both from the standpoint of efficacy and cost of the protection provided. Successful examples from the literature and present and future needs are identified below under headings for the applicable biorational pest control agent.

This paper reviews pesticide application as a process, identifying physical operations within that process, reviews how the literature reflects application equipment in recently reported successful uses of biorational pest control agents, identifies some future equipment and information needs for increased use of biorationals, and considers paths to achieving enhanced use of biorationals and potential societal and individual implications.

## Process of Pesticide Application

Pesticide application has been described and analyzed as a series of discrete steps. Numerous authors (*1, 2, 3*) have expanded and rearranged the process into different discrete steps, but the concept that successful use of a pesticide to attain the desired biological effect requires an unbroken succession of steps or processes, remains constant. That is, in order for a pesticide to successfully control a pest, each successive step must be completed in a timely manner and in a manner compatible with the requirements of the pesticide. Those requirements may be related to thermodynamic, chemical, physical, biological, or any other characteristic of the pesticide.

The process of pesticide application, as used in this paper, will consist of the following steps:

**Mixing.** This process occurs both in the formulation of the product and as the product is diluted in the tank, to the finished volume to be applied per unit surface area or per volume to be treated. Tank diluents are usually water or oil, with perhaps surfactants such as spreader/stickers, drift control agents, or other pesticides. Some products are applied undiluted, or directly as formulated. Mixing in the formulation is attained by the manufacturer or formulator; in many products one of the desired characteristics is the lack of separation over the shelf life of the product.

**Atomization.** Atomization is the process of forming the droplets of product(s) and diluent. It is often accomplished by hydraulic nozzles, rotary atomizers (with either a perforated screen or a spinning disk), or pneumatic or bi-fluid atomization.

Electrodynamic forces and ultrasonic forces have also been used to form liquid droplets. Dry materials are customarily not broken down or atomized at the time of distribution (granules) but may be finely separated and entrained into an airstream as dusts.

**Transport.** Transport of the droplets or granules is most often accomplished by either (a) purposeful entrainment of the droplet or particles into moving airstreams, (b) release of hydraulic pressure, or (c) gravitational forces. Examples of each are aerosol generators, oscillating boom sprayers for citrus, and field crop boom sprayers, respectively. Combinations of these techniques are often used as in aircraft, which use the first and the third, and orchard air carrier or airblast sprayers, which use the first and second.

**Collection.** The process which results in deposition of pesticide on the target field, plant, or animal, is called collection. Modes of collection include impaction, interception, diffusion, gravitational settling, and electrostatic attraction. These are physical processes governed by the fundamental principles of physics. Each may be important under certain circumstances and unimportant under other circumstances. The collection process in any particular application will be some combination of the various modes.

**Deposit Formation and Translocation/Migration.** After the droplet or particle is collected upon the target surface, it may or may not undergo a process of deposit formation. It may also be translocated or transported within the pest or the host. This last sub-step does not always occur, but may play an important role in a successful pesticide application with specific materials.

**Interaction with Pest.** After the pesticide has deposited on the vegetation, the pest, or some other surface, there will be an interaction between the pesticide and the pest. This may be through absorption, ingestion, inhalation, or another mode or modes of action.

**Biological Action.** If all of the previous steps have been successfully completed, and if a sufficient dose has been delivered to the pest in the proper form and in an active state, the desired result, the biological effect on the pest, will likely result. This is the goal of a pesticide application and can be achieved only if all of the previous steps have been successfully completed.

## Examples of Successful Biorational Pest Control Reports

The use of natural compounds, predators, parasites, bacteria, fungi, and other pest control agents for insect, pathogen, and weed control, along with discussions of genetically engineered biorationals and future prospects and needs of the use of biorationals was provided by Lumsden and Vaughn (*4*). Within this volume, the proceedings of a conference held at the USDA ARS Beltsville Agricultural Research Center (Beltsville, MD) in May 1993, several opportunities for biological control of pests are discussed, although most with no reference to application equipment for agents to be released or dispersed throughout the target. Details of selected recently reported successful applications are described below.

**Bacteria.** Successful forestry application of *Bacillus thuringiensis* (B.t.) to control larval stages of several lepdopterous pests has been reported since at least 1975. Recent successes found in the literature include application of *Bacillus thuringiensis* Kurstaki against the gypsy moth, *Lymantria dispar* (L.) by several researchers. Recently, Podgwaite et al. (*5*), reported using 8 Micronair AU5000 Mini Atomizers (Micronair [Aerial], Isle of Wright, England) with Variable Restrictor Unit (VRU) setting = 9, blade angles of 65° and 70° mounted on a 448 kw (600 hp) Grumman AgCat. Flight speed was reported as 161 km/hr at 15 m above the canopy at a swath width of 23 m, and volume median diameter (VMD) for the droplet distribution as observed during characterization trials reported as 229 nm (sic) with a pump pressure of 267 kPa. DuBois et al. (*6*) also used Micronair AU5000 atomizers on three Grumman AgCats at four volumes per hectare and three rates of a.i. per hectare. Their work showed significant differences in deposit characteristics with volume changes, but no differences in post treatment gypsy moth egg mass density between any of the treated plots.

Fleming and van Frankenhuyzen (*7*) reported variations in efficacy of *Bacillus thuringiensis* Berliner, against spruce budworm, *Choristoneura fumiferana* Clemens, in 30 spray block treatments. Applications were made with single engine airplanes using rotary atomizers and larger aircraft equipped with a boom and hydraulic nozzles. No information on the deposit characteristics or application details were presented. They developed a model to predict post-spray larval density based upon pre-spray larval density and 48 hr post-spray mortality. Forty-eight hour mortality was never observed to be over 45% and the authors cited unpublished data indicating that only 50% of larval population received a lethal dose in a field trial in western Ontario. They speculated that either spray distribution was uneven or the doses were too low to achieve higher mortality rates, but did not investigate further.

Ogwang and Matthews (*8*) reported significantly ($P < 0.001$) higher spore counts and *Plutella xylostella* mortality when Brussels sprouts were sprayed with B.t. using a Micron Micro-ulva spinning disc sprayer with electrostatic charging activated as compared with uncharged droplets from the same sprayer. They also suggested a minimum orifice size of 0.8 mm to avoid plugging of the orifice by the B.t. *Bacillus thuringiensis* var. *san diego* was also reported as effective against Colorado Potato Beetle, *Leptinotarsa decemlineata* (Say) by Zehnder et al. (*9*). Applications were made using hollow cone nozzles (no details given) spaced 45 cm apart on a boom, and operated at a pressure of 700 kPa with a volume rate of application of 374 l/ha. Vandenberg and Shimanuki (*10*) assessed the presence of colony forming units of B.t. Berliner following application to beeswax combs by several methods. Colony forming units (CFUs) were sampled from combs after treatment and acceptability or lack of acceptability of the deposit was based upon the authors' previous work where $\log_{10}$ CFUs of $\geq 7$ per 50 $cm^2$ of comb were required for long term control. They found acceptable results with an airless sprayer, a conventional sprayer which sprayed both sides of the combs as it was passed between two nozzles facing each other, and a thermal fogger. Dipping the combs in a dilution of B.t., an aerosol treatment from cans, and a cold fogger application gave unacceptable results. Broza et al. (*11*) used a new isolate of B.t., strain MF-4B-2, which was obtained from soil on an island in Lake Victoria, to control African armyworm, *Spodoptera exempta*, in Kenya.

The use of B.t. for gypsy moth control was described by Smitley and Davis (*12*). The B.t. was applied aerially with both rotary atomizers and conventional hydraulic (fan) nozzles and was equally effective with both atomizers. Later applications of B.t. were made to protect foliage from budworm, and were found to allow increased parasitism of budworm larvae and thus the population of natural enemies was increased (*13*). These applications were made using rotary atomizers (Micronair AU 4000) on a Cessna 188 Agtruck, as were the applications reported by Cadogan and Scharbach (*14*).

The use of a bacterial fungus (Pseudomonus strain 679-2) predator was described by Casida, Jr. (*15*). The material was applied to tomatoes and alfalfa in a evaluation of the ability to control leaf spot diseases and "sprayed to runoff." Control of diseases caused by *Alternaria solani* on tomatoes and *Pseudopeziza medicaginis*, *Phoma medicaginis*, and *Stemphylium botryosum* on alfalfa were all found to be highly significant ($P = 0.01$) when compared to controls sprayed with water.

The November/December 1993 issue of Nut Grower (*16*) magazine lists six formulators (Abbott Labs, Chicago, IL; E.I. duPont de Nemour & Co., Inc., Fresno, CA; Ecogen, Inc., Langhome, PA; Fermone Corporation, Phoenix, AZ; PBI/Gordon,

Kansas City, MO; and Sandoz Crop Protection Corp., Des Plaines, IL) of B.t. registered in California for use in almonds, walnuts, pistachios, or other minor nut crops. These products are commonly applied with conventional orchard sprayers in a total of 100 gallons or more, of water per acre.

**Viruses.** The nucleopolyhedrous virus (baculovirus) Gypchek was reported as reducing egg masses by 81% over counts from control blocks in low and moderate population densities of the gypsy moth when applied with a ground-based hydraulic sprayer (*17*). The application was at a rate of 936 l/ha finished spray using a hydraulic spray gun (FMC 785, FMC Corp., Hoopston, IL) at a pressure of 2112 - 2816 kPa (300 - 400 psi). They reported that "uniform coverage over the lower two-thirds of the canopy was achieved" (no other information supplied). The product was also used to control gypsy moth in Virginia (*5*). Applications were made with a 448 kW (600 hp) Grumman AgCat. Flight speed was reported as 161 km/hr at 15 m above the canopy at a swath width of 23 m, and VMD for the droplet distribution during characterization trials reported as 217 m at a pump pressure of 267 kPa from 8 Micronair AU 5000 Mini Atomizers with VRU setting = 11 and blade angle of 45 and 50.

Young and Yearian (*18*) applied a commercial preparation of *Heliothis* nuclear polyhedrous virus (NPV) for control of *Heliothis* species in soybeans, at low levels using a backpack sprayer with a single TX-10 nozzle at 210 kPa. They found that predators which preyed upon larvae infected with the virus were not important contributors to the spread of the virus throughout the soybean plots, thus implying that the virus might need to be applied at least annually and at appropriate rates over the entire field where control of *Heliothis* is desired.

An NPV isolated from the alfalfa looper was applied to cotton by Vail et al. (*19*). The application was described as being done with a high clearance sprayer at 468 l/ha with up to five nozzles per row. They found differences of up to 141 fold in residue levels of the virus between treated and untreated plots over a several week study period. A wettable powder formulation of an NPV was applied to cotton by Jones et al. (*20*) to control Egyptian cotton leafworm (*Spodoptera littoralis*). The liquid mix of the NPV, inerts, a wetting agent, and water was applied through knapsack sprayers at approximately 207 kPa, using either TY 1.5 or TY 3.0 cone nozzles (Spraying Systems Co., Wheaton, IL). They found that the wettable powder formulation of the virus was capable of controlling the cotton leafworm, with significant differences between the control plots and the doses of $1 \times 10^{12}$ and $5 \times 10^{12}$ polyhedral inclusion bodies (p.i.b.)/hectare in two separate years of testing.

Webb and Shelton (*21*) applied a granulosis virus using a carbon dioxide pressurized backpack sprayer, to control *Pieris rapae* in cabbage. Significant (P = 0.0001) differences were found between untreated check plots and the virus and synthetic chemical treated plots. They were examining the effect of timing, so no information regarding equipment can be extracted beyond the fact that the virus survived, multiplied, and infected pests in the field. Muthiah and Rabindra (22) found a nuclear polyhedrous virus (NPV) to be effective when applied at ultra-low volumes (ULV) with a hand-held, battery powered, spinning disc applicator. Significant (P = 0.05 by Duncan's multiple range test) differences were found between the untreated control and ULV-NPV + crude sugar formulation, ULV-NPV + boric acid and crude sugar, ULV-NPV + endosulfan and crude sugar, and endosulfan treatments. Significant differences were not observed between the control and a high volume NPV formulation and the ULV-NPV formulation with no additives.

**Fungi.** In a summary article, Rogers (*23*) describes research in Australia which controls take-all patch fungus in greenhouses with other fungi. She described the use of *Beauveria bassiana, Metarhizum anisopliae, Verticillium lecanii*, and *Paecilomyces* sp. as insecticides for control of several landscape pests. She also mentioned the difficulty in delivering the fungi to an environment suitable for their survival and effectiveness.

Another summary article (*24*), gives no detail on equipment used for application of fungi to control insects, but describes the strategy of augmentative release as ranging from innoculation to inundation. Baits are also discussed, as well as using an application to adult insects which then return the fungus to nesting areas when laying their eggs. Mention is also made of using additives to protect the fungal pathogen from UV radiation, low humidity, or high temperature.

The fungus *Lagenidium gigantium* continues to be evaluated for its ability to control mosquito larvae. Attempts to gain registration and establish production facilities are currently underway in California as a joint effort of the State Department of Health Services, the University of California, and private companies.

Wright and Chandler (*25*) applied the fungus *Beauveria bassiana* to Boll Weevil (Curculionidae: Colepotera) by dipping at two concentrations (pupae and adults) and feeding adults a sugar solution containing the fungus and cotton squares treated with the fungus. The adult stage was found to have high mortality after topical exposure and feeding on the cotton squares after three days, thus leading the authors to postulate that a foliar application of *B. bassiana* could be an effective control measure.

**Pheromones.** Several companies are involved in the commercial production of pheromones and application systems (including traps) for both monitoring and mating disruption (Phero Tech Inc., Delta, BC, Canada; Hercon Environmental Co., Emigsville, PA; Ecogen Inc., Langhorne, PA; AgriSense Inc., Palo Alto, CA; C & E Enterprises, Mesa, AZ; K & K Aircraft, Inc., Bridgewater, VA and; Scentry Inc., Goodyear, AZ).

Use of the Hercon dispensing system with a variety of synthetic pheromones was reported in Kydonieus and Beroza (*26*) and Quisumbing and Kydonieus (*27*). They described the ability to generate a variety of shapes by cutting or chopping the plastic laminate in the desired length and width dimensions. In all cases the reservoir is between two layers of a plastic barrier which controls the release rate. The polymer used in the outside layers and the thickness of the layers can be used to control the diffusion rate, as can the concentration of the pheromone in the reservoir layer. The larger units are most often used in traps, and thus hand placed, while the smaller units are dispensed most often by aircraft in a proprietary dispenser for complete deposition and mating disruption through diffusion throughout a larger area.

Henneberry et al. (*28*) reported on the efficacy of a variety of pheromones, all dispersed through the use of "flakes" of the laminate of the reservoir and plastic barrier layers. All applications were made by air, through application equipment designed specifically for dispensing the laminate flakes or other similar shapes. Mean swath width was $15.5 \pm 1.1$ m Smith, Baker, and Ninomiya (*29*) stated that for mating disruption, aerial application using conventional equipment was most often the application technique of choice for economic reasons. In order to use conventional equipment, they concentrated on the development of microencapsulated formulations of several pheromones which could be applied with such machinery.

Rothschild (*30*), in a review article, reported on the successful use of microcapsules, chopped hollow fibers, hollow fiber tapes, and rubber tubing dispensers with a number of pheromones for mating disruption of codling moth. He indicated that where any information on the application equipment or details of the dispensers was available, they were made specifically for the product being dispensed or the media (flakes, fibers, etc.) The Hercon fiber was described and potential uses listed in Ashare et al. (*31*). Application of the fibers was by hand placement or an aerial application system under development specifically for the fibers.

Microencapsulated pheromones were applied with a roller pump and conventional hydraulic nozzles to control grape berry moth (*32*). This study also found hollow fibers to be an effective pheromone application system. A single, hollow tube which could be twisted around a portion of the cotton plant, was found to be effective at mating disruption of bollworms for the entire season in Pakistan (*33*). The twist-tie, microencapsulated, and hollow fiber formulations were also evaluated for early season control and delayed pesticide application in Egyptian (*34*) and Pakistani cotton (*35*). Shaver and Brown (*36*) used rubber septa to dispense pheromones and successfully disrupt the mating of the Mexican rice borer in sugarcane field trials, as did Tatsuki (*37*) to control rice stem borer. The length of effective release for rubber septa was found to be at least 10 weeks (*38*). Hercon dispensers were successfully fixed to trees by hand and used to disrupt the mating of oriental fruit moth peaches (*39*). Microcapsules were also used in this trials, applied by a backpack sprayer and described only as a "coarse spray."

The use of insect traps with pheromones for monitoring was reported by Bishop et al. (*40*), Knight and Hull (*41*), Mueller et al. (*42*), Leonhart et al. (*43*) and Grant (*44*). As a monitoring tool, Broza et al. (*11*) reported on the use of pheromone traps to identify areas of high infestation from early generations of African armyworm, *Spodoptera exempta*, and then scouted those areas heavily to track hatch and development of succeeding generations in order to optimize application timing of B.t. Mass trapping for insect control was reported by Beevor et al. (*45*).

**Predators and Parasites.** Several insects including predacious mites, lacewings, and wasps have been shown to be effective biocontrols of aphids, spider mites, pink bollworm, navel orangeworm, leafhoppers, whiteflies, diamondback moth, peach twig borer, and European red mite, among other insect pests. Many of these biocontrol schemes require a balance between predator or parasite and prey, thus implying a non-zero population of the pest. Augmentations may be required early in the growing season or to reduce the pest populations below the natural balance, if the desired level of damage is such that a naturally balanced system is not economically feasible. Many augmentations of predators and parasites have been done with home-built equipment based upon hand held or tractor mounted shakers, spreaders, or even hand releases into fields. VanLenteren (*46*) supplied a thorough overview of the use of predators and parasites worldwide under several insect control strategies, such as inundative control, inoculative control, and seasonal inoculative control.

Thomson (*47*) lists 72 insects, mites, or nematodes which are beneficial to pest control in the farm, garden, or landscape and commercial sources for them. Mracek and Jenser (*48*) reported the presence of two entomogenous species of nematodes in Hungary and suggested that they could be applied to fruit orchards as a biological control for several grubs and larvae.

*Thrichogramma platneri* has been shown to be effective against codling moth in walnut orchards in the Sacramento Valley of California and is being evaluated for use in pears and other crops subject to codling moth damage (*49, 50*). Parasitized *Sitotroga* eggs are glued to cards, which are placed inside fine mesh bags and fastened to tree trunks. Potential problems have arisen from predation of the *Sitotroga* eggs before the *Trichogramma* can emerge and the low densities of codling moth eggs desired for agricultural production systems make repeated "applications" of the parasitized eggs a requirement, thus increasing the cost for parasites and the application costs. Several concepts for releasing parasitized eggs from aircraft are under development.

Andow and Prokrym (*51*) found that 40% of the female *Trichogramma nubilale* disappeared per day through death and dispersal. They used 0.96 l glass jars, covered with screens to allow the Trichogramma to leave, but preventing their predators from entering. Plates were suspended over the jars, which were placed on the ground, to shield the jars from rain and direct sunlight. They concluded that either more *Trichogramma* must survive longer after release, or individuals must be more efficient at finding and parasitizing corn borer eggs in order for this to be an acceptable control method. *Trichogrammatoidea cryptophlebiae* Nagaraja was placed on cards and stapled to trees to control false codling moth in citrus (Newton, *52*).

Bouse et al. (*53*) and Bouse and Morrison (*54*) reported being able to store, transport, and disperse *Trichogramma pretosium* parasitized eggs of *Sitotroga*.

Dispersal was from an aircraft. They concluded that the technology existed for widespread *Trichogramma* release programs for control of *Heliothis spp.* They achieved "application" rates of 125,000 to 370,000 eggs per hectare. A previously developed system for release of *Trichogramma* parasitized *Sitotroga* eggs attached to bran flakes was reported by Bouse et al. (*55*). The application of *Trichogramma* parasitized *Sitotroga* eggs mixed with bran flakes and mucilage—water mixture as a control agent was reported by Jones et al. (*56*). Pickett et al. (*57*) reported on the aerial release of *Phytoseiulus persimilis* in corn, and Pickett et al. (*58*) reported on the storage characteristics of this insect predator, their behavior in a simulated aerial release system, and the "dilution" of the predator by corn cob grits for improved application.

A Bell 47 helicopter and a Brohm aerial seeder were successfully used in a five year period to distribute *Trichogramma minutum* (Riley) parasitized eggs as a control agent against the eastern spruce budworm, *Choristoneura fumiferana* (Clemens) (*59, 60, 61*). Smith et al. (*59*) found that pest egg mass parasitisim patterns followed the distribution patterns of the aerially released parasitized eggs, and that more than 50% of the released parasitized eggs were deposited on the ground. They also found that drift, and thus ineffective use of parasitized eggs, was dependent upon application technique and equipment, particularly poor marking of boundaries and command over the metering and release mechanism. Hope et al. (*60*) described the aerial distribution system as a hopper, a metering device, tubing, and a rotating disc to spread the parasitized eggs over an arc using centrifugal force. They used an orifice plate to meter the parasitized eggs and a rotating brush or wobble plate to agitate the eggs and provide a constant flow to the orifice. They mentioned that metabolic heat buildup inside the hopper containing the parasized eggs and subsequent early *Trichogramma* emergence was a potential problem for large scale releases. Two ground-based application systems were also developed; a) a hand-held leafblower, and b) attachment of the parasitized eggs of a host to cards, which were then distributed in the target areas. Viability of the parasites and parasitism rates were not affected by any of the release methods.

King (*62*) described and compared the advantages and disadvantages of ground and aerial release systems. He stated that ground release systems allowed more precise placement and reduced the likelihood of parasites or predators being placed in inhospitable locations or off target, but that those advantages needed to be compared against the relatively slow and labor intensive nature of ground-based hand release systems. Advantages of aerial release systems cited were the ability to cover large areas quickly and adaptability to various aircraft, while the disadvantages of "drift" and potential damage to the organism being released.

Dicke et al. (*63*) postulated that allelochemicals, which they defined to be "An infochemical that mediates and interaction between two individuals that belong to different species", could be used to enhance the effectiveness of predators. Their concept is that allelochemicals emitted by a plant under attack from pest mite species, could be synthesized, applied to a site and used to initiate a "search mode" in predators. Gross (*64*) suggested that exposing *Trichogramma* parasitized eggs to kairomones as they are released, could help to retain emerging parasites on site.

## Inadequacies of the Current Delivery Systems

**Handling.** Development of delivery systems for biorational agents are often characterized by the trial and error approach, resulting in use of handling systems which were designed for an entirely different purpose. The handling needs of biorational agents can be quite similar or quite different from conventional pest control agents.

In liquid application systems, shear forces within the pump, filtration elements, valves, and fittings have the potential to damage live organisms. The process of continuous pumping and re-circulation, necessary for conventional pesticides formulated as wettable powders, has the potential to seriously impair the viability of biological agents. With some B.t. formulations, a process of "shear thinning" takes place, resulting in a smaller droplet size spectrum than that which would come if the formulation had been sprayed immediately after removal from the packaging. This has potential implications for efficacy under field conditions, as it in turn affects the characteristics of the deposit formed on the target.

Pumps with high mechanical shear characteristics, such as gear and impeller pumps, as well as pressure regulators and control valves, have the potential to actually tear organisms into pieces; obviously an undesirable occurrence when using live control agents.

Although no references to shear damage being a cause of reduced efficacy were found, the introduction of a dry formulation into the carrier, (usually an airstream), has potential for shear damage, as does the metering mechanism chosen. The velocity difference between the agent and the carrier stream should be minimized at the point of introduction, after which the entire stream of agent plus carrier can be accelerated to the desired speed.

**Atomization.** Most current liquid application systems use hydraulic pressure nozzles for atomization. The process of pressuring the liquid, forcing it through a metering orifice, causing it to form a range of droplet sizes, and dispersing those droplets into an initial pattern, generates a variety of internal and external forces. These forces may or may not harm the efficacy of biorationals.

Rotary atomizers are also used for biorationals, primarily for application of B.t.'s from aircraft. Like hydraulic nozzles, rotary atomizers were developed for application of conventional liquid pesticides and have been found to be effective when used to apply some biorationals (*65*).

Broadly stated, the objective of atomization for liquid formulations of biorationals should be to form a viable droplet containing a lethal dose of the pest control agent. Unfortunately, there currently is little information available regarding the characteristics of a lethal dose for many pest—biorational pest control agent systems. Application rates are often developed through trial and error.

**Transport.** Introduction of living organisms into a moving airstream has the potential for damaging the organism through shear induced forces and general shock from extremely rapid changes in humidity, temperature, and other environmental parameters. This has not been reported as a problem to date, but should not be ignored as a potential cause if excessive mortality of the applied organism is observed.

**Collection by Target.** Placement of the biorational deposit on the canopy is a crucial issue. The deposit must be formed at a place where the pest is or will be active. This is in contrast to many conventional pesticides which must only be applied to the site being treated, eg. herbicides. In addition to "collection", the process best described as "retention" on the target must be incorporated for many biorationals. Some must also be protected from their own predators and pesticides during establishment of a population, pupation, or even the entire life cycle.

Rapid deceleration upon capture by the target can potentially damage some living biorational agents, eg. insects. The use of affixing agents such as spreader-stickers or glues will not always be appropriate due to the potential for toxicity or suffocation of the biorational pest control agent by the affixing agent.

## Future Needs for Biorational Application Systems

While predictions of future developments and needs are often inaccurate, several broad categories of needs can be identified as being of likely importance.

**Agent Specificity.** Delivery systems for predators and parasites, in particular, have been demonstrated to have better success when they are developed for the specific agent to be dispersed. Solutions to issues of metering, introduction into the carrier stream, and pattern of initial dispersion are all highly dependent upon the biology of the pest control agent chosen. Such systems for larvae are different than those that are successfully used for parasitized eggs, which are in turn different than systems used for dispersing adults. Different species also can require different systems for dispersion, even within the same life stage.

**Economic Feasibility.** The economics of use for biorational agents are a major issue in adoption rates and the application system is a significant component of the cost for use. The capital and operating costs of such application systems must be competitive with those of application systems for conventional pesticides. This is not meant to imply that a direct comparison must apply, only that the costs must be competitive for pest control with biorational agents when compared to conventional pesticides over the course of an entire growing season.

**Crop Specificity.** Even within the same pest control agent, the application equipment may be different for different crops. Applying a pest control agent to the top of a tree crop canopy is obviously different than applying the same pest control agent to a field or row crop. Pheromones must typically be placed in different crops at different times of the year and in different portions of the canopy and different portions of the planting. Cultural practices, growth habit, time of year, and external factors can combine to require different approaches to dispersing the same agent, either in liquid or dry form, to different crops.

**Operator Feedback.** Current application systems provide operator feedback in such ways as the visible spray, readings on pressure gauges, and the sound of pumps, agitation, and fans. Several biorational agents may be applied at such low rates and in such small quantities that they will no longer be visible upon emission. As discussed earlier, pumps, fans, and other familiar components may no longer be necessary or advisable. Therefore, some mechanism of monitoring must be included to provide feedback to the operator that the biorational pest control agent is indeed being released. The limited time of effectiveness for many biorationals and the economics of control combine to make it imperative that some form of operator feedback be included in all application systems.

## Implications of Possible Future Application Technology for Biorationals

**Limited Number of Applicators.** If application systems for biorationals develop into hardware specific for each control agent and on each crop, the multiplicity of equipment necessary will likely mean that no single applicator can possibly service more than a small number of crops or pest control agents. Thus, for any pest control application needed, there will be a limited number of applicators with the proper equipment available. This can help to develop niche markets for applicators which could have enough demand to insure a reasonable return, but has some potential implications for timeliness of application. This also works against the long term interests of applicators when short term pest infestations likely to be stabilized at low levels by biocontrol techniques are encountered.

**Increased Scouting Becomes Necessary.** Typically, biorational pest control agents are not broad spectrum nor immediate in their action. Scouting of fields gains importance then, for detection of pest population increases before they reach damaging levels, evaluations of the population dynamics between pest and predator, and understanding of the pest life cycle stages found in the field. This information is critical and must be obtained in time to arrange for the acquisition and application of biorational pest control agents while the pest is in a susceptible stage and at suitable population levels. Biorationals require lead time for production in an insectary, or fermentation, transportation and distribution, and lack storage tolerance; this must be factored into scouting to help production facilities identify future demand in a timely manner and respond to demand in sufficient quantities.

**Increased Understanding of Biology of Pest, Control Agent, and Crop is Needed.** Successful use of biorational pest control agents on a large scale will require an increased understanding of the biological relationships between the pest, the host which is to be protected, and the pest control agent. Many biorationals are pest-specific and have no effect on non-target organisms. This makes them attractive to use, but requires specific and accurate identification of the pest in order to achieve control.

Additionally, the life stage affected by the biorational may or may not be the life stage which damages the host crop, product, animal, or landscape. Treatments may be required based on degree days from a previous generation's maturity peak, the beginning of egg hatch or pupation, or biorationals may be applied prophylactically as soon as the environment is hospitable for their survival. Some biorationals are also applied when damage is seen in the field, similar to conventional pesticides.

Widespread use of biorationals will require more education on the part of the public, growers and users, scouts, advisers, and others to maximize their effectiveness. This education will certainly include learning more about the life cycles of the biorational agents and how they may be handled, atomized, and transported by the application systems.

**Application Technology and Pesticide Labels.** Current pesticide labels, when they mention application equipment, often describe a "lowest common technology", i.e. the philosophy is that nearly everyone should be able to apply this product with commonly available equipment. The language on some labels goes so far so as to make illegal the use of innovative technologies. Phrases that specify minimum finished volumes per acre, the use of specific nozzles, prohibiting application by aircraft, and other requirements are common on pesticide labels. They are presumably placed there for legitimate reasons, but inadvertently lock the user into technology in existence at the time the label was produced. As new application technologies are developed, they must either conform to such restrictions or pass through a relabeling or equivalent process. A strategy more open would be to list performance criteria, such as only those handing and atomization requirements necessary to insure the survival and delivery of sufficient viable organisms to achieve pest control. Users, researchers, and regulators must make a more concerted effort to develop and promote the use of appropriate technology through consultation and cooperation at all stages of the research and development.

**Support for Development.** Support for the development of application systems for biorational agents must come from some source. Traditionally, the interested parties such as sprayer manufacturers, product manufacturers, grower organizations, and state and federal governments, including regulators, have been the source of most support for development of pest control application systems. The fractionating of the "sprayer market" into segments by control agent and crop implies that development will likely be required for each combination and there will exist no single market for application systems of the current size. Possible scenarios include; 1) the identification of niche markets by small manufacturers which then aggressively seek to develop, market, and protect through patents, the application equipment for that niche; 2) development of equipment by the producers of the biorational agent; 3) continued development efforts by Universities, the USDA, and other state and federal agencies; 4) growers of a commodity funding research and development of application systems for their commodity and; 5) development of application systems by growers who wish to use the biorational pest control agent and do not want to wait for others to take action.

In all likelihood, all of the above will occur simultaneously, as they have been occurring in the past. Through market forces, much of this progress will come as a result of efforts of individuals, either growers, equipment manufacturers, or those who see an opportunity to develop a business in applying biorational agents, much as custom applicators currently apply conventional pesticides.

## Summary and Conclusions

Development of economically feasible and effective application technology for the use of biorational pest control products will have an even more crucial issue in the years ahead. Research support is often limited to development of the pest control agents themselves, with the method of dispersal as an afterthought. The potential limitations under this scenario, in addition to non-use as a pest control strategy, include damage to the control organism itself, inability to apply the product in a timely manner, and costs that make use of biorationals unattractive to potential users.

Biorationals will likely continue to see increased use in forestry, production agriculture, home gardens and landscapes, and in urban pest control. The success of biorationals has been shown to be linked to use by trained and informed users. Success is also dependent upon effective application systems which incorporate the biology of the host, the pest, and the pest control agent along with engineering in order to develop a system that can place the organism where it needs to be, when it needs to be there, and in a viable form.

In closing, the following questions need to be addressed in order to speed the adoption of biorational pest control agents on a large scale:

*If society wishes to encourage the adoption of biorational pest control technologies as a non-chemical alternative for pest control, how can society be encouraged to support the development of the technology necessary to use and apply biorationals?* There is a perceived societal and environmental benefit to the increased use of biorational pest control agents, from increased environmental safety, worker safety, and food safety. What methods can society, as a whole, use to encourage the more rapid development and adoption of such technology?

*Is the stereotype, independent, family farm still an appropriate intended user of such technology?* The increased need for understanding the biological relationships between the crop, the pest, and the biorational control agent, continue to add expertise needed by a successful grower. Also, a single grower is often diversified into several crops in order to spread out the risk of a crop failure. With specific equipment needed for each pest of each crop, increased scouting requirements, and increased timeliness effect, is it still reasonable or desirable to expect that each grower should be able to use biorational agents without consultative assistance? From where should such assistance come and who should pay for it?

*Can market forces be expected to supply sufficient incentive for adoption of such technology?* For agricultural commodities, the market price is set at the grocery store or processing facility, where by and large, growers using biorational agents must compete against those using conventional pesticides. Some commodities command a premium when grown organically or with sustainable methods, while others do not. Will such a premium develop in other products and will consumers pay the premium, or will other methods be needed? If conventional pesticides are no longer available, will the selection and availability of fresh fruit and produce be reduced? Could tax or other government incentives be used to support the adoption of such technology?

*How can delivery systems for biorationals be made more user-friendly and system-friendly?* This is a technical and design task that must be incorporated into criteria for proper system performance. Will increased competition to supply these devices lead to this result?

## Literature Cited


1. Combellack, J. H. *Weed Research.* 1982, *Vol.* 22, pp 193-204.
2. Young, B. W. *Outlook on Agriculture.* 1986, *Vol.* 15, No. 2, pp 80-87.
3. Hislop, E. C. *Aspects of Applied Biology.* 1987, *Vol.* 14, pp 153-172.
4. *Pest Management: Biologically Based Technologies;* Lumsden, R. D.; Vaughn J. L., Eds.; American Chemical Society Conference Proceedings Series; Washington, DC, 1993.
5. Podgwaite, J. D.; N. R. DuBois; R. C. Reardon; J. Witcosky. *J. Econ. Ent.* 1993, *Vol.* 86, No. 3, pp 730-734.
6. DuBois, N. R.; R. C. Reardon; Karl Mierzejewski. *J. Econ. Entomol.* 1993, *Vol.* 86, No. 1, pp 26-33.
7. Fleming, R. A.; K. van Frankenhuyzen. *The Canadian Entomologist.* 1992, *Vol.* 124, pp 110 -1113.
8. Ogwang, J. A.; G. A. Matthews. *Tropical Pest Management.* 1989, *Vol.* 35, No. 4, pp 362-364.
9. Zehnder, G. W.; G. M. Ghidiu; J. Speese III. *J. Econ. Ent.* 1992, *Vol.* 85, No. 1, pp 281-288.
10. Vandenberg, J. D.; H. Shimanuki. *J. Econ. Ent.* 1990, *Vol.* 83, No. 3, pp 766-771.
11. Broza, M.; M. Brownbridge; B. Sneh. *Crop Protection.* 1991, *Vol.* 10, pp 229-233.
12. Smitley, D. R.; T. W. Davis. *Forest Entomology.* 1993, *Vol.* 86, No. 4, pp 1178-1184.
13. Nealis, V. G.; K. van Frankenhuyzen; B. L. Cadogan. *The Canadian Entomologist.* 1992, *Vol.* 124, pp 1085-1092.
14. Cadogan, B.; R. D. Scharbach. *The Canadian Entomologist.* 1993, *Vol.* 125, pp 479-488.
15. Casida, Jr. L. E. *Plant Disease.* 1992, *Vol.* 76, No. 12, pp 1217-1220.
16. *Nut Grower Magazine.* Western Agricultural Publishing Company, Fresno, CA, 1993, *Vol.* 13, No. 10.
17. Podgwaite, J, D.; R. C. Reardon; D. M. Kolodny-Hirsch; G. S. Walton. *J. Econ. Ent.* 1991, *Vol.* 84, No. 2 , pp 440-444.
18. Young, S. Y.; W. C. Yearian. *J. Entomol. Sci.* 1990, *Vol.* 25, No. 3, pp 486-492.
19. Vail, P. V.; C. Romine; T. J. Henneberry; M. R. Bell. *Southwestern Entomologist.* 1989, *Vol.* 14, No. 4, pp 329-338.
20. Jones, K. A.; N. S. Irving; D. Grzywacz; G. M. Moawad; A. H. Hussein; A. Fargahly. *Crop Protection.* 1994, *Vol. 13*, No. 5, pp 337-340.
21. Webb, S. E.; A. M. Shelton. *Entomophaga.* 1991, *Vol.* 36, No. 3, pp 379-389.
22. Muthiah, C.; R. J. Rabindra. *Indian J. of Agric. Sci.* 1991, *Vol.* 61, No. 6, pp 449-452.
23. Rogers, M. *Grounds Maintenance.* 1994, *Vol.* 29, No. 3, pp 90-94.
24. Hajek, A. E. In *Pest Management: Biologically Based Technologies*; Eds. R. D. Lumsden; J. L. Vaughn; American Chemical Society Conference Proceedings Series; Washington, DC, 1993, pp 54-62.
25. Wright, J. E.; L. D. Chandler. *J. of Invertebrate Pathology.* 1991, *Vol.* 58, pp 448-449.
26. Kydonieus, A. F.; M. Beroza. In *Management of Insect Pests with Semiochemicals Concepts and Practice*, Ed. E. R. Mitchell; Plenum Press, NY, 1981, pp 445-453.
27. Quisumbing, A. R.; A. F. Kydonieus. In *Insect Suppression with Controlled Release Pheromone Systems,* A. F. Kydonieus; M. Beroza, Eds.; CRC Press, Boca Raton, FL, 1982, pp 213-235.

28. Henneberry, T. J.; J. M. Gillespie; L. A. Bariola; H. M. Flint; G. D. Butler, Jr.; P. D. Lindgren; A. F. Kydonieus. In *Insect Suppression with Controlled Release Pheromone Systems*, Eds. A. F. Kydonieus; M. Beroza; CRC Press, Boca Raton, FL, 1982, pp 75-98.
29. Smith, K. L.; R. W. Baker; Y. Ninomiya. In *Controlled Release Delivery Systems*, T. J. Roseman; S. Z. Mansdorf, Eds.; Marcel Dekker, Inc., NY, 1983, pp 325-336.
30. Rothschild, G. H. L. In *Insect Suppression with Controlled Release Pheromone Systems*, A. F. Kydonieus; M. Beroza, Eds.; CRC Press, Boca Raton, FL, 1982, pp 117-134.
31. Ashare, E.; T. W. Brooks; D. W. Swenson. In *Insect Suppression with Controlled Release Pheromone Systems,* Eds. A. F. Kydonieus; M. Beroza; CRC Press, Boca Raton, FL, 1982, pp 237-244.
32. Tashenberg, E. F.; W. L. Roelofs. *Environmental Entomology*. 1976, *Vol.* 5, No. 4, pp 688-691.
33. Chamberlain, D. J.; B. R. Critchley; D. G. Campion; M. R. Attique; M. Rafique; M. I. Arif. *Bull. of Entomological Research. Vol.* 82, pp 449-458.
34. Moawad, G.; A. A. Khidr; M. Zaki; B. R. Critchley; L. J. McVeigh; D. G. Campion. *Tropical Pest Management. Vol.* 37, No. 1, pp 10-16.
35. Critchley, B. R.; D. J. Chamberlain; D. G. Campion; M. R. Attique; M. Ali; A. Ghaffar. *Bull. of Entomological Research.* 1991, *Vol.* 81, pp 371-378.
36. Shaver, T. N.; H. E. Brown. *J. Econ. Ent.* 1993, *Vol.* 86, No. 2, pp 377-381.
37. Tatsuki, S. *Insect Sci. Applic.* 1990, *Vol.* 11, No. 4/5, pp 807-812.
38. Flint, H. M.; L. M. McDonough; S. S. Salter; S. Walters. *J. Econ. Entomol. Vol.* 72, No. 5, pp 798-800.
39. Gentry, C. R.; C. E. Yonce; B. A. Bierl-Leonhart. In *Insect Suppression with Controlled Release Pheromone Systems*, Eds. A. F. Kydonieus; M. Beroza; CRC Press, Boca Raton, FL, 1982, pp 107-115.
40. Bishop, A. L.; H. J. McKenzie; C. P. Whittle. *Entomol. Exp. Appl. Vol.* 67, pp 41-46.
41. Knight, A. L.; L. A. Hull. J. *Econ. Entom.* 1989, *Vol.* 82, No. 4, pp 1019-1026.
42. Mueller, D.; L. Pierce; H. Benezet; V. Krischik. *Journal of the Kansas Entomological Society. Vol.* 63, No. 4, pp 548-553.
43. Leonhart, B. A.; V. C. Mastro; E. D. DeVilbiss. *J. Entomol. Sci.* 1992, *Vol.* 27, No. 3, pp 280-284.
44. Grant, G. G. *Forest Ecology and Management.* 1991, *Vol.* 39, pp 153-162.
45. Beevor, P. S.; H. David; O. T. Jones. *Insect Sci. Applic.* 1990, *Vol.* 11, No. 4/5, pp 787-794.
46. VanLenteren, J. C. In *Pest Management: Biologically Based Technologies*; R. D. Lumsden; J. L. Vaughn, Eds.; American Chemical Society, Washington, DC, 1993.
47. Thomson, W. T. *A Worldwide Guide to Beneficial Animals.* Thomson Publications, Fresno, CA, 1992.
48. Mracek, Z.; G. Jenser. *Acta Phytopathologia et Entomologica Hungarica.* 1988, *Vol.* 23, No. 1-2, pp 153-156.
49. Mills, N. J. Personal communication, 1993.
50. Clark, B. *Ag Consultant.* January, 1994.
52. Newton, P. J. *Bull. Ent. Res.* 1989, *Vol.* 79, pp 507-519.
51. Andow, D. A.; D. R. Prokrym. *Entomophaga.* 1991, *Vol.* 36, No. 1, pp 105-113.
53. Bouse, L. F.; J. B. Carlton; R. K. Morrison. *Transactions of the ASAE.* 1981, *Vol.* 24, No. 5, pp 1093-1098.

54. Bouse, L. F.; R. K. Morrison. *The Southwestern Entomologist.* 1985, Supplement No. 8, pp 36-48.
55. Bouse, L. F.; J. B. Carlton; S. L. Jones; R. K. Morrison; J. R. Ables. *Transactions of the ASAE.* 1980, *Vol.* 23, No. 6, pp 1359-1363, 1368.
56. Jones, S. L.; R. K. Morrison; J. R. Ables; L. F. Bouse; J. B. Carlton; D. L. Bull. *The Southwestern Entomologist.* 1979, *Vol.* 4, No. 1, pp 14-19.
57. Pickett, C. H.; F. E. Gilstrap; R. K. Morrison; L. F. Bouse. *J. Econ. Entom.* 1987, *Vol.* 80, No. 4, pp 906-910.
58. Pickett, C. H.; R. K. Morrison; F. E. Gilstrap; L. F. Bouse. *Internat. J. Acarol.* 1990, *Vol.* 16, No. 1, pp 37-40.
59. Smith, S. M.; D. R. Wallace; J. E. Laing; G. M. Eden; S. A. Nicholson. *Memoirs of the Entomological Society of Canada.* 1990, No. 153, pp 45-55.
60. Hope, C. A.; S. A. Nicholson; J. J. Churcher. *Memoirs of the Entomological Society of Canada.* 1990, No. 153, pp 38-44.
61. Carrow, J. R.; S. M. Smith. *Memoirs of the Entomological Society of Canada.* Ottawa, 1990, No. 153, pp 82-87.
62. King, E. G. In *Pest Management: Biologically Based Technologies*, Eds. R. D. Lumsden; J. L. Vaughn; American Chemical Society Conference Proceedings Series; Washington, DC, 1993, pp 90-100.
63. Dicke, M.; M. W. Sabelis; J. Takabayashi; J. Bruin; M. A. Posthumus. J. Chemical *Ecology.* 1990, *Vol.* 16, No. 11, pp. 3091-3118.
64. Gross, H. R. In *Semiochemicals, Their Role in Pest Control*, Eds. D. A. Nordlund; R. L. Jones; W. J. Lewis. John Wiley and Sons, NY, 1981, pp 137-150.
65. Barry, J. W.; P. J. Skyler; M. E. Teske; J. A. Rafferty; B. S. Grim. *Environ. Tox. and Chem.* 1993, *Vol.* 12, pp 1977-1989.

RECEIVED January 12, 1995

# Chapter 7

# Environmental Fate and Accountancy

**M. E. Teske[1], John W. Barry[2], and H. W. Thistle, Jr.[3]**

[1]Continuum Dynamics, Inc., P.O. Box 3073, Princeton, NJ 08540
[2]Forest Health Protection, Forest Service, U.S. Department of Agriculture, 2121–C Second Street, Davis, CA 95616
[3]Forest Service, U.S. Department of Agriculture, Fort Missoula, Building 1, Missoula, MT 59801

A recently developed analytical model for environmental fate and total accountancy predicts the mass fraction of aerially released material that vaporizes, deposits on canopies, deposits on the ground, or remains aloft, and is part of the FSCBG model, an accepted dispersion prediction system for aerial application of pesticides. In this review the importance of accurate field measurements for model evaluation is discussed in detail, and reference is made to aerial application data collected during 1991 field trials in Utah. Without information on fate (material transport, degradation, and persistence), researchers cannot make predictions, or even reasonable assumptions, about the impact of pesticides on species of concern or ecosystems in general; nor can they weigh the risks and economic benefits of pesticide use. Implications of model use to the environmental fate of pesticides are discussed.

Pesticide use in North American forests developed after World War II with the use of chemicals such as DDT (dichloro diphenyl trichloroethane) to control defoliators of conifers, and 2, 4-D (2, 4-dichlorophenoxyacetic acid) to control vegetation (*1*). The accelerated use of these and other synthetic chemical pesticides was challenged for environmental concerns in the 1960s, with many of them (specifically DDT) withdrawn from use and registration in the decades that followed.

The ban on DDT and similar compounds led to a search for methods to improve the efficacy, economy, safety and accountability of aerial application of pesticides in forests. It soon became apparent that the use of less persistent chemical and biological agents to control defoliators, such as the tussock moth, western spruce budworm, and gypsy moth, would require a higher degree of application precision than previously practiced, with more attention given to application technology and timing, atmospheric conditions, and target physical and behavioral characteristics. Emphasis was placed on developing operational methods which would increase pesticide deposit on the target and reduce spray drift, while also reducing the amount of spray material released.

Concurrent with the need to improve the efficiency and safety of aerial application, needs were expressed for techniques which would assist in accounting environmentally for pesticides released over forests. Information was needed on how much spray reached the tree crown and the forest floor, drifted off target, or remained in the atmosphere (*2*). Understanding how deposition processes interact with canopy

0097–6156/95/0595–0095$12.00/0

architecture, quantifying the collection of drops by foliage elements, and determining how much spray deposits on the forest floor all play a role in environmental fate accountancy within and above a forest canopy. These processes are complex because they include the entire tank mix -- the carrier, dilutent, adjuvant, and active ingredient, and include the transport, deposition, dispersion, and ultimate environmental fate of the spray material.

Consistent with the concern to improve the efficiency of aerial application has been an attempt to determine where spray drops deposit in forests. Numerous studies, summarized and reviewed historically (*3*, *4*), helped to provide an initial understanding of this behavior. The rationale for the referenced studies was based on the assumption that once it could be determined what is deposited in tree canopies, steps might be taken to apply the proper number and size range of drops and spray volume in such a manner so as to increase efficiency of deposition. This approach not only increases efficacy, but also reduces the amount of spray that drifts beyond the intended target. Safe, efficacious, and economical applications are dependent upon information generated by such studies.

Several factors are known to influence deposition of drops on foliage. These factors include drop size and specific gravity, wind speed, target shape and size, foliage density, and velocity of the falling drops. Other factors have a less well-defined role, such as the microenvironment surrounding the target, physical and chemical aspects of the drops, and characteristics of the target surface, including its electrical charge. Research is ongoing to understand the contribution of these factors to canopy penetration, drop deposition, impaction, retention on foliage, and efficacy.

All of these concerns are mirrored in a companion effort, that of developing a predictive model to quantify the behavior of the released spray, through evaporation, transport, diffusion, and deposition processes. The FSCBG (Forest Service Cramer-Barry-Grim) model embodies the latest technology and attempts to recover a consistent picture of the entire dispersion process. The details of this model have been reviewed elsewhere (*5*) and will only be highlighted here.

## FSCBG Model Development

The USDA Forest Service's ban on use of DDT in 1964 was a reaction to the agency's and the public's environmental concerns. In this reaction was the realization that the art of aerial application was primitive at best, having changed little since its inception in the 1920s (*4*). To improve aerial application, the first steps needed were to describe and understand the engineering and physical processes taking place, and to integrate available information data bases, or create new ones. Understanding these processes was basic to improving the safety, economy, efficacy, and environmental acceptance of aerial application. Computer modeling provided the mechanism to use the information data bases, and to understand and quantify the processes that influence spray behavior. With this understanding could come improvements in aerial spray operation techniques, as well as the hope of accounting for most of the spray material released into the atmosphere.

The FSCBG dispersion and deposition model, with its canopy penetration component, is a result of a long-standing USDA Forest Service and U. S. Army partnership to develop a method to predict dispersion, drift, evaporation, canopy penetration, deposition, and total accountancy and environmental fate of aerially released sprays. By the late 1960s provision had been made in the U. S. Army's Gaussian plume modeling techniques to account for the loss of material by gravitational settling of drops from elevated spray clouds, and to predict resulting surface deposition patterns (*6*). Additional work (*7* - *9*) led to the development of algorithms for considering the penetration of drops into canopies and simple expressions for wake effects of spray aircraft. By 1980 the model included an algorithm to consider evaporation of the spray drops as well (*10*).

A prototype model was first applied to forestry use in 1971 to determine application rates in testing of insecticides under consideration at that time for forest insect control in western forests (*11*, *12*). A first reported application of this technology (*13*) estimated the amount of spray material needed to control an outbreak of western spruce budworm. The implications of these early efforts in the use of mathematical models to improve the planning, conducting and subsequent analysis of spray operations and results were noted (*8*) and led to field evaluations (*14*) and further development of the model (*9*). The model was subsequently used (*15*) to determine offset distances in environmentally sensitive areas of Maine. FSCBG was then applied to the development of optimum swath widths, application rates, and aircraft release heights in other projects (*16*) and a pilot project in the Withlacoochee State Seed Orchard in Florida (*17* - *19*) that led to wide acceptance of aerial application in forestry seed orchards in the Southeast.

Continued success in simulating field experiments and control operations led to the inclusion of the near-wake AGDISP model (*20*) in FSCBG (*21*). A personal computer version followed (*22*), succeeded by the development of a more user-friendly interface (*23*). The model has since been applied to the determination of swath widths (*24*) and a complete sensitivity study of parameters affecting aerial application (*25*).

## Spray Accountancy in Forestry

Ecosystem management concerns of forest lands includes wildlife habitat, watershed, recreation, cultural values, and grazing for domestic animals, in addition to silviculture and timber production. Forest lands are critical to maintaining healthy environments and stable climates, unfortunately placing some uses in conflict with each other. For these reasons the informed citizenry is becoming increasingly concerned about threats to forest ecosystems from all uses, especially those with potential risks resulting from the limited use of pesticides on public forest lands. Questions are frequently asked as to the fate of pesticides that are applied to forests -- how much of the spray volatilized, moved off-site, deposited on trees, or contaminated water sources. Detecting and quantifying pesticide deposition, if incorporated in the spray project operations, is often limited to simple procedures such as the use of paper cards placed on the ground to detect spray drop deposits. This method, however, usually accounts for less than thirty percent of the initial spray volume applied to the treatment area (*4*). Responding to concerns about the fate of pesticides applied to forests, regulatory agencies, such as the U. S. Environmental Protection Agency, may soon impose spray accountability requirements on pesticide users, essentially requiring proof that the spray was applied to the treatment area and that off-target drift was avoided. Benefits of such accountability have the potential for further promoting prudent, efficient, and economical use and application of pesticides. Environmental requirements may initiate the development of new technology such as use of on-site samplers to collect, analyze and record pesticide data in real time. With this (and other) technology in place, pesticides can be monitored in the air and water, and on foliage, soil, equipment, and even people. If these requirements result in economic benefit, they will also be more readily accepted (*1*).

## FSCBG and Environmental Fate

FSCBG enables the prediction of the deposition and drift of released material, and when coupled to ground and water fate models, provides a complete environmental fate prediction system. The model includes an option to track the time history of material aloft, evaporated vapor, canopy deposition, and ground deposition; this option provides a total accountancy of the released material. The continuing development and improvement of the model for predicting the fate of released material is a priority need.

The computation process is in fact a straightforward one, once the complete solution to a specified spray problem has been solved with the FSCBG model (*5*). In short, the component elements are obtained as follows:

1. Evaporated vapor. The evaporation model keeps track of the decrease in drop size across the drop size distribution exiting from the nozzles on the aircraft, and compares the amount remaining with what began at the nozzle. If the evaporated fraction is designated $f_E$, then the nonevaporated fraction of material is $1 - f_E = f_R$.

2. Canopy deposition. The canopy penetration model keeps track of the decrease in spray material passing through the canopy, and compares the amount remaining with what entered the canopy. If the fraction lost in the canopy is designated $f_C$, then the canopy deposition fraction is $f_C * f_R$, and the noncanopy-deposited, nonevaporating fraction is $(1 - f_C) * f_R$.

3. Ground deposition. The deposition model present in FSCBG computes the ground deposition from the expression (*20*)

$$f_D = 0.5 * (1 - f_C) * f_R * [1 - erf(z / \sqrt{2} s)]$$

where $erf$ is the commonly computed error function, $z$ is the distance above the ground, and $s$ is the standard deviation of the growth of the dispersion cloud multiplied by the distance of drop fall below the height of release.

4. Material aloft. What is left from $(1 - f_C) * f_R - f_D$ is material still aloft. The solution is time-dependent, as the spray material descends through the forest canopy and toward the ground.

In its present configuration FSCBG takes input data entry from meteorological conditions, aircraft details, nozzle specifications, spray material information, canopy characteristics, and flight path scenario, all through menus managed by the user. FSCBG then predicts the behavior of the released spray material near the wake of the aircraft and into the far downwind environment. FSCBG considers every aspect of the spray process and the significance of atmospheric and aircraft-generated turbulence, leading to ground and canopy predictions. Typical environmental fate results, illustrating the presentation of predictions in four ways, are shown in Figures 1 to 4.

Once on the ground, the deposited material creates a predicted pattern that may be used as input to models that predict environmental fate of material in ground and water. Available pesticide transport models include PRZM (*26*), LEACHM (*27*), and GLEAMS (*28*). Each has been validated in a variety of agricultural applications (*29*). FSCBG is ideally suited for generating the inputs necessary to drive these soil and groundwater models. With FSCBG capability, the entire process of pesticide transport is predictable (*30*).

## Model Evaluation

Field evaluations, mostly involving tree canopies, have always played a critical role in the development of the FSCBG model (*31 - 33*). Many improvements to the model were in fact made possible by the qualitative agreement of model predictions with field data. The model simulates many of the complex processes occurring behind an aircraft and in the atmospheric boundary layer with accuracy and simplicity, but to do so it requires a careful set of field measurements and data before quantitative comparisons can be made. As the usefulness of the model has improved, so has the knowledge of which variables are more important and must be measured accurately (*25*). Of most importance is the accuracy and repeatability of field measurements, particularly in side-by-side sensor comparisons.

Environmental fate accounting will only be as good as the field personnel and instruments taken into the field to measure the residuals, since these measurements will

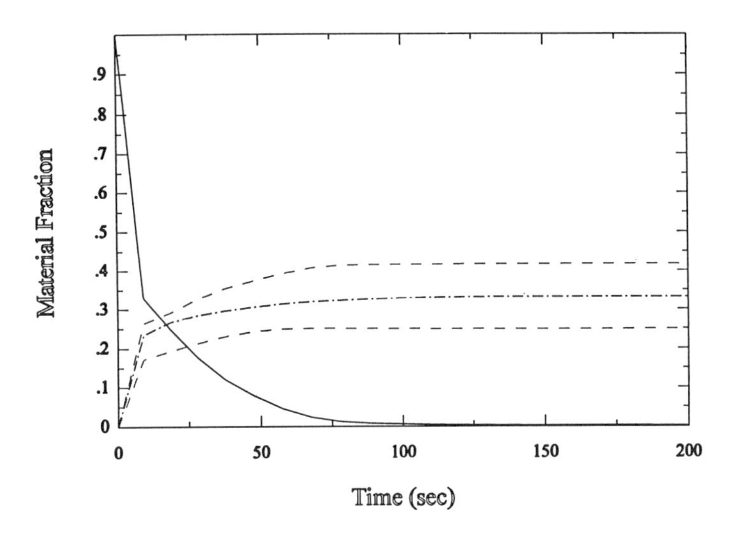


Figure 1. Spray accountancy represented by FSCBG as a time-history plot of the fraction of spray material still aloft (solid curve), deposited on the ground (upper dashed curve), deposited within the canopy (lower dashed curve), or evaporated.

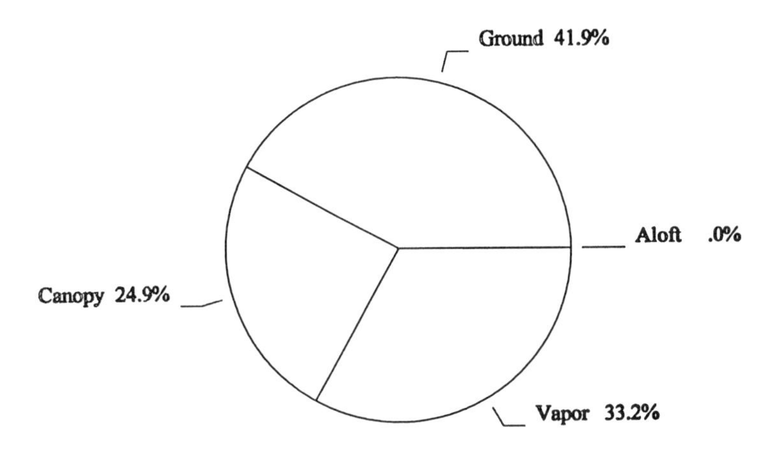


Figure 2. Spray accountancy represented by FSCBG as a pie chart of the final disposition of the spray material.

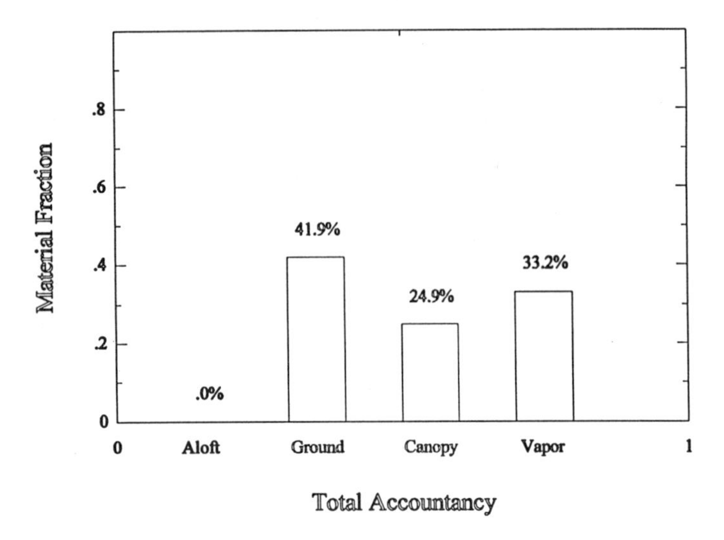


Figure 3. Spray accountancy represented by FSCBG as a bar chart of the final disposition of the spray material.

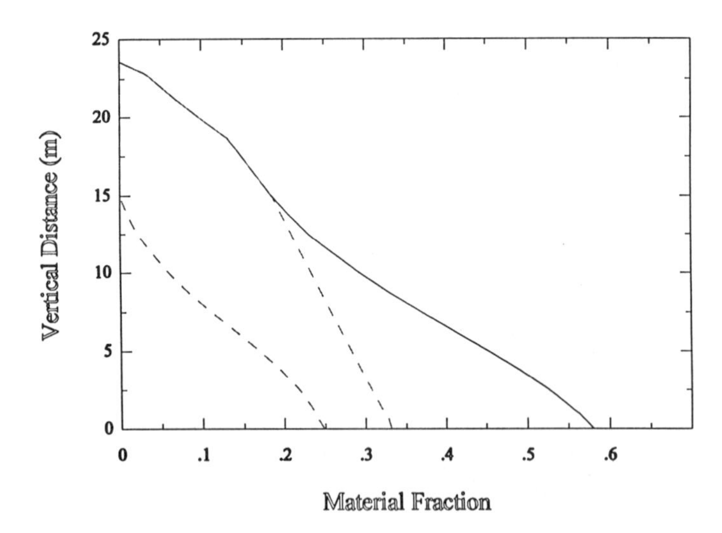


Figure 4. Spray loss accountancy represented by FSCBG as a plot of the fraction of spray material deposited within the canopy (left dashed curve), lost to evaporation (right dashed curve), or loss total, as a function of height.

ultimately be used to validate any predictive model (in this case FSCBG). For the last three years the USDA Forest Service, in cooperation with the U. S. Army and the State of Utah, has conducted aerial spray operations in the Wasatch Mountains above Salt Lake City, Utah, in conjunction with an operational eradication project for gypsy moth infestations in Gambel oak (*Quercus gambelii* Nutt) and big tooth maple (*Acer grandidentatum* Nutt). In 1991 three trials were conducted, and various types of samplers were paired and duplicated to compare recoveries. These results will be reviewed here with an eye toward repeatable field measurements. The complete study is summarized elsewhere (*33*).

## Sampling and Field Measurements

Correct design of the sampling scheme is critical to the goal of total accountancy of material released into the environment. In model validation field experiments it is desirable to collect data for all model variables (which may be a difficult task). In operational aerial spraying, where the model may be called upon to reproduce the material release, a more limited sampling program can be used, and many of the variables can be pre-programmed based upon historical data. In cases where the model is used in planning, hypothetical or pre-existing data are used, often in an effort to simulate worst case conditions.

Two types of sampling and measuring activities are generally conducted in the field: meteorological measurements and pesticide sampling. Meteorological measurements are made with one or more towers instrumented with wind speed and direction, temperature, relative humidity, and solar radiation at one or more heights (especially above any canopy present). To describe turbulence measurements, one-second interval data on wind velocity are needed (*34*). Temperature and relative humidity may be recorded at 10-second intervals.

Two important considerations in pesticide sampling are frequency and duration. Sampling frequency is defined as the number of samples collected in a given space or time period. Duration is the length of time over which samples are collected, or data are taken. Since the goal of pesticide application is often to apply a given amount of material per square area of surface, the data collected are often totals or averages for a given application. These data are essentially independent of time, unless sequencing or time-indexed samplers are used. This field approach usually results in one data point per sampler per test. Unfortunately, this type of low frequency sampling may not be sufficient to answer all of the questions raised in model development and evaluation work, since short-period meteorological events may occur within the overall test and substantially alter test results. Therefore, wherever the technology allows, data are collected at much higher frequencies so that subsequent detailed analyses can be performed.

There are several types of instruments that can be used to collect samples of pesticides applied to forests; and there are a multitude of factors to consider in selecting and using any of them. Samplers are selected to meet the objective of the project with consideration to the need for qualitative (did the spray reach this point) or quantitative (how much spray reached this point) data. Other considerations include cost, ease of handling and assay, dependability, availability, and technician skill. If the objective is to quantify pesticide deposition fallout within or downwind of the treatment area, then samplers such as metal plates, Mylar sheets, or paper cards can be used and positioned horizontally. If the objective is to quantify pesticide in the air, air-drawn samplers such as impingers or filters, static impaction devices such as Rotorods, or vertical impaction samplers like simple soda straws, glass rods, or other cylinders, might be used.

All samplers can be assayed qualitatively and usually quantitatively. Assay methods are dependent upon what will be assayed, the option being the pesticide or a tracer, inherent or intentionally added to the tank mix. The pesticide may be assayed chemically for its toxin, biologically for its bacteria or virus, or physically for its mass, drop size and number of drops. The tracer or other ingredient of the tank mix may be assayed by one or more of the above methods, if applicable, or if the tracer has a fluorescent energy source, the assay may be optical (*35*). In the case of *Bacillus thuringiensis* (Bt) pesticide sprays, the Bt can be assayed chemically for its protein endotoxin, biologically for its bacterial spore, biologically for its effects on susceptible insects, physically for its mass, drop size and drop number, or optically for its fluorescent tracer, if such a tracer is added to the tank mix.

## Sampler Types and Use

The Rotorod sampler (*36*) has proven to be a practical sampler for Bt and one that we have found to produce consistent data. It is a U-shaped rotating arm impaction device capable of obtaining quantitative data of airborne particulates in the size range of 10 to 100 micrometers. The H-shaped Rotorod is designed to sample particles smaller than 10 micrometers, but has not been evaluated as a sampler for Bt. The U-shaped Rotorod spins through a volume of air and the concentration measurement is a function of the ambient wind speed. It samples 120 liters of air per minute rotating at 2400 rpm with a collection efficiency of close to 100 percent (*37*). In a typical field design used by the authors, four Rotorods are co-located as paired duplicates. Rotorods may also be used statically as a qualitative sampler, as its collection efficiency is dependent upon ambient wind energy. Collection efficiency for nonspinning Rotorods may be calculated (*38*) to be nearly 100 percent, given the typically low mean transport winds (1.0 m/s) observed in the Utah tests, and the size of the drops collected.

Other samplers used at various times have been Wagner (filter type) samplers and Reynier (time/concentration type) samplers (*39*). The principal of the Wagner sampler is to draw a known volume of air into the filter holder and across the filter. The Reynier sampler exposes a rotating agar plate to the ambient environment, with only a fraction of the plate exposed at a given time so that the sampler yields a time history of the Bt spray cloud. Wagner samplers are considered 100 percent efficient compared to a standard, the all-glass impinger sampler (*39*). Flat cards not on the ground will always have collection efficiencies less than 100 percent, but recoveries may be quantified with other collection techniques (*40*).

A quantitative problem with most samplers, in a general sense, is the determination of the collection efficiency of the collecting surface. Factors influencing collector efficiency include size and shape of the sampler, nature of the substrate and its surface, and orientation and velocity of the air that carries the agent to be sampled. Physical and chemical properties of the spray drops can also influence collection efficiency. Other approaches are to analyze treated foliage directly for the pesticide residue, or measure the mortality rates of insects fed foliage from the treated area. The latter can provide convincing efficacy data but is often variable due to biological variability. The subject of passive sampling surfaces has received much attention over the last two decades, with the resulting conclusion being that this type of sampler can provide good qualitative, relative information when properly arrayed and collection efficiencies determined, but accurate quantitative information is labor intensive.

Also, since the early work (*41, 42*), there has been considerable focus on the utility and advantage of using natural foliage to collect and quantify the amount of pesticide depositing and impacting on foliage. Bt and most other forest-use insecticides inflict casualties to defoliating insects as they ingest treated foliage; therefore, foliage being the target more closely approaches an ideal sampler. Determining the collection efficiency of foliage is a challenge, with one approach being to use another sampler type and comparing recoveries on a relative basis; and another conducting controlled wind tunnel studies. Assay techniques for foliage are similar to assay of other sampler types, chemical or biological assay, or physically counting and measuring drop stains. Regardless of the sampler type used, one must quantify the surface area or volume of the collecting substrate. The sampling or collection efficiency of the foliage can be calculated by equations (*38*) for cylinders and ribbons.

Sampler duplication, both in numbers and types, is also critical to any field study. When evaluating a physical model that predicts by the laws of physics, it is essential to take great care in the selection and use of samplers and sampling techniques used to collect data for the evaluation. Potential errors from the sampling process include field and laboratory contamination; motor, pump, or battery failure; sample scheduling errors; assay failures; and incorrect sample labeling. To avoid such errors it is prudent to increase sampler density, provide duplicates, and use more than one type of sampler. From a statistical viewpoint fewer stations with a higher sampler density is a superior approach to more stations with lower sampler density for examining sampling errors.

Spatial sampling frequency, more often referred to as sampling density, is usually controlled by test logistics. Typically, a finite number of samplers must be arrayed over a limited number of available sites to yield the best possible spatial representation of the location of the released material. Experience and model simulations may be used *a priori* to attempt to develop the sampling scheme and to optimize sampler location and density. The spatial array of the samplers should reflect the objectives of the specific experimental or operational program, and consider human resources, topography, climatology, and logistical questions of site ownership, around-the-clock accessibility, instrument power and shelter, etc.

The question of sampling duration has in the past been straightforward in spray model validation experiments. Since the material is released over a finite period (typically from minutes to two to three hours), the sampling duration is dictated by the duration of the release. Sampling begins before the cloud arrives and continues until the trailing edge of the cloud has passed the samplers. These times can be predicted given the mean wind speed, distance to the samplers, and the total spraying time. As our understanding advances, more subtle atmospheric questions of long-range drift and material re-entrainment and recirculation will come under investigation and may call for case-by-case decisions regarding sampling duration.

## Utah Studies

The discussion here focuses on a series of model evaluation experiments conducted in the Wasatch Front of Utah during 1991-1993. The Utah studies are presented here as an example of an environmental fate experiment in forest ecosystems. In these experiments model developers took advantage of an existing gypsy moth eradication program conducted by the USDA Forest Service and the State of Utah. This "piggybacking" approach is positive from an economic standpoint and ensures that the experimentalists are participating under actual operational conditions. The disadvantages of this approach are in the loss of flexibility in conducting an experiment due to the logistical (schedule, location, weather, etc.) requirements of the field program. Overall, these experiments yielded a wealth of valuable new model development data.

Each year the experimental effort was conducted in a different location within the eradication area. The objectives and design of the effort varied each year as more data were processed and new questions arose. Among the basic modeling questions being investigated were questions of long-range drift, spray drift in complex terrain, and canopy penetration of released material.

To meet these objectives, the sampling scheme varied from year to year. The sites were all in mountainous terrain of varying severity. Thus, the sampler arrays were generally aligned along valley axes where the maximum spray drift was expected to occur due to topographic control of the wind fields in these circumstances. Samplers were placed at distances ranging from the edge of the spray area to over 10 km downwind. The number of pesticide sampling stations exceeded 25, and various meteorological parameters were measured at several sites during the studies.

Meteorological monitoring required that the basic meteorological inputs to the model be met and that the monitoring be sophisticated enough to answer some of the more difficult questions, especially those surrounding spray drift in and over mountain forests. A basic meteorological station recorded two levels of wind speed and direction, two levels of humidity and two or three levels of temperature. The station also recorded one channel of net radiation and one channel of vertical wind speed. This information was recorded at 1 Hz for subsequent reduction into 10-minute averages. Three of these stations were deployed each year. A number of 1-meter towers with wind speed and direction were also used to enhance spatial coverage. Upper air meteorological information was gathered during the 1991 and 1992 programs using a tethersonde balloon, which raised a package of meteorological instruments into the atmosphere and transmitted the conditions at various heights back to the surface. In the 1993 experiment a SoDAR, which utilizes pulsed sound to determine wind speed and direction in the overlaying atmosphere, was deployed to characterize the upper air flow. To understand meteorological processes within the forest canopy, vertical profiles of wind speed, wind direction and temperature were also collected. In 1993 a sophisticated system of three-dimensional sonic anemometers, which use sound transmitted over a 20-cm path length to measure turbulence in the ambient airstream, was deployed to measure wind speed, direction, and turbulence in the forest canopy.

Various types of droplet and particulate sampling devices have also been used during the three years of spray drift model validation work. The primary types of instruments used are passive surfaces, such as Mylar sheets, placed downwind of the release to measure the material that falls out onto the surface, and Rotorods, vertically oriented rods that collect material impacting on the spinning rods. The passive surface technique has been used for many years with many different types of surfaces ranging from metallic to paper (*35*). Typically, the surfaces are taken to a laboratory and scanned optically to find traces of the released material, or assayed chemically or biologically. In the case of Bt the material on the collector is washed off, diluted and plated. Bt colonies growing on the plate are then counted.

## Statistical Results

Table I summarizes the consistency of the field measurements from the Utah study from four sampler types, using the relative standard deviation (*RSD*) at each of the ten paired stations along the downwind sampling line. The comparison procedure is as follows:

1. The equation used to compute *RSD* of the dosage, total flux and deposition levels $D_i$ recorded by the samplers is

$$RSD = \frac{1}{\overline{D}}\left[\frac{1}{N}\sum(D_i - \overline{D})^2\right]^{1/2}$$

where $\overline{D}$ is the average value $\overline{D} = \frac{1}{N}\sum D_i$ ; the index $i$ denotes each dosage, total flux and deposition data entry; and $N$ denotes the total number of data points considered.

2. The two data points (from the paired instruments) are taken as the values $D_1$ and $D_2$ , averaged to give $\overline{D}$ , and used to evaluate *RSD* from the above equation for each sampler site.

3. The *RSD* values are averaged over the number of sampler sites, to produce the results shown in Table I.

A set of data well-approximated by its average value at each downwind distance might have a relative standard deviation as high as 0.1 . A level of 0.1 would imply that the data exhibited a standard deviation of ten percent from the average value at each downwind distance.

Results in Table I show consistency from trial to trial but indicate some variation between the instruments at each downwind distance. The spinning Rotorods exhibit the least amount of variation, and the nonspinning (or static) Rotorods, the most. This result is understandable, as the spinning Rotorods maintain constant revolutions per minute, whereas the nonspinning Rotorods collect as a function of wind velocity. Given these results and the logistics associated with the aspirated Wagner sampler, Rotorod measurements were favored over the Wagner sampler in follow-up studies.

**Table I. Relative standard deviations at paired duplicate sampling stations for each sampler type, Utah 1991 gypsy moth project**

| Sampler | Trial 1 | Trial 2 | Trial 3 | Combined |
|---|---|---|---|---|
| Spinning Rotorods | 0.143 | 0.174 | 0.196 | 0.172 |
| Wagner Samplers | 0.236 | 0.312 | 0.319 | 0.291 |
| Nonspinning Rotorods | 0.196 | 0.391 | 0.316 | 0.326 |
| Mylar Samplers | 0.250 | 0.224 | 0.121 | 0.206 |

## Summary

This paper has reviewed the FSCBG spray model and aerial application field studies over forested canopies, as they relate to field sampling, measurements, collection efficiencies, pesticide samplers and their applicability in environmental fate and accountancy. The use of appropriate numbers and types of samplers, wisely located and properly operated, to recover field samples of the spray is considered paramount to a consistent understanding of environmental fate of released spray material, and to validate computer model predictions. Continuing field studies and model comparisons to field-collected data will enable improvements in spray accountancy of the model for years to come, and in effect help develop an improved, easier-to-use capability, and a more widely accepted computational tool for both forestry and agricultural application.

## Literature Cited

1. Barry, J. W. In *Application Technology for Crop Protection*; Matthews, G. A.; Hislop, E. C., Eds.; CAB International: 1993; pp 241-273.
2. Boyle, D. G. Report No. FPM 89-3, 1989; USDA Forest Service. Davis, CA.
3. van Liere, J.; Barry, J. W. Study No. 71-152 Phase II Project No. DTC-FR-71-152(II), 1973; Deseret Test Center. Fort Douglas, UT.
4. Barry, J. W. In *Chemical and Biological Controls in Forestry*; Garner, W. Y.; Harvey Jr., J., Eds.; American Chemical Society: Washington, DC, 1984; pp 117-137.
5. Teske, M. E.; Bowers, J. F.; Rafferty, J. E.; Barry, J. W. *Environ. Toxic. and Chem.* **1993**, 12, 453-464.
6. Cramer, H. E.; Bjorklund, J. R.; Record, F. A.; Dumbauld, R. K.; Swanson, R. N.; Faulkner, J. E.; Tingle, A. G. Report No. DTC-TR-72-609-I, 1972; U. S. Army Dugway Proving Ground. Dugway, UT.
7. Grim, B. S.; Barry, J. W. Report No. MEDC-2425, 1975; USDA Forest Service. Missoula, MT.
8. Dumbauld, R. K.; Cramer, H. E.; Barry, J. W. Report No. DPG-TR-M935P, 1975; U. S. Army Dugway Proving Ground. Dugway, UT.

9. Dumbauld, R. K.; Rafferty, J. E.; Bjorklund, J. R. Report No. TR-77-308-01, 1977; H. E. Cramer Co. Salt Lake City, UT.
10. Dumbauld, R. K.; Bjorklund, J. R.; Saterlie, S. F. Report No. 80-11, 1980; USDA Forest Service. Davis, CA.
11. Barry, J. W.; Tysowsky, M.; Ekblad, R. B.; Ciesla, W. M. *Trans. ASAE* **1974**, 17, 645-650.
12. Barry, J. W.; Ekblad, R. B. Paper No. 83-1006, 1983; ASAE. St. Joseph, MI, 1983.
13. Waldron, A. W. Report No. DPG-TN-M605P, 1975; U. S. Army Dugway Proving Ground. Dugway, UT.
14. Boyle, D. G.; Barry, J. W.; Eckard, C. O.; Taylor, W. T.; McIntyre, W. C.; Dumbauld, R. K.; Cramer, H. E. Report No. DPG-DR-C980A, 1975; U. S. Army Dugway Proving Ground. Dugway, UT.
15. Dumbauld, R. K.; Bjorklund, J. R. Report No. TR-77-310-01, 1977; H. E. Cramer Co. Salt Lake City, UT.
16. Dumbauld, R. K.; Bowman, C. R.; Rafferty, J. E. Report No. 80-5, 1980; USDA Forest Service. Davis, CA.
17. Barry, J. W.; Kenney, P. A.; Barber, L. R.; Ekblad, R. B.; Dumbauld, R. K.; Rafferty, J. E.; Flake, H. W.; Overgaard, N. A. Report No. 82-1-23, 1982; USDA Forest Service. Asheville, NC.
18. Rafferty, J. E.; Dumbauld, R. K.; Flake, H. W.; Barry, J. W.; Wong, J. Report No. FPM 83-1, 1982; USDA Forest Service. Davis, CA.
19. Barry, J. W.; Barber, L. R.; Kenney, P. A.; Overgaard, N. A. *Southern J. Appl. Forestry* **1984**, 8, 127-131.
20. Bilanin, A. J.; Teske, M. E.; Barry, J. W.; Ekblad, R. B. *Trans. ASAE* **1989**, 32, 327-334.
21. Bjorklund, J. R.; Bowman, C. R.; Dodd, G. C. Report No. FPM 88-5, 1988; USDA Forest Service. Davis, CA.
22. Curbishley, T. B.; Skyler, P. J. Report No. FPM 89-1, 1989; USDA Forest Service. Davis, CA.
23. Teske, M. E.; Curbishley, T. B. Report No. FPM 91-1, 1991; USDA Forest Service. Davis, CA.
24. Teske, M. E.; Twardus, D. B.; Ekblad, R. B. Report No. 9034-2807-MTDC, 1990; USDA Forest Service. Missoula, MT.
25. Teske, M. E.; Barry, J. W. *Trans. ASAE* **1993**, 36, 27-33.
26. Carsel, R. F.; Smith, C. N.; Mulkey, L. A.; Jowise, P. U. S. Environmental Protection Agency. Athens, GA, 1984; EPA-600/3-84-109.
27. Wagenet, R. J.; Hutson, J. L. *LEACHM: a model for simulating the leaching and chemistry of solutes in the plant root zone* Volume 2; Water Resources Institute. New York, NY, 1989.
28. Leonard, R. A.; Knisel, W. G.; Still, D. A. *Trans. ASAE* **1987**, 30, 1403-1418.
29. Harter, T.; Teutsch, G. *Pesticides in the Next Decade: The Challenges Ahead*; Weigmann, D. L., Ed.; Proceedings of the Third National Research Conference on Pesticides: Richmond, VA, 1990; pp 725-750.
30. Barry, J. W.; Teske, M. E. XIX International Congress of Entomology. Beijing, China, 1992; Session 14S-5.
31. Teske, M. E.; Bentson, K. P.; Sandquist, R. E.; Barry, J. W.; Ekblad, R. B. *J. Applied Meteor.* **1991**, 30, 1366-1375.
32. Anderson, D. E.; Miller, D. R.; Wang, Y.; Yendol, W. G.; Mierzejewski, K; McManus, M. L. *J. Appl ed Meteor.* **1992**, 31, 1457-1466.
33. Barry, J. W.; Skyler, P. J.; Teske, M. E.; Rafferty, J. E.; Grim, B. S. *Environ. Toxic. and Chem.* **1993**, 12, 1977-1989.
34. Haugen, D. A. *J. Applied Meteor.* **1963**, 2, 306-308.
35. Teske, M. E.; MacNichol, A. Z.; Barry, J. W. ASTM 14th Symposium on Pesticide Formulation and Application Systems; Dallas/Fort Worth, TX, 1993.

36. Brown, T. Technical Manual No. 6, 1976; Ted Brown Associates. Las Altos Hills, CA.
37. Edmonds, R. L. *Plant Disease Reporter* **1972**, 56, 704-708.
38. Golovin, M. M.; Putnam, A. A. *Ind. Eng. Fundam.* **1962**, 1, 264-273.
39. Wolf, H. W.; Skaliy, P.; Hall, L. B.; Harris, M. M.; Decker, H. M.; Buchanan, L. M.; Dahlgrem, C. M. *Sampling Microbiological Aerosols*; U. S. Public Health Service: Washington, DC, 1959.
40. Miller, D. R.; Yendol, W. G.; McManus, M. L. *J. Environ. Sci. Health* **1992**, B27, 185-208.
41. Himel, C. M.; Moore, A. D. *Science* **1967**, 156, 1250-1251.
42. Barry, J. W.; Ciesla, W. M.; Tysowsky Jr., M.; Ekblad, R. B. *J. Econ. Entom.* **1977**, 70, 387-388.

RECEIVED August 5, 1994

# Chapter 8

# Factors Affecting Spray Deposition, Distribution, Coverage, and Persistence of Biorational Control Agents in Forest Canopies

**Alam Sundaram**

**Forest Pest Management Institute, Canadian Forest Service, Natural Resources Canada, 1219 Queen Street East, Box 490, Sault Sainte Marie, Ontario P6A 5M7, Canada**

In the past, population suppression of forest insects has been possible in Canada through the judicious use of quick-acting, broad-spectrum synthetic insecticides. Although beneficial, extensive use of these materials could potentially be hazardous to the environment. Currently, intensive research is being conducted to find alternatives with a relatively narrow spectrum of activity. This paper contains a brief overview of research carried out on four biorational insecticides (diflubenzuron, tebufenozide, azadirachtin and *Bacillus thuringiensis*), to study the role of formulation and delivery on field performance. At present, biorational control agents are being viewed as "environmentally safe" compared to conventional chemical pesticides. The data presented here indicate that diflubenzuron and tebufenozide persist in forestry substrates for many months, suggesting the need for a thorough risk-benefit analysis before establishing their "environmental acceptability." In contrast, azadirachtin and *Bacillus thuringiensis* are too short-lived to be consistently effective, thus indicating the need to improve formulation and application methods. This paper also summarizes data for some of the several factors affecting field deposition, distribution, coverage and persistence of pesticides, *viz.,* formulation ingredients and physical properties, application volumes, droplet spectra and droplets per unit area of targets, rain droplet sizes and ageing period of deposits, type of foliage, and nature of target surfaces.

Sustainable development of forest resources is vital to Canada's economic growth. Protection of these resources against unacceptable losses due to various insect pests is essential to provide the basis for such development. Among the strategies available today, the responsible use of environmentally benign and ecologically acceptable insecticides is the most effective way to reduce losses caused by forest insects.

The use of synthetic, broad-spectrum insecticides (i.e., organochlorines, organophosphates and carbamates) for forest protection has been in decline for some

0097–6156/95/0595–0108$13.50/0
Published 1995 American Chemical Society

time, because of their potential for adverse side effects on the environment and on human health (*1*). Considerable effort is being made to find safe and biodegradable alternatives. Research in recent years has been turning more towards selective synthetic chemicals and biorational control agents of a relatively narrow spectrum of activity (e.g., diflubenzuron, tebufenozide), plant-derived pesticides (e.g., azadirachtin), naturally occurring microbial agents (e.g., *Bacillus thuringiensis*, Baculoviruses, Entomopathogenic fungi), entomogenous nematodes and parasites. These classes of agents possess many characteristics that may make them primary candidates as alternatives to broad-spectrum chemical pesticides.

Biorational control agents can be defined as "naturally occurring pesticides obtained from either plants or animals, or those synthesized by man whose geometric and stereochemical structures are identical to their naturally occurring counterparts". Diflubenzuron [Dimilin®, 1-(4-chlorophenyl)-3-(2,6-difluorobenzoyl) urea], though not a biorational pesticide in the true sense, shares many of the qualities of such materials. It is an insect growth regulator that acts on a unique biosynthetic system, chitin formation, in arthropods (*2*). Tebufenozide, known as Mimic®, or RH-5992 [N'-*t*-butyl-N'-(3,5-dimethylbenzoyl)-N-(4-ethylbenzoyl) hydrazine], is a novel type of insect growth regulator interfering with the moulting process of lepidopteran insects (*3,4*). It acts as an agonist or mimic of insect moulting hormone, 20-hydroxyecdysone, causing feeding inhibition, premature ecdysis and eventual death of the exposed insects. Among the botanicals, azadirachtin, a mixture of seven structurally-related tetranortriterpenoids isolated from the seeds of neem tree or Indian lilac [*Azadirachta indica* A. Juss. (Meliaceae)], has attracted the greatest attention (*5*). The compound is an insect repellent, antifeedant and growth regulator, and has a narrow spectrum of activity compared to the organochlorines, organophosphates and carbamates. The insecticidal activity is mainly due to the major isomer (*ca* 85%), azadirachtin-A ($C_{35}H_{44}O_{16}$), present in the extract of neem seeds (*6*).

*Bacillus thuringiensis* var. *kurstaki* (BTK) is a gram-positive, aerobic, spore-forming bacterium that occurs naturally. The bacterium synthesizes an intracellular parasporal glycoprotein crystal which is toxic to lepidopteran insects (*7,8*). The insecticidal activity of the bacterium is mainly associated with the crystalline protein. In highly susceptible insects, only the crystal is needed for control, but in less susceptible insects the synergistic effect of endospores together with the crystal is required (*9*). BTK is usually applied as a suspension in a water- or oil-based formulation, which generally contains additives designed to increase BTK effectiveness. Among these additives are thickeners to provide high viscosity for uniform suspension of spores, parasporal crystals and other ingredients (*10*); wetting agents for enhanced droplet spreading on leaf surfaces (*11*); and non-volatile oils that increase retention and adhesion of spray deposits onto foliage (*12*).

The present paper provides an overview of some of the recent findings from investigations conducted in Forest Pest Management Institute (FPMI) on physical and chemical factors affecting droplet deposition and persistence in target foliage of four biorational control agents, diflubenzuron, tebufenozide, azadirachtin, and *Bacillus thuringiensis* var. *kurstaki*. Two types of investigations will be discussed, laboratory studies on rainfastness of different formulations and field studies on persistence characteristics. The data will be examined to understand the role of formulation ingredients and application volumes on canopy deposition, coverage and persistence in the two types of investigations. The data will also be examined from the standpoint of whether these materials have the potential to be as efficacious as the conventional chemicals, and whether the anticipated "environmental acceptability" of these biorationals can actually be demonstrated under real field-use conditions.

## Diflubenzuron

**Study I - Effect of Carrier Medium on Diflubenzuron Washoff from Foliage.** Environmental persistence of a pest control agent depends largely on how the various weathering processes, e.g., rain-washing, photodegradation and volatilization, affect the persistence of the material on target surfaces. In a laboratory investigation, the rain-washing of diflubenzuron (*13*) was investigated on balsam fir and maple foliage to determine the effect of water- or oil-based carriers on rainfastness. Dimilin® WP-25, a wettable powder formulation, was diluted with water, a light paraffinic oil, a heavy paraffinic oil, or a mixture of canola oil and Cyclosol® 63 (an aromatic solvent). The mixtures were sprayed using mono-sized droplets. Rainfall was applied to determine the role of rain droplet sizes and ageing of deposits on diflubenzuron washoff. The results indicated that in fir foliage, washoff was the most with the water carrier, and the ageing period did not influence the washoff. The three oil-based mixtures showed greater rainfastness depending on the carrier liquid and ageing period. In the maple foliage, however, washoff was the most when the light paraffinic oil was used as the carrier, whereas it was the least with the water carrier; and that ageing of deposits increased rainfastness of all the four mixtures.

From the efficacy standpoint, it appears that no generalized conclusion can be drawn at this point of time as to which formulation would be better, because the water carrier provided greater rainfastness for diflubenzuron on maple foliage, whereas the canola-Cyclosol oil carrier provided greater rainfastness on fir foliage. Nonetheless, with all formulations the rainfall removed > 50% of the initial deposits. Therefore, from the standpoint of environmental contamination after a heavy rainfall, all formulations would contribute to measurable residues in the lithosphere. However, during an aerial treatment the lithosphere would already have received some of the applied material, and therefore, one would expect only a relatively small increase in soil residues as a result of diflubenzuron washoff. If we use the 'best case scenario' of spray deposition, i.e., if all the applied diflubenzuron (at a dosage rate of 70 g /ha) were deposited, the initial deposits on canopy foliage (or on soil in a forest opening) would be 0.7 μg/cm$^2$. If about 50% of 0.7 μg/cm$^2$ were to reach, say, 100 cm$^2$ area of the ground (the rain-washings from a small foliar area would fall on a large area of the ground), the deposit on the lithosphere would only be 0.7035 μg/cm$^2$. Thus, all the four formulations can be considered as 'environmentally acceptable' from the standpoint of environmental contamination due to rain-washing of foliar residues.

**Study II - Effect of Application Volume on Field Deposits and Persistence of Diflubenzuron.** In an aerial application study (*14,15*) using 70 g of active ingredient (AI) in three volume rates, 10, 5.0 and 2.5 L/ha, the droplet size spectrum parameters on artificial samplers decreased gradually, as the application volume decreased. The droplet density (droplets/$cm^2$) and deposit levels were higher on ground samplers and on canopy foliage (pine and maple) at the 10 L/ha volume than at the 5 L/ha rate, which in turn was higher than at the 2.5 L/ha rate (Tables 1 and 2). This behaviour was attributed to the droplet size spectra of sprays.

The disappearance of diflubenzuron from both pine and maple foliage was gradual, and followed first order kinetics according to equation (1):

$$Y = A + B\ e^{-Ct} \qquad (1)$$

where Y is the percentage residues remaining at time 't'; and A, B and C are constants. Using equation (1), values of $DT_{50}$, i.e., the time required for dissipation of 50% of the initial deposits, were also calculated (Table 2). The data indicated similar $DT_{50}$ values for pine and maple foliage at the three volume rates used, except for the markedly higher values for maple foliage in the 10 L/ha block. This could be due to the high initial deposits obtained. The rate of pesticide loss from the environment is known (*16*) to be inversely proportional to the initial deposit obtained.

Table 1. **Diflubenzuron:** Spray application details and meteorological parameters

| Parameter | 70 g AI in 10.0 L/ha | 70 g AI in 5.0 L/ha | 70 g AI in 2.5 L/ha |
|---|---|---|---|
| Date of application | June 5, 1986 | June 5, 1986 | June 5, 1986 |
| Time of application (EDT) | 0645 | 0745 | 1215 |
| Wind speed (av.) | 7.2 km/h | 5.7 km/h | 6.6 km/h |
| Temperature (av.) | 13.6°C | 13.4°C | 10.6°C |
| Relative humidity (av.) | 85% | 87% | 86% |
| Aircraft type | ←------Piper Pawnee Brave------→ | | |
| Aircraft speed | ←------160 km/h------→ | | |
| Atomizer used | ←------4 x Micronair® AU4000------→ | | |
| Blade angle setting | ←------35°------→ | | |
| Spray height | ←------20 m above canopy------→ | | |
| Swath width | ←------45 m------→ | | |

The present findings on the long persistence of diflubenzuron in pine and maple foliage agree with those observed in aquatic plants (*17,18*), grass (*19*), cotton leaves (*20,21*), citrus foliage (*22*), hardwood and conifer foliage (*23*), spruce needles (*24*) and oak foliage (*25*). Diflubenzuron is not a systemic insecticide because little translocation occurs (*26*) after foliar deposition. Therefore, the dissipation process can only involve loss of surface deposits by physicochemical means such as sloughing, volatilization, co-distillation, photolysis, hydrolysis, rain-washing etc. (heavy rain occurred between 24 and 48 h after treatment), during the initial stages. During the later stages, however, the dissipation could have been partly due to the slow loss of deposit (a cumulative rainfall of about 427 mm occurred during the 120-d sampling period), and partly due to the metabolic and/or hydrolytic attack (*27*) of the chemical.

The dissipation pattern of diflubenzuron from pine and maple foliage is quite different from the biphasic pattern of disappearance of aminocarb (*28*), and fenitrothion (*29,30*) from balsam fir foliage when water-based formulations were used. These two chemicals dissipated very rapidly within a few hours initially, but

Table 2. **Diflubenzuron:** Droplet spectra, foliar deposits and persistence characteristics of diflubenzuron after aerial application over a forest ecosystem

| Parameter | 70 g AI in 10.0 L/ha | 70 g AI in 5.0 L/ha | 70 g AI in 2.5 L/ha |
|---|---|---|---|
| Droplet size spectra: | | | |
| Minimum diameter | 5 µm[a] | 5 µm[a] | 5 µm[a] |
| Maximum diameter | 490 µm | 420 µm | 327 µm |
| Number median diameter | 132 µm | 95 µm | 77 µm |
| Volume median diameter | 250 µm | 195 µm | 150 µm |
| Diflubenzuron deposits (µg/g fresh weight) (mean): | | | |
| *Pine foliage:* | | | |
| 1 h (after spray) | 13.1 | 6.90 | 4.67 |
| 6 h | 12.2 | 5.23 | 3.62 |
| 1 d | 11.1 | 4.61 | 3.52 |
| 3 d | 6.51 | 3.57 | 2.71 |
| 5 d | 5.41 | 2.95 | 2.23 |
| 7 d | 4.93 | 1.61 | 1.40 |
| 10 d | 3.81 | 1.05 | 0.82 |
| 15 d | 3.50 | 0.72 | 1.02 |
| 20 d | 3.41 | 0.40 | 0.34 |
| 30 d | 2.76 | 0.29 | 0.17 |
| 120 d | 0.24 | ND[b] | ND |
| $DT_{50}$ (d) | 4.60 | 3.3 | 3.9 |
| *Maple foliage:* | | | |
| 1 h (after spray) | 19.1 | 4.10 | 3.36 |
| 6 h | 18.1 | 3.95 | 3.20 |
| 1 d | 17.2 | 2.90 | 2.81 |
| 3 d | 16.0 | 2.01 | 1.95 |
| 5 d | 12.7 | 1.72 | 1.45 |
| 7 d | 9.23 | 1.41 | 1.52 |
| 10 d | 8.75 | 1.01 | 0.63 |
| 15 d | 7.54 | 0.66 | 0.34 |
| 20 d | 5.46 | 0.42 | 0.26 |
| 30 d | 4.67 | 0.21 | 0.11 |
| 120 d | 1.04 | ND | ND |
| $DT_{50}$ (d) | 8.70 | 3.5 | 4.3 |

a: Detection limit of droplet sizing technique = 5 µm; b: ND = not detectable.

the residues tapered off very slowly during later stages. In the case of diflubenzuron, however, the dissipation was gradual, and the chemical stayed in foliage at low levels even up to 120 d after treatment at the high volume (10 L/ha) and up to 30 d at the two lower volume rates (Table 2).

The present findings indicated that the higher the initial deposits, the longer the duration of persistence in foliage. This means that a knowledge of the initial deposits alone would be inadequate, because the target insects are exposed to a pesticide on a continuous basis throughout the duration of its persistence. This implies that the cumulative exposure levels (i.e., a function of concentration x duration of persistence) are more important than the initial concentrations alone. The data also indicated that from the field efficacy standpoint, high volume rates would be preferable to low volume rates in aerial application studies, because of the high deposit recovery on foliage. Nevertheless, the correspondingly high ground deposits obtained were not desirable because of the potential for environmental contamination (*15*).

## Tebufenozide

**Study I - Effect of Formulation Ingredients on Rainfastness of Tebufenozide under Laboratory Conditions.** A laboratory study was conducted (*31*) to examine the effect of formulation type, i.e., aqueous flowable *versus* emulsion-suspension on rainfastness of tebufenozide (Mimic®, or RH-5992). Spray mixtures of the flowable and emulsion-suspension concentrates (described as the 2F and ES mixtures respectively) were sprayed in a laboratory chamber, onto balsam fir branch tips collected from field trees and greenhouse-grown seedlings, at dosage rates ranging from 35 to 150 g AI in volumes of 2.0 to 5.0 L/ha (Tables 3 and 4). Droplet spectra and mass deposit were determined on artificial samplers. Simulated rainfall of two different intensities was applied at different rain-free periods, and rain droplet sizes were determined. Foliar washoff of RH-5992 was assessed after application of different amounts of rain, and the increase in soil residues was evaluated.

The results of the investigation indicated (Tables 3 and 4) a direct relationship between the amount of rainfall and RH-5992 washoff. The larger the rain droplet size, the greater the amount washed off. This can be attributed to the higher impact velocity of the large rain droplets compared to the small ones, resulting in a greater probability of the RH-5992 particles being knocked off. Longer rain-free periods made the deposits more resistant to rain. Regardless of the amount of rainfall, rain droplet size and rain-free periods, foliar deposits of the 2F mixtures were washed off to a greater extent than those of the ES mixtures. The increase in soil residues due to foliar washoff was greater for the 2F than for the ES mixtures. The deposits of the emulsion-suspension were consistently more resistant to rain-washing than those of the aqueous flowable formulation. This could be due to the presence of a lipophilic material in the emulsion-suspension formulation, which might have caused better droplet retention and adhesion to foliar surfaces (*12*).

**Study II - Field Studies on Single Tree Treatment of Tebufenozide Using Ground Application Equipment.** In a ground application study (*32*), young trees (2.3 to 2.5 m high) of white spruce [*Picea glauca* (Moench Voss)] were sprayed using a spinning disc atomizer (Flak®, Micron Agri Sprayers Canada, Walkerton, Ontario) with the 2F mixtures at 35, 70 or 140 g AI/ha. Droplet size spectra were assessed on Kromekote® cards (Table 5). Initial deposits and persistence of RH-5992 were

Table 3. **Tebufenozide:** Rain-washing of tebufenozide following application of an aqueous flowable 2F spray mixture

| Parameter | 35 g AI in 2.0 L/ha | 70 g AI in 2.0 L/ha | 140 g AI in 4.0 L/ha |
|---|---|---|---|
| Initial deposits (μg/branch) | 8.8 ± 0.8 | 18.7 ± 1.7 | 43.1 ± 2.9 |
| 1. Influence of 4 mm and 8 mm cumulative rainfall on tebufenozide washoff: | | | |
| Rain droplet size spectra (for rainfall intensity of 5 mm/h): | | | |
| Maximum diameter | 1100 μm | 1100 μm | 1100 μm |
| Minimum diameter | 80 μm | 80 μm | 80 μm |
| Number median diameter | 445 μm | 445 μm | 445 μm |
| Volume median diameter | 780 μm | 780 μm | 780 μm |
| Tebufenozide washoff (mean): | | | |
| Washoff by 4 mm (μg) | 5.2 | 12.9 | 34.0 |
| Percent washoff | 59.1 | 69.0 | 78.9 |
| Washoff by 8 mm (μg) | 7.2 | 16.6 | 40.9 |
| Percent washoff | 81.8 | 88.8 | 94.9 |
| 2. Influence of 1 mm/h and 5 mm/h rainfall intensities on tebufenozide washoff: | | | |
| Rain droplet size spectra (for rainfall intensity of 1 mm/h): | | | |
| Maximum diameter | 340 μm | 340 μm | 340 μm |
| Minimum diameter | 30 μm | 30 μm | 30 μm |
| Number median diameter | 155 μm | 155 μm | 155 μm |
| Volume median diameter | 315 μm | 315 μm | 315 μm |
| Rain droplet size spectra (for rainfall intensity of 5 mm/h): | | | |
| Maximum diameter | 1100 μm | 1100 μm | 1100 μm |
| Minimum diameter | 80 μm | 80 μm | 80 μm |
| Number median diameter | 445 μm | 445 μm | 445 μm |
| Volume median diameter | 780 μm | 780 μm | 780 μm |
| Tebufenozide washoff (mean): | | | |
| Washoff by 1 mm/h (μg) | 1.55 | 5.52 | 17.2 |
| Percent washoff | 17.6 | 29.5 | 39.9 |
| Washoff by 5 mm/h (μg) | 6.5 | 15.3 | 38.1 |
| Percent washoff | 73.9 | 81.8 | 88.4 |
| 3. Influence of 8 h and 72 h rain-free periods on tebufenozide washoff (mean): | | | |
| Washoff after 8 h (μg) | 5.8 | 14.3 | 37.0 |
| Percent washoff | 65.9 | 76.5 | 85.9 |
| Washoff after 72 h (μg) | 1.7 | 5.4 | 15.7 |
| Percent washoff | 19.3 | 28.9 | 36.4 |

Table 4. **Tebufenozide:** Rain-washing of tebufenozide following spray application of aqueous flowable (2F) and emulsion-suspension (ES) mixtures

| Parameter | 50 g AI in 2.5 L/ha | 100 g AI in 2.5 L/ha | 150 g AI in 5.0 L/ha |
|---|---|---|---|
| Rainfall intensity | ←— | 5 mm/h | —→ |
| Cumulative rainfall | ←— | 5 mm | —→ |
| Number median diameter | ←— | 445 μm | —→ |
| Volume median diameter | ←— | 780 μm | —→ |
| Initial deposits of 2F mix | 7.6 μg/branch | 15.3 μg/branch | 25.5 μg/branch |
| Washoff of 2F mix | 5.1 μg | 11.8 μg | 22.2 μg |
| Percent washoff | 67.1 | 77.1 | 87.1 |
| Initial deposits of ES mix | 10.5 μg/branch | 21.3 μg/branch | 31.8 μg/branch |
| Washoff of ES mix | 1.4 μg | 4.66 μg | 9.03 μg |
| Percent washoff | 13.3 | 21.9 | 28.4 |

Table 5. **Tebufenozide:** Droplet spectra, foliar deposits and persistence characteristics after ground application over white spruce trees

| Parameter | 35 g AI in 2.0 L/ha | 70 g AI in 2.0 L/ha | 140 g AI in 2.0 L/ha |
|---|---|---|---|
| Droplet size spectra: | | | |
| Maximum diameter | 116 μm | 132 μm | 155 μm |
| Minimum diameter | 20 μm | 20 μm | 20 μm |
| Number median diameter | 56 μm | 63 μm | 67 μm |
| Volume median diameter | 72 μm | 86 μm | 100 μm |
| Tebufenozide deposits (μg/g fresh weight) (mean): | | | |
| *One-year-old spruce needles:* | | | |
| 1 h (after spray) | 2.25 | 4.78 | 8.54 |
| 4 d | 1.99 | 4.42 | 8.14 |
| 16 d | 1.66 | 3.94 | 7.03 |
| 31 d | 1.34 | 2.90 | 5.44 |
| 85 d | 0.592 | 1.220 | 2.761 |
| 135 d | 0.415 | 0.691 | 1.666 |
| *New growth (spruce shoots):* | | | |
| 1 h (after spray) | 1.59 | 3.34 | 6.13 |
| 4 d | 1.41 | 3.09 | 6.00 |
| 16 d | 0.95 | 2.12 | 4.29 |
| 31 d | 0.49 | 1.06 | 2.08 |
| 85 d | 0.115 | 0.234 | 0.450 |
| 135 d | 0.050 | 0.114 | 0.238 |

determined on spruce foliage. The data indicated persistence of tebufenozide residues in foliage for more than 408 d (data are given in Table 5 for only up to 135 d after spray application).

An aquatic study and soil persistence study (*32*) to determine residues in several substrates, showed similar prolonged persistence in water, sediment, soil and other substrates (Table 6). Thus, the potential for cumulative exposure for the aquatic and soil organisms is considerable when the persistence aspects are taken into account. Although tebufenozide is claimed to be selectively toxic only to lepidopterous insects and to be non-toxic to crustacea, arachnida and several other insect orders (beetles, aphids, flies etc.), the prolonged exposure to the myriad other types of biota present in the aquatic and soil ecosystems should not be taken lightly, especially in view of the lack of data establishing safety on them.

Table 6. **Tebufenozide:** Initial concentrations and persistence characteristics in water and sediment samples in aquatic enclosures at three dosage levels

| | ←Concentration spiked→ | | |
|---|---|---|---|
| Parameter | 0.10 mg/L | 0.26 mg/L | 0.50 mg/L |
| *Concentration in water samples (mg/L) (mean):* | | | |
| 0.33 d (after spray) | 0.133 | 0.331 | 0.659 |
| 3 d | 0.119 | 0.304 | 0.547 |
| 12 d | 0.103 | 0.232 | 0.363 |
| 35 d | 0.101 | 0.210 | 0.311 |
| 49 d | 0.090 | 0.202 | 0.298 |
| 92 d | 0.069 | 0.189 | 0.293 |
| 154 d | 0.008 | 0.033 | 0.052 |
| 393 d | 0.003 | 0.012 | 0.030 |
| *Concentration in sediment samples (mg/kg) (mean):* | | | |
| 0.33 d (after spray) | 1.57 | 6.13 | 8.99 |
| 3 d | 2.42 | 5.71 | 16.53 |
| 12 d | 2.73 | 7.68 | 21.09 |
| 35 d | 2.97 | 11.44 | 26.97 |
| 49 d | 2.41 | 8.03 | 23.40 |
| 92 d | 1.90 | 4.19 | 22.16 |
| 154 d | 0.85 | 4.46 | 28.13 |
| 393 d | 0.39 | 2.77 | 13.20 |

## Azadirachtin

**Study I - Persistence of Azadirachtin in Balsam Fir and Oak Foliage:** Foliar deposits and persistence of azadirachtin (AZ) were investigated after spraying a commercial non-aqueous formulation, Margosan-O® (W.R. Grace and Co.-Conn., Cambridge, MA, USA), at three dosage rates (i.e., 7, 15 and 40 g AZ in 2.8, 5.9 and

15.8 L/ha respectively) onto balsam fir [*Abies balsamea* (L.) Mill.] and oak (*Quercus rubra* L.) seedlings in a laboratory chamber (*33*). Droplet size spectra and mass deposit were assessed on artificial samplers. Initial deposits were measured on canopy foliage by high-performance liquid chromatography (HPLC) together with the residual concentrations at different intervals of time after treatment to assess persistence characteristics. The data indicated that azadirachtin disappeared from both types of foliage relatively rapidly, although the duration of persistence was generally longer in oak foliage than in fir, probably because of high initial deposits (Table 7). The half-life values ranged from 17 to 22 h. The initial concentrations varied widely from 4 to 96 µg/g (fresh weight), but the rate constants were not significantly different.

Table 7. **Azadirachtin**: Droplet spectra, foliar deposits and persistence characteristics after application of Margosan-O over balsam fir and oak seedlings

| Parameter | 7 g AI in 2.8 L/ha | 15 g AI in 5.9 L/ha | 40 g AI in 15.8 L/ha |
|---|---|---|---|
| Droplet size spectra: | | | |
| Maximum diameter | 138 µm | 145 µm | 164 µm |
| Minimum diameter | 4 µm[a] | 4 µm[a] | 4 µm[a] |
| Number median diameter | 76 µm | 72 µm | 74 µm |
| Volume median diameter | 94 µm | 97 µm | 98 µm |
| Azadirachtin deposits (µg/g fresh weight) (mean): | | | |
| *Balsam fir needles:* | | | |
| 0.25 h (after spray) | 4.3 | 12.8 | 28.8 |
| 5.5 h | 3.5 | 10.4 | 16.6 |
| 10 h | 2.7 | 7.9 | 12.8 |
| 24 h | 2.2 | 6.7 | 8.6 |
| 48 h | 1.1 | 3.2 | 6.2 |
| 72 h | 0.2 - 0.4 | 0.2 - 0.4 | 2.8 |
| 120 h | ND[b] | ND | 0.2 - 0.4 |
| 192 h | ND | ND | ND |
| $DT_{50}$ (h) | 22 | 22 | 17 |
| *Oak foliage:* | | | |
| 0.25 h (after spray) | 10.5 | 31.4 | 96.2 |
| 5.5 h | 8.3 | 24.8 | 57.3 |
| 10 h | 6.3 | 18.9 | 46.2 |
| 24 h | 4.3 | 12.8 | 30.9 |
| 48 h | 2.3 | 6.9 | 20.9 |
| 72 h | 1.3 | 4.0 | 16.4 |
| 120 h | ND | 1.6 | 12.9 |
| 192 h | ND | ND | 6.2 |
| $DT_{50}$ (h) | 20 | 20 | 19 |

a: Detection limit of droplet sizing technique = 4 µm; b: ND = not detectable.

The influence of spray droplet size and cuticular wax content of foliage on persistence of azadirachtin was investigated using both foliar types, and glass microscope slides without and with the foliar wax coating. Regardless of the presence or absence of cuticular wax, and the sizes of droplets applied, the half-life of persistence was very similar (Tables 8 and 9). However, the rate of azadirachtin loss from both foliar types was slower than from the glass slides, and was unaffected by the amount of cuticular wax present (Table 8). The fir foliage contained higher cuticular wax content than the oak foliage, but the persistence of azadirachtin was similar in both foliar types.

Margosan-O® formulation is registered for use on nonfood crops and ornamentals, and the approval for use on food crops is in progress (*34,35*). Thomas et al. (*36*) have tested this formulation against spruce budworm (*Choristoneura fumiferana* Clem.), a serious defoliator of fir and spruce forests in eastern North America. Schmutterer and Hellpap (*37*) found that neem seed extracts were effective against gypsy moth (*Lymantria dispar* L.). At present, researchers in Canadian Forest Service (*38*) are investigating the efficacy of some azadirachtin formulations against spruce budworm in field applications. In view of the low persistence of this chemical in the environment, further research is being carried out to identify suitable additives that can provide increased persistence of AZ in foliage. Jacobson (*39*) summarizes the extensive research that has been carried out on the pharmacology and toxicology of neem, including the effect on several nontarget organisms. However, these were mainly laboratory investigations or small scale field investigations under controlled conditions. There is a need for field investigations under the actual forestry application situation, so that the environmental behaviour of azadirachtin and its formulations can be evaluated in forest ecosystems.

### ***Bacillus thuringiensis* var. *kurstaki* (BTK)**

Rainfall, exposure to sunlight, and spray droplet size are the three major factors contributing to the rapid disappearance of *Bacillus thuringiensis* (BTK) activity in foliar deposits (*12,40-42*). Sunlight-mediated inactivation of pure crystals (delta-endotoxin protein) under laboratory conditions (*43*) showed that within 40 h after irradiation, the crystals became non-toxic. Previous studies on field persistence of spore viability (*44-46*) indicated a complete loss of spore viability within 24 h after treatment. The degree of inactivation was dependent, to some extent, on formulation ingredients and tree species. Spray droplets of aqueous BTK formulations, depending on the concentration of ingredients present, are known to partially evaporate in-flight forming moist spheres. These impinge on target surfaces either as spheres or spherical segments. They can also bounce off from foliage without being retained, if the spherical surface of the droplet was hardened enough during drying in-flight (*47*). All of these processes have a profound effect on the amount of foliar deposits obtained, and also on the duration of persistence of BTK activity (*48*). In view of these findings, laboratory studies were conducted to investigate the role of formulation type (i.e., water-based or oil-based) on rainfastness of BTK deposits and biological activity, on physical properties and droplet size spectra, and on spreading and adhesion.

Table 8. **Azadirachtin:** Initial concentrations of AZ per g foliage and persistence during the first 48 h after application (Reproduced with permission from reference 33. Copyright 1994 Pesticide Science.)

| Droplet diameter 'd' (μm) | Number of droplets per $cm^2$ | Initial concn. of AZ present in foliage (μg/g fresh weight) | Mass of AZ (μg/g) at | | | | Percent loss at | | | | $DT_{50}$ (h) |
|---|---|---|---|---|---|---|---|---|---|---|---|
| | | | 8 h | 18 h | 30 h | 48 h | 8 h | 18 h | 30 h | 48 h | |
| | | | Fir Foliage[a] (cuticular wax content = 9.2 g per 100 g foliage) | | | | | | | | |
| 95 | 30 | 2.176 | 1.71 | 1.26 | 0.88 | 0.51 | 21 | 42 | 60 | 76 | 23[x] |
| 164 | 15 | 5.632 | 4.21 | 2.92 | 1.88 | 0.98 | 25 | 48 | 67 | 83 | 19[x] |
| 255 | 9 | 12.67 | 9.31 | 6.33 | 3.99 | 1.99 | 27 | 50 | 69 | 84 | 18[x] |
| 345 | 6 | 20.86 | 14.75 | 9.56 | 5.69 | 2.61 | 29 | 54 | 73 | 88 | 16[x] |
| 445 | 6 | 44.86 | 34.45 | 24.76 | 16.66 | 9.20 | 23 | 45 | 63 | 79 | 21[x] |
| | | | Oak Foliage[b] (cuticular wax content = 6.2 g per 100 g foliage) | | | | | | | | |
| 95 | 30 | 2.720 | 2.03 | 1.41 | 0.91 | 0.47 | 25 | 48 | 67 | 83 | 19[x] |
| 164 | 15 | 7.040 | 4.98 | 3.23 | 1.92 | 0.88 | 29 | 54 | 73 | 88 | 16[x] |
| 255 | 9 | 15.84 | 12.45 | 9.21 | 6.41 | 3.73 | 21 | 42 | 60 | 76 | 23[x] |
| 345 | 6 | 26.08 | 19.16 | 13.04 | 8.21 | 4.11 | 27 | 50 | 69 | 84 | 18[x] |
| 445 | 6 | 56.08 | 41.21 | 28.04 | 17.66 | 8.83 | 27 | 50 | 69 | 84 | 18[x] |

a: Fir needle: average surface area = 0.39 $cm^2$; average mass = 6.14 mg; number of needles per g foliage = 163; surface area per g foliage = 64 $cm^2$.
b: Oak leaf: average surface area = 20 $cm^2$; average mass per leaf = 0.24 g; number of leaves per g foliage = 4; surface area per g foliage = 80 $cm^2$.
x: The $DT_{50}$ values were not significantly different from one another (ANOVA $P > 0.05$).

*Table 9.* ***Azadirachtin:*** *Initial* concentrations (µg/ml) of AZ in the glass slide extracts and persistence during the first 40 h after treatment (Reproduced with permission from reference 33. Copyright 1994 Pesticide Science.)

| Droplet diameter 'd' (µm) | Initial concn. of AZ present in extract (µg/ml) | Mass of AZ (µg/ml of extract) at | | | | Percent loss at | | | | $DT_{50}$ (h) |
|---|---|---|---|---|---|---|---|---|---|---|
| | | 10 h | 16 h | 30 h | 40 h | 10 h | 16 h | 30 h | 40h | |
| | | | | Glass slides without wax coating | | | | | | |
| 95 | 3.192 | 1.95 | 1.45 | 0.72 | 0.44 | 39 | 55 | 77 | 86 | 14[x] |
| 164 | 4.096 | 2.40 | 1.75 | 0.83 | 0.49 | 41 | 57 | 80 | 88 | 13[x] |
| 255 | 3.080 | 1.54 | 1.02 | T[a] | ND[b] | 50 | 67 | > 87 | > 94[c] | 10[x] |
| 345 | 3.808 | 2.32 | 1.72 | 0.86 | 0.53 | 39 | 55 | 77 | 86 | 14[x] |
| 445 | 8.176 | 4.09 | 2.70 | 1.02 | 0.51 | 50 | 67 | 88 | 94 | 10[x] |
| | | | | Glass slides with fir wax coating | | | | | | |
| 95 | 3.192 | 1.94 | 1.45 | 0.72 | 0.44 | 39 | 55 | 77 | 86 | 14[x] |
| 164 | 4.096 | 2.30 | 1.63 | 0.72 | 0.41 | 44 | 60 | 82 | 90 | 12[x] |
| 255 | 3.080 | 1.81 | 1.31 | 0.62 | T | 41 | 57 | 80 | > 87 | 13[x] |
| 345 | 3.808 | 1.90 | 1.26 | 0.48 | ND | 50 | 67 | 88 | > 94 | 10[x] |
| 445 | 8.176 | 4.59 | 3.24 | 1.44 | 0.81 | 44 | 60 | 82 | 90 | 12[x] |
| | | | | Glass slides with oak wax coating | | | | | | |
| 95 | 3.192 | 1.87 | 1.36 | 0.64 | T | 41 | 57 | 80 | > 87 | 13[x] |
| 164 | 4.096 | 2.30 | 1.63 | 0.72 | T | 44 | 60 | 82 | > 87 | 12[x] |
| 255 | 3.080 | 1.64 | 1.12 | 0.46 | ND | 47 | 64 | 85 | > 94 | 11[x] |
| 345 | 3.808 | 1.90 | 1.26 | 0.48 | ND | 50 | 67 | 88 | > 94 | 10[x] |
| 445 | 8.176 | 3.78 | 2.38 | 0.81 | T | 54 | 71 | 90 | > 87 | 9[x] |

a: 'T' = Trace amounts (i.e., between 0.2 and 0.4 µg/mL). b: ND = not detectable (i.e., ≤ 0.2 µg/mL).
c: Calculated from 'T' and 'ND' values. x: Values were not significantly different from one another (ANOVA $P > 0.05$).

**Study I - Rainfastness of *Bacillus thuringiensis* Formulations Using Force-Feeding Bioassay and Total Protein Assay Methods.** Eight oil-based and four aqueous formulations of BTK (Tables 10 and 11) obtained from Abbott laboratories (Long Grove, Illinois) were sprayed in a laboratory chamber equipped with a spinning disc atomizer (Flak®, Micron Agri Sprayers Canada, Walkerton, Ontario), over potted seedlings of white spruce and balsam fir (*12*). Spray deposit on foliage was assessed

Table 10. ***Bacillus thuringiensis***: Droplets per needle, spray volume deposit, percent mortalities after force-feeding bioassay of foliar extracts, and BTK deposits (IU/g foliage and IU/cm$^2$ of foliar area) of the twelve formulations prior to 3 mm rainfall (Reproduced with permission from reference 12. Copyright 1993 ASTM.)

| Formulation abbreviation | Drops per needle | Spray vol. deposit (nL/cm$^2$) | Mortality (%) (corrected)[a] | IU per larva | IU/g foliage | IU/cm$^2$ foliar area |
|---|---|---|---|---|---|---|
| | | | Spruce Foliage | | | |
| ABG-6158A (O)[b] | 5.3 | 12.4 | 66 ± 4 | 4.13 | 15488[c] | 215[c] |
| ABG-6192A (O) | 4.0 | 10.5 | 25 ± 3 | 2.96 | 11100 | 154 |
| ABG-6167 (W)[b] | 2.6 | 5.2 | 15 ± 3 | 1.89 | 7088 | 98 |
| ABG-6167A (W) | 2.5 | 7.2 | 24 ± 3 | 2.26 | 8475 | 118 |
| ABG-6282 (W) | 5.9 | 12.9 | 16 ± 2 | 2.06 | 7725 | 107 |
| ABG-6158F (O) | 4.7 | 13.3 | 56 ± 5 | 4.86 | 18225 | 253 |
| ABG-6192B (O) | 5.9 | 12.2 | 63 ± 3 | 3.81 | 14828 | 206 |
| ABG-6158B (O) | 4.8 | 10.1 | 42 ± 4 | 3.77 | 14138 | 196 |
| ABG-6281 (W) | 2.4 | 5.9 | 6 ± 2 | 1.51 | 5663 | 79 |
| ABG-6222 (O) | 4.1 | 12.2 | 67 ± 3 | 3.77 | 14138 | 196 |
| ABG-6222D (O) | 3.6 | 14.9 | 41 ± 3 | 4.18 | 15675 | 218 |
| ABG-6158D (O) | 5.6 | 12.1 | 43 ± 2 | 4.96 | 18600 | 258 |
| | | | Balsam fir Foliage | | | |
| ABG-6158A (O)[b] | 3.9 | 12.8 | 61 ± 3 | 3.72 | 13950[c] | 215[c] |
| ABG-6192A (O | 4.4 | 10.4 | 31 ± 2 | 3.47 | 13013 | 200 |
| ABG-6167 (W)[b] | 2.4 | 6.9 | 16 ± 2 | 1.96 | 7350 | 113 |
| ABG-6167A (W) | 2.3 | 5.9 | 26 ± 3 | 2.42 | 9075 | 140 |
| ABG-6282 (W) | 4.9 | 11.3 | 15 ± 2 | 1.89 | 7088 | 109 |
| ABG-6158F (O) | 5.4 | 12.1 | 61 ± 5 | 5.79 | 21713 | 334 |
| ABG-6192B (O) | 4.6 | 11.1 | 57 ± 4 | 3.47 | 13013 | 200 |
| ABG-6158B (O) | 5.6 | 12.2 | 55 ± 4 | 4.92 | 18450 | 284 |
| ABG-6281 (W) | 2.3 | 6.2 | 8 ± 2 | 1.76 | 6600 | 102 |
| ABG-6222 (O) | 3.4 | 14.1 | 61 ± 4 | 3.27 | 12263 | 189 |
| ABG-6222D (O) | 4.6 | 15.2 | 51 ± 3 | 5.86 | 21975 | 338 |
| ABG-6158D (O) | 4.2 | 12.5 | 39 ± 3 | 4.36 | 16350 | 252 |

a: Mortality values were corrected for the control mortality.

b: 'O' = oil-based formulation; 'W' = aqueous formulation.

c: The standard deviations (not illustrated) ranged from 18 to 28%.

before and after 3 mm of simulated rainfall applied at an intensity of 5 mm/h, using a bioassay method (*49*). Sprayed foliage was soaked in an alkaline buffer, and the toxic protein was extracted. Four-µL aliquots of the extracts were force-fed directly into the mid-gut of the sixth instar spruce budworm larvae using a microsyringe. In addition to the bioassay, a total protein assay method was used to quantify the total protein (active + inactive proteins) to compare the two sets of results (Table 12).

Table 11. ***Bacillus thuringiensis*:** Percent mortalities after force-feeding bioassay of foliar extracts, and BTK deposits (IU/g foliage and IU/cm$^2$ of foliar area) of the twelve formulations, and percent deposit lost in 3 mm rainfall (Reproduced with permission from reference 12. Copyright 1993 ASTM.)

| Formulation abbreviation | Mortality (%)[a] | IU per larva | IU/g foliage | IU/cm$^2$ foliar area | Percent lost in rain |
|---|---|---|---|---|---|
| | | Spruce Foliage | | | |
| ABG-6158A (O)[b] | 31 ± 2 | 1.86 | 6970[c] | 97[c] | 55[c] |
| ABG-6192A (O) | None | None | None | None | 100 |
| ABG-6167 (W)[b] | None | None | None | None | 100 |
| ABG-6167A (W) | None | None | None | None | 100 |
| ABG-6282 (W) | None | None | None | None | 100 |
| ABG-6158F (O) | 39 ± 3 | 2.67 | 10024 | 139 | 45 |
| ABG-6192B (O) | 12 ± 2 | 1.19 | 4448 | 62 | 70 |
| ABG-6158B (O) | 6 ± 1 | 1.32 | 4948 | 69 | 65 |
| ABG-6281 (W) | None | None | None | None | 100 |
| ABG-6222 (O) | 23 ± 2 | 1.32 | 4948 | 69 | 65 |
| ABG-6222D (O) | 22 ± 3 | 1.88 | 7054 | 98 | 55 |
| ABG-6158D (O) | 27 ± 2 | 2.98 | 11160 | 155 | 40 |
| | | Balsam Fir Foliage | | | |
| ABG-6158A (O)[b] | 24 ± 3 | 1.58 | 5925[c] | 91[c] | 58[c] |
| ABG-6192A (O) | 5 ± 1 | 1.24 | 4650 | 72 | 64 |
| ABG-6167 (W)[b] | None | None | None | None | 100 |
| ABG-6167A (W) | 5 ± 1 | 0.76 | 2850 | 44 | 69 |
| ABG-6282 (W) | None | None | None | None | 100 |
| ABG-6158F (O) | 33 ± 2 | 2.16 | 8100 | 125 | 63 |
| ABG-6192B (O) | 10 ± 1 | 1.10 | 4125 | 63 | 69 |
| ABG-6158B (O) | 7 ± 1 | 1.38 | 5175 | 80 | 72 |
| ABG-6281 (W) | None | None | None | None | 100 |
| ABG-6222 (O) | 17 ± 1 | 1.09 | 4088 | 63 | 66 |
| ABG-6222D (O) | 20 ± 2 | 1.69 | 6338 | 98 | 71 |
| ABG-6158D (O) | 21 ± 2 | 2.30 | 8625 | 133 | 47 |

a: Mortalities were corrected for the control mortality.
b: 'O' = oil-based formulation; 'W' = aqueous formulation.
c: The standard deviations (not illustrated) ranged from 18 to 28%.

Table 12. ***Bacillus thuringiensis:*** Total protein content of the twelve formulations, and of foliar extracts before and after 3 mm rainfall, and deposit lost in the rain (Reproduced with permission from reference 12. Copyright 1993 ASTM.)

| Formulation abbreviation | Total protein content (g/L) | Protein deposit prior to rainfall (ng protein/cm$^2$) | Protein deposit after rainfal (ng protein/cm$^2$) | Percent deposit lost in rain (mean) |
|---|---|---|---|---|
| | | Spruce Foliage | | |
| ABG-6158A (O)[a] | 75 ± 3 | 640 ± 26 | 305 ± 31 | 52 |
| ABG-6192A (O) | 92 ± 4 | 496 ± 22 | 245 ± 15 | 51 |
| ABG-6167 (W)[a] | 67 ± 3 | 230 ± 10 | 77 ± 9 | 67 |
| ABG-6167A (W) | 71 ± 4 | 297 ± 17 | 85 ± 8 | 71 |
| ABG-6282 (W) | 45 ± 3 | 288 ± 19 | 95 ± 9 | 67 |
| ABG-6158F (O) | 88 ± 3 | 718 ± 24 | 338 ± 29 | 53 |
| ABG-6192B (O) | 98 ± 5 | 773 ± 39 | 374 ± 29 | 52 |
| ABG-6158B (O) | 91 ± 3 | 633 ± 21 | 335 ± 33 | 47 |
| ABG-6281 (W) | 70 ± 4 | 343 ± 20 | 124 ± 15 | 64 |
| ABG-6222 (O) | 92 ± 7 | 673 ± 51 | 365 ± 28 | 46 |
| ABG-6222D (O) | 82 ± 5 | 707 ± 43 | 316 ± 45 | 55 |
| ABG-6158D (O) | 97 ± 4 | 575 ± 24 | 285 ± 25 | 50 |
| | | Balsam Fir Foliage | | |
| ABG-6158A (O)[a] | 75 ± 3 | 575 ± 23 | 275 ± 21 | 52 |
| ABG-6192A (O) | 92 ± 4 | 425 ± 18 | 183 ± 15 | 57 |
| ABG-6167 (W)[a] | 67 ± 3 | 335 ± 15 | 114 ± 14 | 66 |
| ABG-6167A (W) | 71 ± 4 | 375 ± 21 | 104 ± 9 | 72 |
| ABG-6282 (W) | 45 ± 3 | 405 ± 27 | 123 ± 11 | 70 |
| ABG-6158F (O) | 88 ± 3 | 678 ± 23 | 339 ± 14 | 50 |
| ABG-6192B (O) | 98 ± 5 | 595 ± 30 | 268 ± 17 | 55 |
| ABG-6158B (O) | 91 ± 3 | 555 ± 18 | 261 ± 19 | 53 |
| ABG-6281 (W) | 70 ± 4 | 505 ± 29 | 157 ± 21 | 69 |
| ABG-6222 (O) | 82 ± 7 | 775 ± 59 | 357 ± 22 | 54 |
| ABG-6222D (O) | 82 ± 5 | 675 ± 41 | 344 ± 24 | 49 |
| ABG-6158D (O) | 97 ± 4 | 615 ± 25 | 295 ± 19 | 52 |

a: 'O' = oil-based formulation; 'W' = aqueous formulation.

The bioassay method showed that generally, experiments with high initial deposits (in units of IU/cm$^2$) on foliage, showed some residual activity after the rainfall, whereas experiments with low deposits showed no activity after the rain (Table 11). This trend was observed regardless of the formulation type, oil-based or aqueous. The total protein method was more sensitive than the force-feeding bioassay method. With all formulations, some protein deposits (ng/cm$^2$) remained on foliage after the rain. When the total protein assay method was used, the oil-based formulations showed greater rainfastness than the aqueous formulations, a finding that was not as evident in the force-feeding bioassay method. The differences in the two sets of data were due to the fact that the bioassay method depended on the biological response of an insect, thus resulting in high variability in deposits, whereas the total protein method provided a direct estimation of protein that were less variable.

**Study II - Droplet Size Spectra and Deposits of Four *Bacillus thuringiensis* Formulations on Simulated and Natural Fir Foliage.** In aerial spray studies reported in the literature using concentrated aqueous formulations, droplets that remained on the foliar surface were measured after impaction. The droplet volume that could have penetrated into the foliar cuticle was not taken into account. The objectives of this study (*11*) were to spray mono-sized droplets over natural and simulated balsam fir needles, to determine droplet spreading, to measure the sizes of droplet segments above the foliage, and to compare the droplet volume that penetrated into the foliage. Four aqueous formulations of *Bacillus thuringiensis*, var. *kurstaki*, Futura® XLV, Futura® XLV-HP, Thuricide® 48LV and Foray® 48B, containing a dye and a chemical tracer, (triethyl phosphate), were sprayed in a laboratory chamber over balsam fir branch tips. Spray was also applied to aluminum fir branch simulators with and without a coating of the cuticular wax extracted from natural fir foliage. Droplet size spectra, droplets/$cm^2$ and mass deposits were assessed on the natural foliage and wax-coated aluminum foliar simulators, but only droplets/$cm^2$ and mass deposits could be measured on the uncoated aluminum foliar simulators. Both natural foliage and foliar simulators received similar droplet sizes, droplets/$cm^2$, and mass deposits, but the latter two parameters were higher on the uncoated aluminum foliage. This was attributed to the tiny crevices present on the bare metal surface which acted as an excellent collector of a large number of fine droplets. The investigation indicated the importance of assessing droplet penetration into foliage, otherwise considerable errors could be encountered in droplet sizing. The investigation also provided a new method to determine the actual sizes of droplets that deposited on a foliar simulator. The simulator not only had similar size and shape, but also the same surface characteristics. Deposit quantification was also faster on the simulator than on the natural foliage.

**Study III - Droplet Spectra, Spreading and Adhesion and Physical Properties of *Bacillus thuringiensis* Formulations after Spray Application under Laboratory Conditions.** Three oil-based formulations, Dipel® 6L, Dipel® 8L and Dipel® 12 L, and five aqueous formulations, Thuricide® 48LV, Thuricide® 64B, Futura® XLV, Dipel® 6AF and Dipel® 8AF, of BTK, were sprayed at a dosage rate equivalent to 30 billion international units (BIU) per ha in a laboratory chamber using a spinning disc atomizer (*10*). The formulations were mixed with Day-Glo® fluorescent dyes to facilitate droplet detection. Spray deposit was collected on Kromekote® cards and glass plates. Droplets were counted on balsam fir foliage of potted seedlings. Physical properties, viz., viscosity, surface tension, volatility and pseudoplastic behaviour, were determined for each of the formulations. The data are given in Tables 13 and 14.

Spray droplets of the three oil-based formulations underwent complete spreading on the Kromekote cards and fir foliage because no three-dimensional segments were observed on the surface. The degree of spreading was influenced by viscosity and

Table 13. ***Bacillus thuringiensis*:** Dosage and volume rates, droplet spectra and mass deposits of three oil-based Dipel formulations and Thuricide 64B after atomization in a chamber, and physical properties SOURCE: Reproduced with permission from reference 10. Copyright 1989 ASTM.

| Parameters | Dipel 6L | Dipel 8L | Dipel 12L | Thuricide 64B[p] |
|---|---|---|---|---|
| Dosage rate | ←---------- | 30 BIU/ha | ---------- | ----------→ |
| Volume rate | 2.36 L/ha | 1.78 L/ha | 1.18 L/ha | 1.78 L/ha |
| Droplet size spectra on Kromekote cards: | | | | |
| Droplets/cm$^2$ | 110 | 70 | 76 | 96 |
| Maximum diameter | 125 µm | 125 µm | 105 µm | 100 µm |
| Minimum diameter | 7 µm[q] | 7 µm[q] | 7 µm[q] | 20 µm |
| Number median diameter | 58 µm | 58 µm | 58 µm | 66 µm |
| Volume median diameter | 81 µm | 84 µm | 74 µm | 76 µm |
| Droplet data on fir foliage: | | | | |
| a. Droplets/needle | 12.8 | 11.6 | 8.8 | 7.6 |
| b. Droplets/cm$^2$ | 32.0 | 29.0 | 22.0 | 19.0 |
| Spray volume deposit on glass plate by fluorometric analysis of the tracer dye.: | | | | |
| a. Volume deposit | 2.36 L/ha | 1.78 L/ha | 1.18 L/ha | 1.78 L/ha |
| b. Percent deposition | 100 | 100 | 100 | 100 |
| Physical properties of formulations: | | | | |
| a. Viscosity at 2000/s shear rate (mPa.s) | 33.3 | 35.2 | 47.4 | 75.0 |
| b. Surface tension (mN/m) | 31.7 | 32.8 | 33.9 | 35.2 |
| c. Non-volatiles (%)[s] | 100 | 100 | 100 | 67.5 |
| d. Evap. rate (%)[t] | NA[u] | NA | NA | 1.01 |

p: Thuricide 64B is a water-based formulation containing a lipophilic agent.
q: Detection limit of the droplet sizing technique = 7 µm.
s: The non-volatile components are expressed in w/w%.
t: Evap. rate represents the rate of evaporation, i.e., decrease of w% per min.
u: NA = not applicable.

surface tension values (Tables 13 and 14). Droplets of the five aqueous formulations were either totally spherical or hemispherical on all sampling surfaces, and the degree of spreading was related to surface tension and pseudoplastic behaviour. The data on droplet size spectra indicated little influence from the physical properties of formulations. This is because of the differences in the application volumes and emission rates used, which influenced the droplet size spectra. All three oil-based Dipel formulations were non-volatile and provided 100% recovery of the spray volume on the glass plates. The four aqueous formulations, except Thuricide 64B, provided lower percent recovery of the spray volume on glass plates.

Table 14. ***Bacillus thuringiensis***: Dosage and volume rates, droplet spectra and mass deposits of four aqueous formulations after atomization in a chamber, and physical properties SOURCE: Reproduced with permission from reference 10. Copyright 1989 ASTM.

| Parameters | Dipel 6AF | Dipel 8AF | Thuricide 48LV | Futura XLV |
|---|---|---|---|---|
| Dosage rate | ←------------ | 30 BIU/ha | ------------ | ------------→ |
| Volume rate | 2.36 L/ha | 1.78 L/ha | 2.36 L/ha | 2.10 L/ha |
| Droplet size spectra on Kromekote cards: | | | | |
| Droplets/cm$^2$ | 60 | 36 | 90 | 100 |
| Maximum diameter | 160 µm | 170 µm | 90 µm | 80 µm |
| Minimum diameter | 20 µm[q] | 30 µm[q] | 30 µm[q] | 40 µm |
| Number median diameter | 63 µm | 80 µm | 56 µm | 54 µm |
| Volume median diameter | 80 µm | 102 µm | 66 µm | 64 µm |
| Droplet data on fir foliage: | | | | |
| a. Droplets/needle | 4.8 | 4.0 | 3.6 | 5.0 |
| b. Droplets/cm$^2$ | 12.0 | 10.0 | 9.0 | 12.5 |
| Spray volume deposit on glass plate by fluorometric anaylsis of the tracer dye: | | | | |
| a. Volume deposit | 1.47 L/ha | 1.62 L/ha | 1.28 L/ha | 1.17 L/ha |
| b. Percent deposition | 63 | 91 | 54 | 56 |
| Physical properties of formulations: | | | | |
| a. Viscosity at 2000/s shear rate (mPa.s) | 48.5 | 69.5 | 15.1 | 27.2 |
| b. Surface tension (mN/m) | 38.9 | 43.5 | 37.5 | 42.5 |
| c. Non-volatiles (%)[s] | 55.2 | 58.4 | 15.0 | 17.7 |
| d. Evap. rate (%)[t] | 1.71 | 1.75 | 1.76 | 1.70 |

q: Detection limit of the droplet sizing technique = 15 µm.
s: The non-volatile components are expressed in w/w%.
t: Evap. rate represents the rate of evaporation, i.e., decrease of w% per min.

**Study IV - Persistence of *Bacillus thuringiensis* Deposits in a Hardwood Forest, after Aerial Application of a Commercial Formulation at Two Dosage Rates.** *Bacillus thuringiensis* var. *kurstaki* is presently being viewed as an effective biorational alternative to conventional chemical insecticides for the control of several defoliating insects in forestry (*50*). Nonetheless, several researchers have reported the need to improve field efficacy (*51-55*). BTK can present its own ecological risks on nontarget arthropods (*56*). Some studies have reported the effect of field applications on the natural enemies (parasitoids) of gypsy moth (*57*), on the nontarget leaf-feeding lepidoptera (*50*), and on the monarch butterfly in its over-wintering habitat (*58*).

Therefore, from the view point of forest managers who wish to achieve maximum field efficacy against target lepidopterous insects with minimum BTK dosages, and the environmentally conscious public who demand minimum ecological risks to nontarget organisms, the knowledge on the duration of exposure, i.e., persistence of BTK in foliage would be valuable.

A commercial formulation of BTK, Foray® 48B, was sprayed aerially over four blocks (referred to as B1, B2, B3 and B4) in an oak forest in Wayne County, Pennsylvania during May, 1990 (*48*). B1 and B2 were sprayed at 75 BIU in 5.91 L/ha, and B3 and B4 at 50 BIU in 3.94 L/ha. Droplet spectra were assessed at canopy and ground levels on water-sensitive paper strips (Table 15). Mass deposits were determined using glass micro-fiber filters. Oak foliage was collected at different time intervals after treatment. Three types of bioassays were conducted (Tables 16 and 17) using 4th instar gypsy moth larvae, *viz.,* direct feeding on sprayed foliage, feeding on diet containing homogenized foliage and force-feeding on foliar extracts. Larval mortalities were converted into IU/$cm^2$ of foliage. Foliar extracts were also subjected to enzyme-linked immunosorbent assay (ELISA) (Table 18) to determine the concentration of delta-endotoxin protein (*59*). Regardless of the type of bioassay used, bioactivity persisted in foliage for about a week in all the blocks. The half-life of inactivation, $DT_{50}$, ranged from 12 to 22 h. The immunoassay data indicated a shorter duration of persistence (i.e., about 2 d) of the delta-endotoxin protein, with the $DT_{50}$ values ranging from 10 to 15 h. Formulation ingredients present in Foray 48B played a role in the toxicity of BTK to gypsy moth larvae.

Table 15. ***Bacillus thuringiensis***: Droplet size spectra on water-sensitive paper at ground and mid-canopy levels in aerial spray trials of Foray 48B in Pennsylvania in 1990

| Block number | B1 | B2 | B3 | B4 |
|---|---|---|---|---|
| Dosage rate (BIU/ha) | 75 | 75 | 50 | 50 |
| Volume rate (L/ha) | 5.91 | 5.91 | 3.94 | 3.94 |
| Data at ground level: | | | | |
| Droplets/$cm^2$ (mean ± SD) | 9.0 ± 3.2[a] | 14.2 ± 3.6[a] | 19.9 ± 3.9[a] | 17.6 ± 4.2[a] |
| Maximum diameter | 380 μm | 380 μm | 395 μm | 320 μm |
| Minimum diameter | 15 μm[b] | 15 μm[b] | 15 μm[b] | 15 μm[b] |
| Number median diameter | 52 μm | 69 μm | 70 μm | 89 μm |
| Volume median diameter | 224 μm | 180 μm | 167 μm | 159 μm |
| Data at canopy level: | | | | |
| Droplets/$cm^2$ (mean ± SD) | 15.7 ± 3.3[a] | 22.6 ± 5.0[a] | 21.5 ± 5.2[a] | 20.3 ± 4.7[a] |
| Maximum diameter | 395 μm | 395 μm | 355 μm | 340 μm |
| Minimum diameter | 15 μm[b] | 15 μm[b] | 15 μm[b] | 15 μm[b] |
| Number median diameter | 85 μm | 74 μm | 70 μm | 82 μm |
| Volume median diameter | 200 μm | 203 μm | 157 μm | 169 μm |

a: Mean ± SD of values obtained from 36 water-sensitive papers used in each block.
b: The minimum detection limit of the droplet sizing technique was 15 μm.

Table 16. ***Bacillus thuringiensis***: Initial mortality and persistence of BTK activity from foliar feeding and diet bioassays using sixth instar gypsy moth larvae

| Time after spray | Foliar feeding | Foliar feeding[b] | Foliage plus diet | Time after spray | Foliar feeding | Foliar feeding[b] | Foliage plus diet |
|---|---|---|---|---|---|---|---|
| Mortality[a] (%) (mean ± SD) in B1 (75 BIU in 5.91 L/ha) | | | | Mortality[a] (%) (mean ± SD) in B2 (75 BIU in 5.91 L/ha) | | | |
| 1.0 h | 28.8 ± 2.3 | 36.0 ± 2.9 | 30.3 ± 2.1 | 1.0 h | 49.5 ± 3.7 | 62.0 ± 4.5 | 55.0 ± 5.2 |
| 3.5 h | 20.0 ± 1.8 | 25.0 ± 2.5 | 22.0 ± 1.9 | 4.0 h | 41.0 ± 4.5 | 51.3 ± 6.7 | 46.0 ± 5.2 |
| 12.0 h | 10.3 ± 1.5 | 13.0 ± 2.5 | 10.3 ± 2.5 | 12.0 h | 29.0 ± 3.7 | 37.2 ± 4.5 | 31.4 ± 3.5 |
| 1.0 d | 8.5 ± 0.8 | 10.6 ± 1.6 | 8.5 ± 1.7 | 1.0 d | 25.4 ± 4.2 | 32.3 ± 3.2 | 26.3 ± 4.2 |
| 2.0 d | 5.3 ± 0.7 | 6.6 ± 1.4 | 5.3 ± 2.1 | 2.0 d | 17.4 ± 2.7 | 22.0 ± 1.2 | 16.2 ± 3.7 |
| 4.0 d | 3.2 ± 0.5 | 4.0 ± 0.8 | 4.0 ± 1.0 | 4.0 d | 8.3 ± 0.5 | 10.4 ± 1.2 | 7.2 ± 2.7 |
| 6.0 d | ND[c] | ND | ND | 6.0 d | ND | ND | ND |
| Mortality[a] (%) (mean ± SD) in B3 (50 BIU in 3.94 L/ha) | | | | Mortality[a] (%) (mean ± SD) in B4 (50 BIU in 3.94 L/ha) | | | |
| 1.0 h | 60.0 ± 6.2 | 78.0 ± 7.0 | 65.3 ± 4.5 | 1.0 h | 59.2 ± 5.2 | 68.7 ± 8.0 | 62.7 ± 6.3 |
| 4.5 h | 49.6 ± 6.3 | 64.5 ± 6.0 | 53.2 ± 3.5 | 4.0 h | 50.7 ± 6.7 | 60.8 ± 6.3 | 55.0 ± 7.2 |
| 12.5 h | 28.8 ± 6.2 | 37.4 ± 6.0 | 34.2 ± 3.0 | 12.0 h | 32.9 ± 5.8 | 39.9 ± 7.3 | 32.4 ± 4.3 |
| 1.0 d | 24.7 ± 5.2 | 32.1 ± 4.7 | 30.0 ± 2.5 | 1.0 d | 32.2 ± 4.0 | 38.0 ± 6.9 | 30.5 ± 4.1 |
| 2.0 d | 19.7 ± 3.8 | 26.0 ± 2.8 | 22.6 ± 1.3 | 2.0 d | 21.6 ± 2.9 | 24.4 ± 4.6 | 20.5 ± 4.5 |
| 4.0 d | 18.6 ± 2.7 | 23.3 ± 2.7 | 18.6 ± 3.7 | 4.0 d | 15.0 ± 3.2 | 26.7 ± 4.2 | 15.3 ± 3.2 |
| 6.0 d | 16.9 ± 1.5 | 21.3 ± 1.8 | 13.5 ± 1.2 | 6.0 d | 9.2 ± 2.1 | 12.3 ± 4.2 | 9.2 ± 2.8 |
| 8.0 d | 5.9 ± 0.7 | 7.8 ± 1.2 | 5.8 ± 1.0 | 8.0 d | 3.8 ± 0.8 | 7.9 ± 2.7 | 4.0 ± 1.2 |

a: Percent mortality was corrected for the control mortality ( which was ≤ 5%).
b: Corrected for differences in feeding behaviour of foliage stored in the freezer, and fresh foliage
c: ND = Not detectable (i.e., mortality was the same as in the control group).

Table 17. ***Bacillus thuringiensis:*** Initial deposits and persistence (mean ± SD) of BTK at intervals of time after spray, computed from bioassay[a] using 4th instar larvae force-fed with foliar extracts, and extracts of micro-fiber filters

| Time after spray | Foliar extract (% mortality) | IU/cm$^2$ of foliage | Micro-fiber filter extract (% mortality) | IU/cm$^2$ of micro-fiber filter | Time after spray | Foliar extract (% mortality) | IU/cm$^2$ of foliage | Micro-fiber filter extract (% mortality) | IU/cm$^2$ of micro-fiber filter |
|---|---|---|---|---|---|---|---|---|---|
| Block B1 (75 BIU in 5.91 L/ha) | | | | | Block B2 (75 BIU in 5.91 L/ha) | | | | |
| 1.0 h | 25.2 ± 2.2 | 32.8 ± 5.1 | 37.3 ± 3.8 | 53.2 ± 7.2 | 1.0 h | 34.8 ± 3.3 | 48.2 ± 6.5 | 38.9 ± 2.8 | 56.4 ± 5.9 |
| 3.5 h | 18.7 ± 1.5 | 24.2 ± 4.1 | 28.1 ± 3.0 | 37.3 ± 6.2 | 4.0 h | 27.7 ± 3.5 | 36.5 ± 6.8 | 31.1 ± 0.9 | 42.1 ± 3.2 |
| 12.0 h | 12.5 ± 2.0 | 16.9 ± 4.8 | 19.9 ± 2.7 | 25.9 ± 5.8 | 12.0 h | 20.3 ± 4.2 | 26.2 ± 7.6 | 22.2 ± 2.1 | 28.9 ± 5.0 |
| 1.0 d | 9.9 ± 1.2 | 13.9 ± 3.7 | 17.2 ± 3.2 | 22.5 ± 6.4 | 1.0 d | 17.9 ± 3.0 | 23.1 ± 6.1 | 19.5 ± 2.4 | 25.4 ± 5.4 |
| 2.0 d | 7.3 ± 0.9 | 11.1 ± 3.1 | 12.3 ± 1.8 | 16.7 ± 4.6 | 2.0 d | 11.3 ± 2.2 | 15.5 ± 5.1 | 13.3 ± 1.5 | 17.9 ± 4.1 |
| 4.0 d | 2.3 ± 0.5 | 5.2 ± 2.4 | 5.3 ± 1.5 | 8.9 ± 4.1 | 4.0 d | 7.5 ± 1.5 | 11.3 ± 4.1 | 7.2 ± 0.9 | 11.0 ± 3.2 |
| 6.0 d | ND[b] | ND | ND | ND | 6.0 d | ND | ND | ND | ND |
| Block B3 (50 BIU in 3.94 L/ha) | | | | | Block B4 (50 BIU in 3.94 L/ha) | | | | |
| 1.0 h | 46.7 ± 3.3 | 73.7 ± 6.5 | 51.2 ± 5.2 | 86.7 ± 8.8 | 1.0 h | 44.3 ± 4.7 | 67.8 ± 8.2 | 49.8 ± 6.7 | 82.6 ± 10.5 |
| 4.5 h | 36.1 ± 4.5 | 50.6 ± 7.9 | 43.1 ± 4.3 | 65.5 ± 7.8 | 4.0 h | 36.8 ± 5.4 | 51.9 ± 9.0 | 40.1 ± 7.2 | 58.8 ± 11.0 |
| 12.5 h | 26.4 ± 2.8 | 34.6 ± 5.9 | 29.8 ± 3.7 | 40.0 ± 7.0 | 12.0 h | 28.8 ± 6.6 | 38.2 ± 10.3 | 32.7 ± 5.2 | 44.8 ± 8.8 |
| 1.0 d | 23.1 ± 3.0 | 29.9 ± 6.1 | 26.4 ± 3.0 | 34.8 ± 6.2 | 1.0 d | 25.1 ± 3.2 | 32.7 ± 6.4 | 29.3 ± 4.9 | 39.2 ± 8.4 |
| 2.0 d | 18.2 ± 2.9 | 23.6 ± 6.0 | 20.4 ± 2.4 | 26.5 ± 5.4 | 2.0 d | 20.1 ± 1.8 | 26.0 ± 4.6 | 22.4 ± 2.1 | 29.2 ± 5.0 |
| 4.0 d | 12.9 ± 1.8 | 17.3 ± 4.6 | 15.4 ± 1.9 | 20.3 ± 4.7 | 4.0 d | 12.1 ± 2.2 | 16.4 ± 5.1 | 14.3 ± 1.5 | 19.0 ± 4.1 |
| 6.0 d | 6.7 ± 1.5 | 10.4 ± 4.1 | 9.8 ± 0.8 | 13.9 ± 3.0 | 6.0 d | 7.8 ± 1.9 | 11.6 ± 4.7 | 9.9 ± 1.0 | 14.0 ± 3.3 |
| 8.0 d | 2.3 ± 0.4 | 5.2 ± 2.1 | 5.9 ± 0.6 | 9.6 ± 2.6 | 8.0 d | 2.7 ± 1.0 | 5.8 ± 3.3 | 4.9 ± 0.3 | 8.4 ± 1.9 |

a: Percent mortality was corrected for the control mortality (which was ≤ 5%).
b: ND = Not detectable (i.e., mortality was the same as in the control group).

From the forest managers standpoint, the bioactivity of *Bacillus thuringiensis* is short-lived in sprayed foliage, and there is a need to increase field persistence to achieve a longer half-life of up to 2 or 3 days. However, from the viewpoint of the environmentally conscious public, no conclusion can be made from the findings of this study because no biological data were collected on nontarget biota.

Table 18. ***Bacillus thuringiensis:*** Enzyme-linked immunosorbent assay (ELISA) to quantify BTK delta-endotoxin protein (ng/cm$^2$) on foliage and micro-fiber filters collected at different intervals of time after application

| Time after spray | B1 (75 BIU/ha) | | B2 (75 BIU/ha) | | B3 (50 BIU/ha) | | B4 (50 BIU/ha) | |
|---|---|---|---|---|---|---|---|---|
| | Foliage | Filter | Foliage | Filter | Foliage | Filter | Foliage | Filter |
| 1.0 h | 77[a] | 105 | 83 | 114 | 98 | 132 | 97 | 122 |
| 6.0 h | 57 | 86 | 64 | 94 | 75 | 93 | 73 | 102 |
| 9.0 h | 40 | 69 | 47 | 70 | 55 | 78 | 52 | 82 |
| 0.5 d | 29 | 52 | 36 | 53 | 38 | 57 | 41 | 61 |
| 1.0 d | 20 | 37 | 23 | 39 | 27 | 36 | 29 | 45 |
| 1.5 d | NQ[b] | 27 | NQ | 29 | NQ | 25 | NQ | 31 |
| 2.0 d | NQ | NQ | NQ | NQ | NQ | NQ | NQ | NQ |

a: The standard deviations (not given here) ranged from 15 to 25% of the means.
b: NQ = Not quantifiable. Minimum quantification limit (MQL) of delta-endotoxin in Foray 48B was 8 ng/mL. Quantifiable amount of delta-endotoxin applied to field foliage exposed to natural weather conditions was 1400 ng/g or 18 ng/cm$^2$.

**Study V - Droplet Size Spectra and Deposits of Two Aerially Sprayed *Bacillus thuringiensis* Formulations on Artificial Samplers and Live Foliage.** Two commercial formulations of BTK, Foray® 48B and Thuricide® 48LV, were applied aerially over nine spray blocks in a hardwood forest in West Virginia in May, 1991 (*60*). Droplet size spectra were determined using water-sensitive papers and castor oil. Mass deposits were assessed on glass micro-fiber filters, and glass plates. Mass deposits on natural foliage were assessed by two bioassay methods, i.e., feeding of homogenized foliage containing a starch-sucrose solution and force-feeding bioassay using foliar extracts containing re-dissolved protein precipitate. Deposits on canopy foliage and ground samplers were also evaluated by total protein assay and enzyme-linked immunosorbent assay (ELISA) (*59*). The data indicated that the droplet spectra on the water-sensitive paper were different from those on castor oil. Droplets at the ground level on the horizontal water-sensitive paper were larger than those on the vertical water-sensitive paper. Similar samplers placed at the canopy level collected more droplets than those at the ground level. The total protein deposits (determined by using a modified bicinchoninic acid method, unpublished data of the author), expressed in ng/cm$^2$ were consistently higher on all blocks than the delta-endotoxin protein deposits. Spray mass deposit on the ground samplers were low, and ranged from 2.9 to 8.0% of the applied material.

## Conclusions

Research has thus far indicated that biorational control agents have the potential to become effective alternatives to conventional chemical insecticides. However, extensive work is needed to develop and improve the field performance of some of these products. The botanicals and bacterial pathogens show very short duration of persistence in foliage of host trees. Formulation research is being conducted to increase their field stability. Research is also being focused on rainfastness and photostability of the formulated products. Biorational pesticides that show prolonged persistence both in crops and in the environment should be viewed with caution. A thorough risk/benefit analysis will be necessary before establishing the "environmental acceptability' of these materials.

## Acknowledgment

The author thanks K.M.S. Sundaram for providing the necessary data on tebufenozide and azadirachtin to include in this paper.

## Literature Cited

1. Pimentel, D.; Acquay, H.; Biltonen, M.; Rice, P.; Silva,M.; Nelson, J.; Lipner, V.; Giordano, S.; Horowitz, A.; D'Amore.M. *BioScience,* **1992,** *42,* 750-760.
2. Anon. *Dimilin®: An Insecticide Interfering with Chitin Deposition,* Technical Information, 9th edition, Duphar, B.V., Weesp, Holland, 1985, 36 pp.
3. Wing, K.D.; Slawecki, R.A.; Carlson, G.R. *Science,* **1988,** *241,* 470-472.
4. Hurt, S.S. *Bulletin on RH-5992 Toxicology,* Rohm and Haas Company, Independence Mall West, Philadelphia, PA, 1990, p. 2.
5. Rembold, H. In *Insecticides of Plant Origin,* J.T. Arnason, B.J.R. Philogene and P. Morand (Eds.), Amer. Chem. Soc. Symp. Series No. 387, American Chemical Society, Washington, D.C., 1989, pp. 150-163.
6. Schmutterer, H. *Annual Rev. Entomol.,* **1990,** *35,* 271-297.
7. Bechtel, D.B.; Bulla, L.A., Jr. *J. Bacteriol.,* **1976,** 127, 1472-1481 (1976).
8. Bulla, L.A., Jr.; Kramer, K.J.; Davidson, L.I. *J. Bacteriol.,* **1977,** *130,* 375-383.
9. Ghassemi, M.; Fargo, L.; Painter, P.; Painter, P.; Quinlivan, S.; Scofield, R.; Takata, A. *Environmental Fate and Impacts of Major Forest Use Pesticides,* TRW Environmental Division, Report No. PB83 - 124552, Redondo Beach, CA 90278; EPA Contract No. 68-02-3174, US-EPA Office of Pesticides and Toxic Substances, Washington, D.C., 20460, 1981, pp. A-398-A410.
10. Sundaram, A. In *Pesticide Formulations and Application Systems: International Aspects, Nineth Volume,* ASTM STP 1036, J.L. Hazen and D.A Hovde, (Eds.), American Society for Testing and Materials, Phildelphia, PA, 1989, pp. 129-141.
11. Sundaram, A. *Trans. ASAE,* **1994,** *37,* 9-17.
12. Sundaram, A.; Leung, J.W.; Devisetty, B.N. In *Pesticide Formulations and Application Systems: Thirteenth Volume,* ASTM STP 1183, P.D. Berger, B.N. Devisetty and F.R. Hall, (Eds.), American Society for Testing and Materials, Philadelphia, PA, 1993, pp. 227-241.
13. Sundaram, K.M.S.; Sundaram, A. *J. Environ. Sci. Health,* **1994,** *B29,* 757-783.
14. Sundaram, K.M.S. *Pestic. Sci.,* **1991,** *32,* 275-293.

15. Sundaram, K.M.S.; Holmes, S.B.; Kreutzweiser, D.P.; Sundaram, A.; Kingsbury, P.D. *Arch. Environ. Contam. Toxicol.*, **1991,** *20,* 313-324.
16. Lichtenstein, E.P. In *Fate of Pesticides in the Environment,* A.S. Tahori, Ed., Gordon and Breach Science, New York, NY, 1972, pp. 1-22.
17. Metcalf, R.L.; Lu, P.Y.; Bowlus, S. *J. Agric. Food Chem.*, **1975,** *23,* 359-364.
18. Booth, G.M.; Ferrell, D. In *Pesticides in Aquatic Environments,* M.A.Q. Khan, Ed., Plenum Press, New York, NY, 1977, pp. 221-243.
19. Schaefer, C.H.; Dupras, E.F., Jr. *J. Agric. Food Chem.*, **1977,** *25,* 1026-1030.
20. Bull, D.L.; Ivie, G.W. *J. Agric. Food Chem.*, **1978,** *26,* 515-520.
21. Mansager, E.R.; Still, G.G.; Frear, D.S. *Pestic. Biochem. Physiol.*, **1979,** *12,* 172-182.
22. Nigg, H.N.; Cannizzaro, R.D.; Stamper, J.H. *Bull. Environ. Contam. Toxicol.*, **1986,** *36,* 833-838.
23. van den Berg, G. *Dissipation of Difluebenzuron Residues after Application of Dimilin WP-25 in a Forestry Area in N. Carolina (USA) and Some Ecological Effects,* Duphar, B.V., Crop Protection Division, Report No. 56637/47/1986, The Netherlands, 1986, 11 pp.
24. Sundaram, K.M.S. *Persistence and Degradation of Diflubenzuron in Conifer Foliage, Forest Litter and Soil, Following Simulated Aerial Application,* Govt. Can., For. Serv., For. Pest Manage. Inst., Inf. Rept. FPM-X-74, Sault Ste. Marie, Ontario, Canada, 1986, 19 pp.
25. Martinat, P.J.; Christman, V.; Cooper, R.J.; Dodge, K.M.; Whitmore, R.C.; Booth, G.; Seidel, G. *Bull. Environ. Contam. Toxicol.*, 1987, *39,* 142-149.
26. Rabenort, B.; de Wilde, P.C.; de Boer, F.G.; Korver, P.K.; Diprima, S.J.; Canizzaro, R.D. *In Analytical Methods for Pesticides and Plant Growth Regulators,* Volume 10, Zweig, G., Academic Press, New York, NY, 1987, pp. 57-72.
27. Mian, L.S.; Mulla, M.S. *Residue Rev.*, **1982,** *84,* 27-112.
28. Sundaram, K.M.S.; Szeto, S.Y. *J. Agric. Food Chem.*, **1984,** *32,* 1138-1141.
29. Sundaram, K.M.S.; Sundaram, A. *Environ. Can., For. Serv. Res. Notes,* **1982,** *2,* 2-5.
30. Sundaram, K.M.S.; Sundaram, A. *Pestic. Sci.*, **1987,** *18,* 259-271.
31. Sundaram, K.M.S. *J. Environ. Sci. Health,* **1994,** *B29*, 541-579.
32. Sundaram, K.M.S. *Persistence and Fate of RH-5992 in Aquatic and Terrestrial Components of a Forest Environment,* A Special Report to Rohm and Haas Company, Natural Resources Canada, Canadian Forest Service, Forest Pest Management Institute, 1219 Queen Street East, P.O. Box 490, Sault Ste. Marie, Ontario, Canada, 1993, 50 pp.
33. Sundaram, K.M.S.; Curry, J. *Pestic. Sci.*, **1994,** *41,* 129-138.
34. Saxena, R.C. *Insecticides of Plant Origin,* J.T. Arnason, B.J.R. Philogene and P. Morand, (Eds.), Amer. Chem. Soc. Symp. Series No. 387, Amer. Chem. Soc., Washington, D.C., 1989, pp. 110-135.
35. Locke, J.C.; Lawson, R.H. *Neem's Potential in Pest Management Programs,* Proceedings of the USDA Neem Workshop, USDA Agric. Res. Serv., ARS-86 1990, 136 pp.
36. Thomas, A.W.; Strunz, G.M.; Chaisson, M.; Chan., T.N. *Entomol. Exp. Appl.*, **1992,** *62,* 37-46.
37. Schmutterer, H.; Hellpap, C. In *Focus on Phytochemical Pesticides - Volume I:*

*The Neem Tree,* M. Jacobson, (Ed.), CRC Press, Inc., Boca Raton, Florida, 1988, pp. 69-86.

38. Helson, B.V. *For. Chron.,* **1992,** *68,* 349-354.
39. Jacobson, M. In *Focus on Phytochemical Pesticides, Volume I: The Neem Tree,* M. Jacobson, (Ed.), CRC Press, Inc., Boca Raton, Florida, 1988, pp. 133-153.
40. Morris, O.N.; Moore, A. *Studies on the Protection of Insect Pathogens from Sunlight Inactivation. II. Preliminary Field Trials,* Environ. Can., For. Serv., Chem Cont. Res. Inst. Inf. Rept. CC-X-113, 1975, 34 pp.
41. Bryant, J.E.; Yendol, W.G. *J. Econ. Entomol.,* **1988,** *81,* 130-134.
42. Sundaram, K.M.S.; Sundaram, A. In *Proceedings: IX Simposium Pesticide Chemistry, Degradation and Mobility of Xenobiotics.* A.A.M. Del Re, E. Capri, S.P. Evans, P. Natali and M. Trevisan, (Eds.), Istituto di Chimica Agraria ed Ambientale - Sez. Vegetale Facolta di Agraria, Vis Emilia Parmanse, 84 - 29100, Piacenza, Italy, 1993, pp. 497-507.
43. Pozgay, M.; Fast, P.; Kaplan, H.; Carey, P.R. *J. Invertebr. Pathol.,* **1987,** *50,* 246-253.
44. Morris, O.N.; McErlane, B. *Studies on the Protection of Insect Pathogens from Sunlight Inactivation. I. Preliminary Tests,* Environ. Can., For. Serv., Chem Cont. Res. Inst. Inf. Rept. CC-X-112, 1975, 45 pp.
45. Pinnock, D.E.; Brand, R.J.; Milstead, J.E.; Jackson, K.L. *J. Invertebr. Pathol.,* **1975,** *25,* 209-214.
46. Leong, K.L.H.; Cano, R.J.; Kubinski, A.M. *Environ. Entomol.,* **1980,** *9,* 593-599.
47. Sundaram, K.M.S.; Sundaram, A.; Zhu, J.S.; Nott, R.; Curry, J.; Leung, J.W. *J. Environ. Sci. Health,* **1993,** *B28,* 243-273.
48. Sundaram, K.M.S.; Sundaram, A.; Hammock, B.D. *J. Environ. Sci. Health,* **1994,** *B29,* 999-1052.
49. Sundaram, K.M.S.; Sundaram, A. *J. Environ. Sci. Health,* **1992,** *B27,* 73-112.
50. Miller, J.C. *Field Assessment of the Effects of a Microbial Pest Control Agent on Nontarget Lepidoptera,* The American Entomologist, **1990,** pp. 135-139.
51. Cadogan, B.L.; Zylstra, B.F.; Nystrom, C.; Ebling, P.M.; Pollock, L.B. *Proc. Ent. Soc. Ont.,* **1986,** *117,* 59-64.
52. Cadogan, B.L.; Scharbach, R.D. *Can. Ent.,* **1993,** *125,* 479-488.
53. Cadogan, B.L. *Crop Protection,* **1993,** *12,* 351-356.
54. Dubois, N.R.; Reardon, R.C.; Mierzejewski, K. *J. Econ. Entomol.,* **1993,** *86,* 26-33.
55. Smitley, D.R.; Davis, T.W. *J. Econ. Entomol.,* **1993,** *86,* 1178-1184.
56. Pimentel, D.; Glenister, C.; Fast, S.; Gallahan, D. *Oikos,* **1984,** *42,* 283-290.
57. Webb, R.E.; Shapiro, M.; Podgwaite, J.D.; Reardon, R.C.; Tatman, K.M.; Venables, L.; Kolodny-Hirsch, D.M. *J. Econ. Entomol.,* **1989,** *82,* 1695-1701.
58. Brower, L.P. *Atala,* **1986,** *14,* 17-19.
59. Sundaram, K.M.S.; Sundaram, A.; Gee, S.J.; Harrison, R.O.; Hammock, B.D. In *Pesticide Formulations and Application Systems*, ASTM STP 1234, F.R. Hall, P.D. Berger and H.M. Collins, (Eds.), American Society for Testing and Materials, Philadelphia, PA, 1994, (in press).
60. Sundaram, A.; Sundaram, K.M.S.; Leung, J.W.; Sloane, L. *J. Environ. Sci. Health,* **1994,** *B29,* 697-738.

RECEIVED January 12, 1995

# Chapter 9

# Photostability and Rainfastness of Tebufenozide Deposits on Fir Foliage

**Kanth M. S. Sundaram**

**Forest Pest Management Institute, Canadian Forest Service, Natural Resources Canada, 1219 Queen Street East, Box 490, Sault Sainte Marie, Ontario P6A 5M7, Canada**

Two formulation concentrates of the insecticide, tebufenozide, [Mimic®, also known as RH-5992, N'-t-butyl-N'-(3,5-dimethylbenzoyl)-N-(4-ethylbenzoyl) hydrazine], an aqueous flowable (2F) and an emulsion-suspension (ES), were diluted with water and sprayed onto balsam fir branch tips at 140 to 150 g of active ingredient (AI) in 4.0 to 5.0 L/ha. Simulated rainfall was applied onto treated branch tips after different ageing periods of deposits. Foliar washoff of RH-5992 was assessed after application of different amounts of rain. A direct relationship existed between the amount of rainfall and AI washoff. The larger the rain droplet size, the greater the washoff. Longer rain-free periods made the deposits more resistant to rain. Regardless of the amount of rainfall, rain droplet size and ageing period, foliar deposits of the 2F mixture were washed off more than those of the ES mixture.

Another set of branch tips was exposed to simulated sunlight at two different radiation-free periods, and the emission-intensity spectra were measured. The amount of AI disappeared from foliage after exposure to radiation was measured. A direct relationship existed between radiation intensity and AI disappearance from foliage. The longer the duration of exposure, the greater the disappearance. Unlike the rain-washing, the ageing of foliar deposits had little influence on photo-induced disappearance of the AI. Regardless of the amount and intensity of radiation, and radiation-free period, AI deposits of the ES mixture disappeared more than those of the 2F mixture.

Rain-washing of foliar deposits of sprays reduces pesticidal activity (*1-3*), depending upon how much of the active ingredient (AI) is washed off (*4-9*). Pesticide residues are also susceptible to photodegradation from treated surfaces (*10-11*), depending upon the emission spectrum and duration of radiation. Residues must persist intact in foliage at a concentration at or above the threshold level throughout the critical period of pest development, to ensure adequate pest control. In the case of pesticides

0097–6156/95/0595–0134$12.00/0

that are systemic in action, the longer the 'ageing' (i.e., rain-free or radiation-free) period, the greater the amount undergoing penetration and translocation into the untreated parts of plants, thereby removing the pesticide residues from the site of exposure (*12*). However, information is sparse on the rain-washing and photo-disappearance of insecticides that do not translocate into the untreated parts of plants, and little is known about how much of the spray deposited on foliage will be lost in rain, or after exposure to sunlight. Protection of foliar deposits from rain and sunlight becomes even more important for those insecticides that need to be orally ingested for pesticidal activity.

Field studies have shown that RH-5992 [N'-t-butyl-N'-(3,5-dimethylbenzoyl)-N-(4-ethylbenzoyl) hydrazine], an ecdysone agonist, persisted in the lithosphere for over 5 months (*13*). The higher the initial residues, the longer the persistence. However, any increase in soil residues as a result of rain-washing of foliar deposits would only be small. For example, if we assume the 'worst case scenario' for deposit washoff, i.e., even if all the deposited material (maximum deposit would be 140 g of RH 5992 per ha area of leaves, or 1.4 µg per $cm^2$) was washed off, the increase in soil residues would only be a small fraction of 1.4 µg per $cm^2$, because the rain-washings from a small foliar area would fall on a very large ground area). If the deposit on 1.0 $cm^2$ of foliage would reach, for example, 100 $cm^2$ of the ground, the increase in soil residues would only be 0.014 µg per $cm^2$. This value is too small to influence persistence. Nonetheless, from the viewpoint of the field operators to achieve maximum efficacy against the target pests, the information on rain-washing of foliar deposits is useful for understanding the field behaviour of the material.

The objectives of the present investigation were to determine the loss of RH-5992 from foliage after spray application onto balsam fir [*Abies balsamea* (L.) Mill] branch tips, in order to examine the effect of: (i) different amounts of simulated rainfall; (ii) rain droplet sizes; (iii) rain-free periods; (iv) formulation type on rainfastness; (v) variable amounts of sunlight radiation; (vi) radiation intensity; (vii) radiation-free periods; and (viii) formulation type on photostability.

## Experimental Procedures

Two types of formulation concentrates, aqueous flowable (2F) and emulsion-suspension (ES) each containing 240 g AI/liter, were obtained from Rohm and Haas (Westhill, Ontario, Canada). Each formulation was diluted with water of moderate hardness (0.75 mM of $Ca^{++}$ + 0.25 mM of $Mg^{++}$, or 1.00 mM/L) (*14*) to provide a spray mixture containing 140 g AI in 4.0 L, or 150 g AI in 5.0 L.

The investigation consisted of eight studies, I to VIII, corresponding to the eight objectives listed above. The temperature, relative humidity (RH) and photoperiod were, respectively, 22 ± 2°C, 70 ± 5% and 16h light:8h darkness for all studies. Branch tips (15 ± 1.5 cm long containing 76 ± 7 needles), without any open buds, were collected from the field (tree height, *ca* 1 m). The stems of the branches were placed in tap water and used within two days. Fresh branch tips were collected whenever needed in order to avoid storage requirements. The surface area available for droplet impaction on a branch tip was measured by separating the needles, stalk and petiole. From the average surface area of a needle (0.4 $cm^2$), and the area of the stalk and petiole (2.25 $cm^2$), the total surface area (76 needles x 0.4 $cm^2$ + 2.25 $cm^2$ = 32.65 $cm^2$) was computed.

**Study I - Washoff of RH-5992 by 4 mm and 8 mm rainfall at a rain-free period of 8 h after treatment with the 2F mixture.** The 2F mixture was sprayed at the rate of 140 g AI in 4.0 L/ha in a chamber equipped with a spinning disc atomizer (Flak®, Micron Agri Sprayers Canada, Walkerton, Ontario) mounted on a central rail. The rail facilitated the to-and-fro movement of the atomizer (*15*). The chamber was calibrated (*16*) with no branch tips inside, by using water-sensitive paper strips (WSPS, Ciba Geigy Ltd., Basle, Switzerland), and a glass plate (GP, 10 cm x 10 cm). The application parameters (Table 1) were optimized by repeated trials (conducted in still air) to get similar droplet stains on the WSPS for the replicate treatments. The WSPS-GP assemblies (referred to as 'samplers') were removed 15 min post-spray, and the stain sizes were measured microscopically. By using a spread factor for droplet spreading (*17*), the number and volume median diameters ($D_{N0.5}$ and $D_{V0.5}$ respectively) and the maximum and minimum diameters ($D_{max}$ and $D_{min}$ respectively) were evaluated (*18*). The glass plates were eluted with acetonitrile and the AI deposits were determined (Table 1) by high-performance liquid chromatography (HPLC) (*19*).

Table 1. **Study I** - Spray application parameters, droplet size spectra, droplets/$cm^2$ and mass deposit of RH-5992 on samplers during the calibration and actual trials of the 2F mixture

| Spray delivery parameters used in the calibration and actual trials: | |
|---|---|
| Power supply (volts) | 6 |
| Flow rate (mL/min) | 0.77 |
| Number of mL delivered | 1.54 |
| Number of passes | 16 |
| Track speed (m/s) | 0.57 |
| Dosage rate (g AI/ha) | 140 |
| Volume rate (L/ha) | 4.0 |
| **Droplet spectra and AI deposits on samplers after spraying 2F mixture:** | |
| Number median diameter (µm) | 70 ± 3[a] |
| Volume median diameter (µm) | 114 ± 5[a] |
| Maximum diameter (µm) | 189 |
| Minimum diameter (µm) | 15[b] |
| Droplets/$cm^2$ | 85 ± 15[a] |
| Deposit on glass plates (g AI/ha) | 132 ± 4.3[a] |
| Percent recovery | 94.3 ± 3.1[a] |

a: Values are given in mean ± SD.
b: The minimum detection limit (MDL) of the droplet sizing technique was 15 µm.

Sixteen fir branch tips were equally divided into two sets of eight. One set was used for the 4 mm cumulative rain and the other, for the 8 mm rain. For RH-5992 treatment, four branch tips were used at a time (i.e., 2 sets x 4 branch tips x 2 replicate treatments = 16 branch tips). Two samplers were used (Figure 1a) to assess droplet sizes and AI deposits. Spray was applied using the parameters selected in the calibration trials (Table 1). The branches and the samplers were removed 15 min post-spray. The branches were taken out of the conical flasks, the water contaminated with the RH-5992 was replaced with fresh water, and the branches were returned to the flasks. Of the 8 branch tips used in each set, two were randomly selected to determine the initial deposits of RH-5992, and the other six were kept aside for rain-washing studies. The entire branch tips were analyzed for initial deposits by HPLC as described by Sundaram et al. (*19*) (2 branch tips x 2 replicate measurements of each branch tip = 4 measurements in total). The deposits on the artificial samplers were also analyzed to determine the droplet spectra and the AI deposited (Table 1).

Prior to the actual investigation, test trials were conducted to determine the persistence of RH-5992 residues in fir foliage after treatment, without any rainfall. No loss was detected even up to 72 h after treatment. Therefore, in the actual investigation, the pre-rain foliar residues (i.e., 8 h after spray) were assumed to be the same as those measured at 15 min after spray (i.e., initial deposits).

To determine rain-washing of RH-5992 from foliage, the water in each conical flask was decanted, a glass funnel was placed over the flask, and the branch tip was placed through the funnel (Figure 1b). This arrangement facilitated collection of the rain-washing directly into the flasks for residue analysis. The branch tips in each set were treated with rain in two replicate rainfall treatments, using 3 branch tips at a time (Figure 1b). Simulated rainfall was generated in the spray chamber using the Veejet® 8002 nozzle (Spraying Systems Co., Wheaton, Illinois, USA). The spinning disc atomizer was removed from the central rail of the chamber, and the 8002 nozzle was mounted. The rainfall application parameters (Table 2) were chosen by repeated trials to provide 4 and 8 mm amounts of cumulative rain. A rain gauge was used to measure the cumulative rainfall (Figure 1b). The rain droplet size spectra were determined as described by Sundaram (*20*), and the $D_{max}$, $D_{min}$, $D_{N0.5}$, and $D_{V0.5}$ were obtained (Table 2). After the rainfall, the conical flasks were removed from the chamber and the branch tips were discarded. The rain water in the flasks containing the washoff of RH-5992 was transferred into amber-colored glass bottles and stored at –20°C until analysis by HPLC (*19*). The amount of RH-5992 washed off from foliage is given in Table 3.

**Study II - Effect of 1 mm/h and 5 mm/h rainfall intensity on washoff of RH-5992 from foliage after application of the 2F mixture.** Fir branch tips were treated with the 2F mixture using the application parameters described in Study I. To apply rainfall of different intensities, two types of Veejet nozzles, 8001 and 8002, were used (Table 4). The duration of rainfall was adjusted to provide the same cumulative rainfall (6 mm) from both types of nozzles, and the rain droplet sizes (Table 4) were measured. At 15 min after the rainfall, the conical flasks were removed, the branch tips were discarded, and the RH-5992 residues (Table 5) in the rain water were analyzed by HPLC.

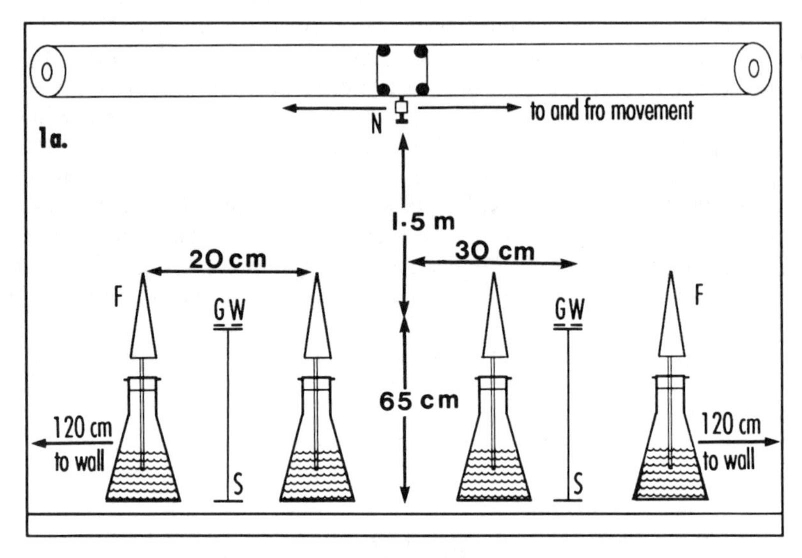


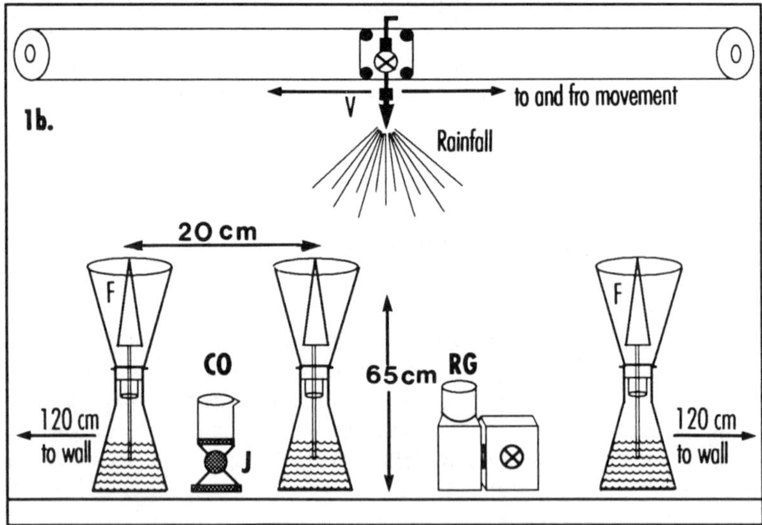


Figure 1a. Positions of balsam fir branches (F) in conical flasks, and samplers (G, Glass plates, and W, water–sensitive paper on stands, S) during spray application of 2F end-use mixes. N: Spinning disc atomizer.

Figure 1b. Postions of the balsam fir branches (F) in funnels in flasks, castor oil (CO) in a beaker on a jack (J), and rain gauge (RG) during rainfall application. V: Veejet 8002 nozzle.

Table 2. **Study I** - Details on rainfall simulation to determine rain-washing of RH-5992 from foliar deposits of the 2F mixture

| | | |
|---|---|---|
| Cumulative rainfall | 4 mm | 8 mm |
| Nozzle type | Veejet 8002 | Veejet 8002 |
| Nozzle orifice | 0.9 mm | 0.9 mm |
| Spray angle | 74° | 74° |
| Spray pressure | 180 kPa | 180 kPa |
| Flow rate | 0.58 L/min | 0.58 L/min |
| Intensity of rain | 5 mm/h | 5 mm/h |
| Duration of rain | 48 min | 96 min |
| Rain droplet size spectra:[a] | | |
| Minimum diameter | 80 μm | 80 μm |
| Maximum diameter | 1100 μm | 1100 μm |
| Number median diameter[b] | 445 ± 85 μm | 445 ± 85 μm |
| Volume median diameter[b] | 780 ± 95 μm | 780 ± 95 μm |

a: Rain droplet sizes were measured by collecting the droplets for 20 s in castor oil in a 400 mL beaker. Sizes were assessed by microscopy during sedimentation of the water droplets towards the bottom of the beaker.

b: The data represent mean ± SD of three replicate measurements.

Table 3. **Study I** - Initial deposits of RH-5992, the amount washed off by 4 mm and 8 mm of simulated rainfall, and the percent washoff from fir foliage sprayed with the 2F mixture

| | |
|---|---|
| Application rate | 140 g of RH-5992 in 4.0 L/ha |
| Surface area of a branch tip[p] | 32.65 $cm^2$ (needles + petiole + stalk) |
| Initial deposits[q] | 43.1 ± 2.9 (μg per branch tip) |
| RH-5992 washoff after 4 mm rainfall at a rain-free period of 8 h: | |
| Amount (μg) washed off in rain[q] | 34.0 ± 2.3 |
| Mean residues per branch tip (μg)[r] | 9.1 |
| Percent washoff[p] | 78.9 |
| RH-5992 washoff after 8 mm rainfall at a rain-free period of 8 h: | |
| Amount (μg) washed off in rain[q] | 40.9 ± 2.0 |
| Mean residues per branch tip (μg)[r] | 2.2 |
| Percent washoff[p] | 94.9 |

p: The data represent average values.

q: Values represent mean ± SD.

r: Calculated by subtracting the amount washed off from initial residues.

Table 4. **Study II** - Rainfall application details to determine RH-5992 washoff at two rainfall intensities from foliar deposits of the 2F mixture

| | | |
|---|---|---|
| Cumulative rainfall | 6 mm | 6 mm |
| Rainfall intensity | 1 mm/h | 5 mm/h |
| Nozzle type | Veejet 8001 | Veejet 8002 |
| Nozzle orifice | 0.65 mm | 0.90 mm |
| Spray angle | 72° | 74° |
| Spray pressure | 160 kPa | 180 kPa |
| Flow rate | 0.24 L/min | 0.58 L/min |
| Duration of rain | 6.0 h | 1.2 h |
| Rain droplet size spectra:[a] | | |
| Maximum diameter | 340 µm | 1100 µm |
| Minimum diameter | 30 µm | 80 µm |
| Number median diameter[b] | 155 ± 25 µm | 445 ± 85 µm |
| Volume median diameter[b] | 315 ± 25 µm | 780 ± 95 µm |

a and b: See footnotes of Table 2.

Table 5. **Study II** - Initial deposits (mean ± SD), the amount washed off by 6 mm rainfall and the percent washoff from foliage sprayed with 2F mixture

| | |
|---|---|
| Application rate | 140 g in 4.0 L/ha |
| Initial deposits[q] | 43.1 ± 2.9 (µg per branch tip) |
| RH-5992 washoff at 1.0 mm/h rainfall intensity and 6 mm of cumulative rain after a rain-free period of 8 h: | |
| Amount (µg) washed off in rain[q] | 17.2 ± 1.1 |
| Mean residues per branch tip (µg)[r] | 25.9 |
| Percent washoff[p] | 39.9 |
| RH-5992 washoff at 5.0 mm/h rainfall intensity and 6 mm of cumulative rain after a rain-free period of 8 h: | |
| Amount (µg) washed off in rain[q] | 38.1 ± 3.3 |
| Mean residues per branch tip (µg)[r] | 5.0 |
| Percent washoff[p] | 88.4 |

p, q and r: See footnotes of Table 3.

**Study III - Effect of 8 h and 72 h rain-free periods on RH-5992 washoff from foliage after application of the 2F mixture.** Fir branch tips were treated with the 2F mixture as described in Study I, and rain was applied at 8 h and 72 h after RH-5992 treatment. The data on rain-washing of RH-5992 at the two different rain-free periods are given in Table 6.

Table 6. **Study III** - Initial deposits of RH-5992, washoff by a 5 mm cumulative rainfall at 8 h and 72 h rain-free periods, and the percent washoff from foliage sprayed with 2F end-use mixes

| | |
|---|---|
| Application rate | 140 g in 4.0 L/ha |
| Initial deposits[q] | 43.1 ± 2.9 (μg per branch tip) |
| <u>RH-5992 washoff by a 5 mm cumulative rainfall after a rain-free period of 8 h:</u> | |
| Amount (μg) washed off in rain[q] | 37.0 ± 2.4 |
| Mean residues per branch tip (μg)[r] | 6.1 |
| Percent washoff[p] | 85.9 |
| <u>RH-5992 washoff by a 5 mm cumulative rainfall after a rain-free period of 72 h:</u> | |
| Amount (μg) washed off in rain[q] | 15.7 ± 3.3 |
| Mean residues per branch tip (μg)[r] | 27.4 |
| Percent washoff[p] | 36.4 |

p, q and r: See footnotes of Table 3.

**Study IV - Role of formulation type [flowable (2F) versus emulsion-suspension (ES)] on the amount of RH-5992 washed off.** Ten fir branch tips were divided into two sets of five. One set was sprayed with the 2F mixture, and the other, with the ES mixture. The two formulation concentrates, RH-5992 2F and RH-5992 ES, were diluted with moderately hard water (as in Study I). Each mixture was sprayed onto five branch tips (two branches in one treatment, and three branches in the replicate treatment) at 150 g AI in 5.0 L/ha (see Table 7 for application parameters, droplet spectra and mass deposits on samplers). Two branches were randomly selected to assess initial deposits (2 branches x 2 measurements of each branch = 4 measurements in total), and the remaining 3 branches were used for rainfall application. The data on rainfall intensity, cumulative rainfall, rain droplet spectra, initial deposits of RH-5992 on foliage and the amount washed off are given in Table 8.

**Study V - Loss of RH-5992 after 8-d and 16-d exposure to sunlight of intensity 437 W/m$^2$, at 48h after application of the 2F mixture.** Prior to the start of the investigation, the intensity and emission spectra of the natural sunlight were measured outside of the Sault Ste. Marie laboratory in the open sky, between 0900 to 1500 h every day for a total period of 140 h using a portable Spectroradiometer (Model LI-1800, LI-COR, Inc., Lincoln, Nebraska). The wavelength ranged from 320 to 1100 nm (Figure 2), the average intensity was 678 W/m$^2$ and the mean temperature was 16.5°C. Similar measurements inside the greenhouse, where the branches were held, indicated the same wavelength range, but with an intensity of 437 W/m$^2$ (*ca* 64% of the value in the open sky) at a location 10 cm above the branches, and at an average temperature of 22°C. Fifteen fir branches were divided into two sets , one of five and the other one of ten. Both sets were treated with the 2F mixture at 140 g AI in 4.0 L/ha. Spray application was made as described in Study IV. After 15 min, the branches were removed, transported to the greenhouse, and kept in darkness for

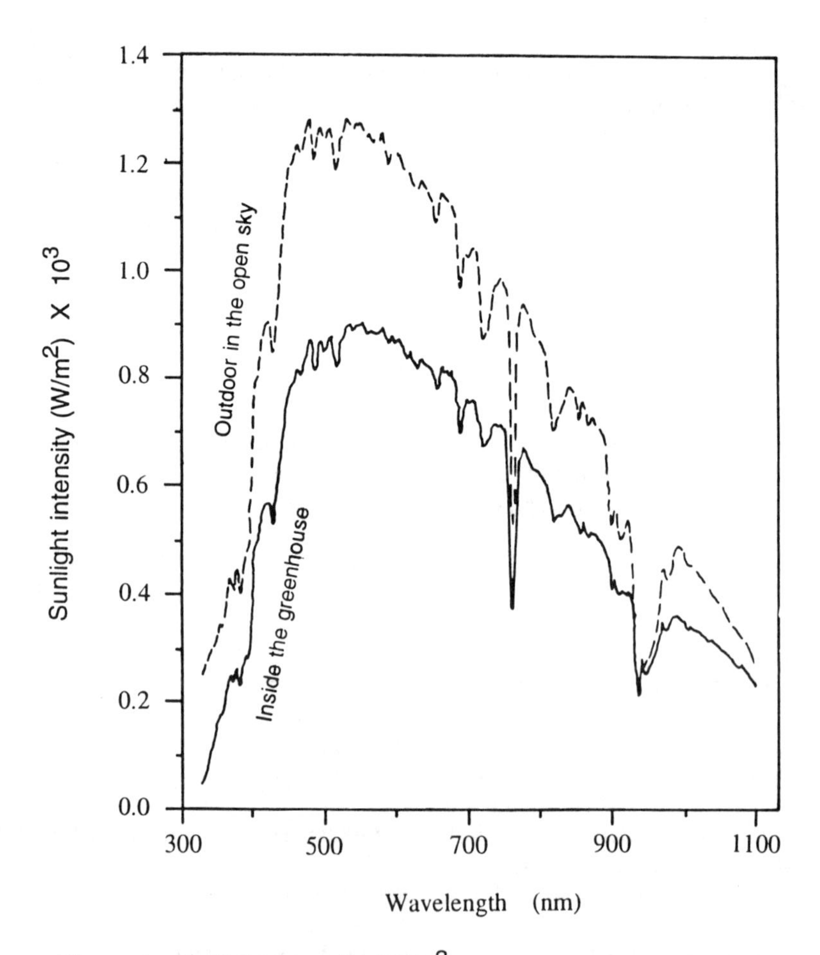


Figure 2. Sunlight intensity (W/m²) *versus* wavelength (nm) of radiation during the photostability studies

Table 7. **Study IV** - Spray parameters, droplet size spectra[a], droplets/unit area[a] and deposit recovery[a] of RH-5992 on artificial samplers during the calibration and actual trials of the 2F and ES mixtures

| | |
|---|---|
| Spray delivery parameters used in the calibration and actual trials: | |
| Power supply | 7 volts |
| Flow rate | 0.65 mL/min |
| Volume delivered | 1.94 mL |
| Number of passes | 20 |
| Track speed | 0.44 m/s |
| Application rate | 150 g AI in 5.0 L/ha |
| Droplet size spectra and AI deposits on samplers after application of 2F mixes: | |
| Number median diameter (μm) | 68 ± 4 |
| Volume median diameter (μm) | 102 ± 7 |
| Minimum - maximum diameter (μm) | 10[b] - 157 |
| Droplets/$cm^2$ | 102 ± 15 |
| Deposit on glass plates (g AI/ha) | 127 ± 6.2 |
| Percent recovery of deposit | 84.7 ± 4.1 |
| Droplet size spectra and AI deposits on samplers after application of ES mixes: | |
| Number median diameter (μm) | 76 ± 4 |
| Volume median diameter (μm) | 113 ± 6 |
| Minimum - maximum diameter (μm) | 10[b] - 190 |
| Droplets/$cm^2$ | 96 ± 8 |
| Deposit on glass plates (g AI/ha) | 138 ± 4.3 |
| Percent recovery of deposit | 92.0.± 2.9 |

a: Values are given in mean ± SD.

b: The MDL of the droplet sizing technique was 10 μm in Studies IV to VI, as opposed to 15 μm in Studies I to III. This was because there were more small droplets in Studies IV to VI. Therefore, the droplet stains were measured at a higher magnification of the microscope with an MDL of 10 μm.

48 h (22°C, RH 70%). One set of five branches was then analyzed individually for RH-5992 residues (5 replicates). The second set of 10 branches was divided equally into two sub-sets, for exposure to sunlight in the greenhouse (16 h light: 8 h darkness) at an intensity of 437 W/$m^2$, one sub-set for 8 days, and the other one for 16 days. Post-exposure residues of RH-5992 in the branches are given in Table 9.

**Study VI - Loss of RH-5992 after 12-d exposure to sunlight of intensity 437 and 752 W/$m^2$ at 48 h after application of the 2F mixture.** Fifteen fir branches were divided into two sets, sprayed with the 2F mixture at 140 g AI in 4.0 L/ha, and kept in darkness for 48 h as described in Study V. The first set was then analyzed for RH-5992 residues. The other set was divided into two sub-sets. One sub-set of five

Table 8. **Study IV** - Initial deposits of RH-5992, rainfall application details and washoff by 5 mm rainfall from foliage sprayed with 2F and ES mixtures

| | |
|---|---|
| Application rate | 150 g in 5.0 L/ha |
| Rainfall application details: | |
| Nozzle type | Veejet 8002 |
| Intensity of rain (mm/h) | 5.0 |
| Duration of rain (min) | 60 |
| Cumulative rainfall (mm) | 5.0 |
| Minimum - maximum diameter (μm) | 80 - 1100 |
| Number median diameter (μm)[b] | 445 ± 85 |
| Volume median diameter (μm)[b] | 780 ± 95 |
| End-use mixes of the 2F formulation concentrate: *RH-5992 washoff by a 5 mm cumulative rainfall after a rain-free period of 8 h:* | |
| Initial deposits (μg/branch tip)[q] | 25.5 ± 2.4 |
| Amount (μg) washed off in rain[q] | 22.2 ± 2.2 |
| Mean residues per branch tip (μg)[r] | 3.3 |
| Percent washoff[p] | 87.1 |
| End-use mixes of the ES formulation concentrate: *RH-5992 washoff by a 5 mm cumulative rainfall after a rain-free period of 8 h*: | |
| Initial deposits (μg/branch tip)[q] | 31.8 ± 1.9 |
| Amount (μg) washed off in rain[q] | 9.03 ± 1.3 |
| Mean residues per branch tip (μg)[r] | 22.77 |
| Percent washoff[p] | 28.4 |

b: See footnote of Table 2. p, q and r: See footnotes of Table 3.

branches was exposed to sunlight at 437 W/m$^2$. To expose the other sub-set to a higher intensity, two household table lamps (each containing 2 x 100 W bulbs) were used for 4 h every day during 0900 to 1300 h. Measurements indicated the same wavelength and temperature ranges as before, but with an average intensity of 752 W/m$^2$. Both sub-sets were exposed to light for 12 days. The branches were analyzed individually for residues of RH-5992, prior to exposure and after exposure to the simulated sunlight. The data are given in Table 10.

**Study VII - Loss of RH-5992 after 8-d exposure to sunlight of intensity 752 W/m$^2$ at 48 h and 168 h after application of the 2F mixture.** Twenty fir branches were divided into two sets and sprayed with the 2F mixture at 140 g AI in 4.0 L/ha. One set of ten was kept in darkness for 48 h and the other set for 168 h. Each set was divided into two sub-sets of five, and one sub-set was analyzed for RH-5992 residues. The other sub-set was exposed to simulated sunlight at 752 W/m$^2$ as in Study V for 8 days. The branches were analyzed individually for RH-5992, prior to and after exposure to simulated sunlight. The data are given in Table 11.

Table 9. **Study V** - Deposits of RH-5992 before exposure, and the loss after exposure to sunlight at 48 h radiation-free period after application of the 2F mixture

| | |
|---|---|
| Application rate | 140 g in 4.0 L/ha |
| Number of branch tips used | 5 for initial deposits |
| Deposits before exposure to sunlight[q] | 42.9 ± 2.1 (μg per branch tip) |
| <u>RH-5992 loss after 8-d exposure to sunlight at 437 W/m$^2$ intensity, after a radiation-free period of 48 h:</u> | |
| Number of branch tips used | 5 after exposure |
| Residual amount per branch tip[q] | 30.7 ± 3.2 μg |
| Amount lost after exposure (mean)[r] | 12.2 μg |
| Percentage loss (mean) | 28.4 |
| <u>RH-5992 loss after 16-d exposure to sunlight at 437 W/m$^2$ intensity, after a radiation-free period of 48 h:</u> | |
| Number of branch tips used | 5 after exposure |
| Residual amount per branch tip[q] | 19.0 ± 2.1 μg |
| Amount lost after exposure (mean)[r] | 23.9 μg |
| Percentage loss (mean) | 55.7 |

q: Values represent mean ± SD.
r: Calculated by subtracting the residual amount per branch tip (mean of 5 branches).

Table 10. **Study VI** - Deposits of RH-5992 before exposure, and the loss after 12-d exposure to sunlight at 48 h radiation-free period after application of 2F mixture

| | |
|---|---|
| Application rate | 140 g in 4.0 L/ha |
| Number of branch tips used | 5 for initial deposits |
| Deposits before exposure to sunlight[q] | 42.9 ± 2.1 (μg per branch tip) |
| <u>RH-5992 loss after 12-d exposure to sunlight at 437 W/m$^2$ intensity, after a radiation-free period of 48 h:</u> | |
| Number of branch tips used | 5 after exposure |
| Residual amount per branch tip [q] | 25.6 ± 3.1 μg |
| Amount lost after exposure(mean) [r] | 17.3 μg |
| Percentage loss (mean) | 40.3 |
| <u>RH-5992 loss after 12-d exposure to sunlight at 752 W/m$^2$ intensity, after a radiation-free period of 48 h:</u> | |
| Number of branch tips used | 5 after exposure |
| Residual amount per branch tip[q] | 17.0 ± 1.9 μg |
| Amount lost after exposure (mean)[r] | 25.9 μg |
| Percentage loss (mean) | 60.4 |

q and r: See footnotes of Table 9.

**Study VIII - Loss of RH-5992 after 16-d exposure to sunlight of intensity 752 W/m$^2$ at 48 h after application of the 2F and ES mixtures.** Twenty fir branches were divided into two sets of ten. One set was sprayed with the 2F mixture, and the other with the ES mixture, both at 150 g AI in 5.0 L/ha. The branches were kept in darkness for 48 h. Each set was divided into two sub-sets of five, and one sub-set was analyzed for RH-5992 residues. The other sub-set was exposed to sunlight of intensity 752 W/m$^2$ for 16 days. The branches were analyzed for RH-5992, and the data are given in Table 12.

Table 11. **Study VII** - Deposits of RH-5992 before exposure, and the loss after exposure to sunlight at two radiation-free periods after application of 2F mixture

| | |
|---|---|
| Application rate | 140 g in 4.0 L/ha |
| Deposits before exposure to sunlight[q] | 42.9 ± 2.1 (μg per branch tip) |
| RH-5992 loss after 8-d exposure to sunlight at 752 W/m$^2$ intensity, after a radiation-free period of 48 h: | |
| Residual amount per branch tip[q] | 24.2 ± 1.8 μg |
| Amount lost after exposure (mean)[r] | 18.7 μg |
| Percentage loss (mean) | 43.6 |
| RH-5992 loss after 8-d exposure to sunlight at 752 W/m$^2$ intensity, after a radiation-free period of 168 h: | |
| Mean residues per branch tip[q] | 23.0 ± 2.3 μg |
| Amount lost after exposure (mean)[r] | 19.9 μg |
| Percentage loss (mean) | 46.4 |

q and r: See footnotes of Table 9.

Table 12. **Study VIII** - RH-5992 deposits before exposure, and the loss after exposure to sunlight after application of the 2F and ES mixtures

| | |
|---|---|
| Application rate | 150 g in 5.0 L/ha |
| End-use mixture of the 2F formulation concentrate *RH-5992 loss after 16-d exposure to sunlight at 752 W/m$^2$ intensity, after a radiation-free period of 48 h*: | |
| Deposits before exposure to sunlight[q] | 49.3 ± 3.3 (μg per branch tip) |
| Mean residues per branch tip[q] | 11.6 ± 1.9 μg |
| Amount lost after exposure (mean)[r] | 37.7 μg |
| Percentage loss (mean) | 76.5 |
| End-use mixture of the ES formulation concentrate *RH-5992 loss after 16-d exposure to sunlight at 752 W/m$^2$ intensity, after a radiation-free period of 48 h*: | |
| Deposits before exposure to sunlight[q] | 53.1 ± 2.1 (μg per branch tip) |
| Mean residues per branch tip[q] | 0.2 ± 0.1 μg |
| Amount lost after exposure[r] | 52.9 μg |
| Percentage loss (mean) | 99.6 |

q and r: See footnotes of Table 9.

## Results and Discussion

**Study I - RH-5992 washoff by 4 mm and 8 mm cumulative rainfall at 5 mm/h intensity at a rain-free period of 8 h after application of 2F mixture.** The data in Table 3 indicate that the average initial deposit (i.e., pre-rain residue) of the 2F flowable mixture on foliage was about the same (43.1 μg) on all branch tips. After a 4 mm rainfall, about 79% of the initial deposit was washed off, but the washoff increased to about 95% after the 8 mm rainfall. Thus, the data clearly indicated that the higher the amount of rainfall, the greater the washoff of RH-5992.

**Study II: - RH-5992 washoff by 6 mm cumulative rainfall at 1 mm/h and 5 mm/h intensities at a rain-free period of 8 h after application of 2F mixture.** The use of two different rainfall intensities with the same cumulative rainfall, provides information on the role of rain droplet size on washoff of RH-5992 particles from the foliar surface. For example, at an intensity of 1.0 mm/h the $D_{V0.5}$ of the rain droplet size spectrum was 315 ± 25 μm (Table 4), and a 6 h continuous rainfall (6 mm cumulative rain) washed off only *ca* 40% of the foliar deposits (Table 5). However, at the 5 mm/h intensity, the $D_{V0.5}$ of the rain droplet spectrum was 780 ± 95 μm (Table 4), and the same 6 mm of rain washed off *ca* 88% of the deposits (Table 5). The data thus indicate that the larger the rain droplet size spectrum, the greater the amount of RH-5992 washed off. This finding is in contrast with the one reported for glyphosate washoff from white birch foliage (*20*), where the amount washed off was independent of the rain droplet size. The reason is that the foliar deposits of RH-5992 consisted of particles that were susceptible to being knocked off by the impact velocity of the rain droplets. The larger the rain droplet size, the greater its impact velocity, and the higher the probability of RH-5992 particles being knocked off. In contrast, the glyphosate deposits were not in the particulate form but were present as a homogeneous solution in the study reported by Sundaram (*20*). Therefore, the deposits were not susceptible to being knocked off by the impact velocity of the rain droplets.

**Study III - RH-5992 washoff by 5 mm rain at an intensity of 5mm/h rainfall at rain-free periods of 8 h and 72 h after application of 2F mixture.** Table 6 presents data on RH-5992 washoff at two rain-free periods. The data clearly demonstrate that at the 8 h rain-free period after RH-5992 treatment, the washoff by a 5 mm rainfall, *ca* 86%, was much greater than the washoff of only *ca* 36% at the 72 h rain-free period. The data thus indicate the importance of ageing of foliar deposits on rain-washing characteristics, and are in agreement with the findings reported in the literature. With pesticides that undergo translocation into plants, the greater the rain-free period, the greater the time available for translocation, thus offering rain-protection for the amount removed from the washoff site (*12*). Even with the diflubenzuron insecticide that had no systemic action, Sundaram and Sundaram (*21*) found that the longer the rain-free period, the lesser the washoff of diflubenzuron from foliage, probably because of interaction of the deposits with foliar cuticle. With permethrin, despite its liquid state, Willis et al. (*8*) found that foliar residues became increasingly resistant to washoff with time. The reason is probably that permethrin, being highly lipophilic, might gradually interact with the waxy cuticle of foliage, undergoing absorption and penetration into the subsurface regions. As a result, the deposit might become increasingly resistant to rain-washing with time. In the present study, despite the lack of systemic action of RH-5992, the finding that the deposits have become increasingly resistant to washoff, suggests possible interaction of the RH-5992 particles with cuticular waxes.

**Study IV - Role of formulation type, the 2F flowable versus the ES emulsion-suspension, on RH-5992 washoff from foliage at 8 h after treatment.** Tables 7 and 8 provide data for the 2F and ES end-use mixtures, on the dosage rate applied, size spectra of the RH-5992 sprays and rain droplets, and washoff of RH-5992 by a 5 mm cumulative rainfall at an intensity of 5 mm/h and a rain-free period of 8 h after treatment. The results indicate the role of formulation ingredients on initial deposits on foliage. The ES mixture provided higher initial deposits on foliage than the 2F mixture. Two reasons could be offered for this finding. Firstly, the ES mixture, probably because of different physicochemical properties, provided larger droplets in the vicinity of the target site than the 2F mixture, despite the use of same application parameters. This is evident from the larger $D_{max}$, $D_{N0.5}$ and $D_{V0.5}$ values obtained for the ES mixture (Table 7). Larger droplets (of the sizes reported here) are known to have greater impaction efficiency (*22*) and to provide greater initial deposits on target surfaces, than smaller droplets. Secondly, the presence of a lipophilic substance in the ES mixture (information provided by Rohm and Haas company), probably caused better retention of the spray droplets on the foliar surface with minimum 'bounce off'. In contrast, the 2F formulation concentrate contained little lipophilic materials, and as a result, spray droplets probably could not be retained as much as those of the ES mixture on the foliar surface. Moreover, the presence of the lipophilic material is known to make spray droplets spread more (*17,23*) and to adhere better to the waxy cuticle of the foliage than the aqueous 2F mixture, thus enhancing droplet retention (*24-26*).

Similar to the enhanced droplet deposition and retention, the deposits of the ES mixture were also more resistant to washoff from foliage by rain than the 2F mixture. For example, the washoff was *ca* 28% for the ES mixture, compared to 87% for the 2F mixture (Table 8). The reason appears to be the interaction of the lipophilic materials in the ES mixture with the foliar waxes, thus contributing to enhanced adhesion of the RH-5992 particles to the foliar cuticle (*21*). The present finding is in agreement with that of Pick et al. (*4*), who noted that at a rain-free period of 2.5 h after application of two formulations of parathion, one as an emulsifiable concentrate (EC) and the other as the wettable powder (WP), a 3 mm rainfall washed off only *ca* 50% of initial deposits of the EC mixes (containing lipophilic ingredients) from cotton leaves, as opposed to *ca* 66% washoff of the WP mixes that contained little lipophilic ingredients. Similarly, Sundaram et al. (*27*) found that oil-based *Bacillus thuringiensis* formulations showed greater rainfastness than aqueous ones.

**Study V - Loss of RH-5992 after 8-d and 16-d exposure to sunlight of intensity 437 W/m$^2$, at 48h after application of 2F mixture.** The data in Table 9 indicate that after keeping the branch tips in darkness for 48 h post-spray (i.e., after a radiation-free period of 48 h), about 28% of the initial deposits (i.e. residues prior to exposure) of RH-5992 were lost following exposure to sunlight at an intensity of 437 *W/m$^2$* for 8 d. However, when the exposure period increased to 16 days, the amount lost increased to twice the value, i.e., about 56%. Thus, the data clearly indicated that the longer the exposure period the greater the disappearance of RH-5992. The seemingly linear increase in loss of RH-5992 with duration of exposure is in disagreement with the log-linear relationship between intensity of radiation and photolytic loss of materials (*28*). However, the present study provides data only for two exposure periods. Unless data are generated for several exposure periods, it would be difficult to conclude whether the seemingly linear relationship would hold for a wide range of exposure periods.

**Study VI - Loss of RH-5992 after 12-d exposure to sunlight of intensity 437 and 752 W/m$^2$ at 48h after application of 2F mixture.** Table 10 presents results of RH-5992 loss after a 12-d exposure to sunlight of two different intensities after a radiation period of 48 h. The data show that the higher the intensity of radiation, the greater the loss. For example, after exposure to sunlight of intensity 437 *W/m$^2$* the loss was about 40%, whereas the corresponding value was 60% for the 752 *W/m$^2$*. Thus, the data has demonstrated that the loss of RH-5992 was proportional to the intensity of radiation.

**Study VII - Loss of RH-5992 after an 8-d exposure to sunlight of intensity 752 W/m$^2$at 48h and 168 h after application of 2F mixture.** Table 11 presents results of RH-5992 disappearance after exposure to simulated sunlight of 752 W/m$^2$ following radiation-free periods of 48 and 168 h. The data clearly demonstrated that when the radiation-free period increased from 48 and 168 h, there was little change in the amount of RH-5992 disappeared as a result of ageing of deposits. These results are in contrast to that observed in the rainfastness study (Study III), where the longer the ageing period, the greater the rainfastness of deposits. The reason could be that any interaction that might have occurred between the RH-5992 particles and the foliar cuticle, did not help in providing photo-protection for the material.

**Study VIII - Loss of RH-5992 after a 16-d exposure to sunlight of intensity 752 W/m$^2$ at 48 h after application of the 2F and ES mixtures.** Table 12 provides data on the amount of RH-5992 disappeared from balsam fir foliage after a 16-d exposure to sunlight of intensity 752 W/m$^2$, for the 2F and ES end-use mixtures. Unlike the rain-washing study where the deposits of the ES mixture were more rainfast than those of the 2F mixture, the photostability study indicated that the deposits of the 2F mixture were more photostable than those of the ES mixture. For example, after a 16-d exposure, the deposits of the 2F mixture were decreased by *ca* 77%, whereas those of the ES mixture were decreased by nearly 100%. This behaviour was probably due to the presence of a lipophilic material in the ES mixture which caused droplets to spread more than the 2F mixture, thus providing a greater surface area of exposure to sunlight for the RH-5992 particles. This could have been the cause for the greater loss of RH-5992 from deposits of the ES mixture than from those of the 2F mixture.

## Discussion and Conclusions

Understanding pesticide washoff from foliage of host plants is important from the standpoint of field efficacy of foliar applied pesticides. The presence of oils in the end-use mixes has been reported to provide increased rain-protection for pesticide deposits on foliage (*29-32*). Nonetheless, if high concentrations of surfactants are present in the mixes together with oils, spray droplets can be emulsified during rainfall, and be washed off as readily as water soluble pesticides. In fact, Hatfield et al. (*33*) found a slight decrease in rainfastness of oil-based permethrin, compared to that of an emulsion. Baker and Shiers (*34*) found little difference in washoff of water-based and water/oil-based herbicides. Sundaram et al. (*27*) found that the presence of oil carriers in *Bacillus thuringiensis* var. *kurstaki* formulations provided only a slight increase in rainfastness of foliar deposits. In this investigation, the presence of a lipophilic ingredient in the emulsion-suspension spray mixture provided definite advantages over the aqueous flowable mixture, from the standpoint of greater rainfastness for RH-5992 deposits. The increased rainfastness was evident in all stages of the investigation regardless of the amount of rainfall, rain droplet size, and rain-free period. The lipophilic materials and other ingredients present in the droplets appear to enhance interaction with the foliar cuticle, resulting in better retention and adhesion to foliar surfaces. The longer the rain-free period, the greater the rainfastness, thus indicating the time-dependent interaction of the droplet with the foliar cuticle.

Unlike the rainfastness studies, literature information on photo-induced disappearance of pesticides from plant surfaces is sparse, although extensive research data are available for photolysis of pesticides in water (*35*). Also, some researchers have reported data on photolytic loss of pesticides from glass surfaces (*36,37*). Ford and Salt (*38*) indicated that although the natural sunlight has little energy below 290 nm, sufficient energy is still available above this wavelength to deactivate pesticide deposits on plant surfaces if the quantum efficiency of the reaction is greater than 10%. The presence of unsaturated lipids in the plant cuticle can also shift the absorption maxima of organic materials by charge transfer and other electronic interactions, thereby protecting the pesticides from photodegradation (*39*).

The present study provided data on relationships between the intensity of radiation, duration of exposure and the effect of formulation ingredients on photo-induced disappearance of RH-5992. The higher the intensity of radiation, the greater the loss of RH-5992 from foliage. The longer the duration of exposure, the greater the disappearance. Unlike the rain-washing, the radiation-free period (i.e., the ageing period of deposits) failed to influence the disappearance of the chemical. Regardless of the amount of cumulative radiation, radiation intensity, and radiation-free period, deposits of the emulsion-suspension mixture disappeared to a greater extent than the aqueous flowable mixture. This behaviour, as mentioned earlier, could be due to the greater spreading of droplets of the emulsion-suspension mixture than that of the flowable mixture, thus providing a greater surface area of exposure for the RH-5992 particles.

## Acknowledgment

The author thanks Johanna Curry, John W. Leung, Linda Sloane, Martin J. Reed and Derek A. Tesky for their technical assistance in the investigation.

## References

1. Taylor, N.; Matthews, G.A. *Crop Protection,* **1986,** *5,* 250-253.
2. Thonke, K.E.; Kudsk, P.; Streibig, J.C. In *Adjuvants and Agrochemicals,* Volume, II, P.N.P. Chow, C.A. Grant, A.M. Hinshalwood and E. Simundson, Eds., CRC Press, Inc., Boca Raton, Florida, USA, 1989, pp. 103-110.
3. Mashaya, N. *Crop Protection,* **1993,** *12,* 55-58.
4. Pick, F.E.; van Dyk, L.P.; de Beer, P.R. *Pestic. Sci.,* **1984,** *15,* 616-623.
5. McDowell, L.L.; Willis, G.H.; Smith, S.; Southwick, L.M. *Trans. ASAE,* **1985,** *28,* 1896-1900.
6. McDowell, L.L.; Willis, G.H.; Southwick, L.M.; Smith, S. *Pestic. Sci.,* **1987,** *21,* 83-92.
7. Willis, G.H.; McDowell, L.L.; Smith, S.; Southwick, L.M. *Trans. ASAE,* **1988,** *31,* 86-90.
8. Willis, G.H.; McDowell, L.L.; Smith, S.; Southwick, L.M. *J. Agric. Food Chem.,* **1992,** *40,* 1086-1089.
9. Willis, G.H.; McDowell, L.L.; Southwick, L.M.; Smith, S. *J. Environ. Qual.,* **1992,** *21,* 373-377.
10. Silk, P.J.; Unger, I. *Internat. J. Environ. Anal. Chem.,* **1973,** *2,* 213-220.
11. Hapeman-Somich, C.J. In *Pesticide Transformation Products: Fate and Significance in the Environment,* L. Somasundaram; J.R. Coats, Eds., ACS Symposium Series No. 459, American Chemical Society, Washington, D.C., 1991, pp. 133-147.
12. Caseley, J.C.; Coupland, D. *Ann. appl. Biol.,* **1980,** *96,* 111-118.
13. Sundaram, K.M.S. *Persistence and Fate of RH-5992 in Aquatic and Terrestrial Components of a Forest Environment,* A Special Report to Rohm and Haas Company, Natural Resources Canada, Canadian Forest Service, Forest Pest Management Institute, 1219 Queen Street East, P.O. Box 490, Sault Ste. Marie, Ontario, Canada, 1993, 50 pp.
14. Hem, J.D. *Study and Interpretation of the Chemical Characteristics of Natural Water,* Second Edition, U.S. Department of the Interior, U.S. Geological Survey, Virginia, USA, 1983, pp. 224-229.
15. Sundaram, K.M.S.; Sundaram, A. *Pestic. Sci.,* **1987,** *18,* 259-271.
16. Sundaram, K.M.S. *J. Environ. Sci. Health,* **1994,** *B29,* 541-579.
17. Sundaram, A.; Sundaram, K.M.S.; Leung, J.W. *Trans. ASAE,* **1991,** *34,* 1941-1951.

18. Johnstone, D.R. In *Proceedings BCPC Symposium: Controlled Drop Application 1978,*" The British Crop Protection Council, BCPC Monograph No. 22, BCPC, Croydon, England, 1978, pp. 35-42.
19. Sundaram, K.M.S.; Zhu, J.S.; Nott, R. *J. AOAC Int.,* **1993,** *76,* 668-673.
20. Sundaram, A. *J. Environ. Sci. Health,* **1991,** *B26,* 37-67.
21. Sundaram, K.M.S.; Sundaram, A. *J. Environ. Sci. Health,* **1993,** *B29,* 757-783.
22. Ford, R.E.; Furmidge, C.G.L. In *Pesticidal Formulations Research: Physical and Colloidal Chemical Aspects,* R.F. Gould, Ed., American Chemical Society Publications, Washington, D.C., 1969, pp. 155-182.
23. Sundaram, A.; Sundaram, K.M.S.; Zhu, J.S.; Nott, R.; Curry, J.; Leung, J.W. *J. Environ. Sci. Health,* **1993,** *B28,* 243-273.
24. Brunskill, R.T. In *Proceedings of the 3rd British Weed Control Conference,* Volume 2, 1956, pp. 593-603.
25. Becher, P.; Becher, D. In *Pesticidal Formulations Research, Physical and Colloidal Chemical Aspects,* R.F. Gould, Ed., American Chemical Society Publications, Washington, D.C., 1969, pp. 15-23.
26. Johnstone, D.R. In *Pesticide Formulations,* W. Van Valkenburg, Ed., Marcel Dekker, Inc., New York, USA, 1973, pp. 343-386.
27. Sundaram, A.; Leung, J.W.; Devisetty, B.N. In *Pesticide Formulations and Application Systems: 13th Volume, ASTM STP 1183,* Paul D. Berger, Bala N. Devisetty and Franklin R. Hall, Eds., American Society for Testing and Materials, Philadelphia, 1993, pp. 227-241.
28. Wayne, R.P. *Photochemistry,* Butterworths and Co. (Publishers) Ltd., London, England, 1970, pp. 25-28.
29. Wheeler, H.G.; Smith, F.F.; Yeomans, A.H.; Fields, E. *J. Econ. Entomol.,* **1967,** *60,* 400-402.
30. Nemec, S.J.; Adkisson, P.L. *J. Econ. Entomol.,* **1969,** *62,* 71-73.
31. Mayo, Z.B. *J. Econ. Entomol.,* **1984,** *77,* 190-193.
32. Omar, D.; Matthews, G.A. *Trop. Pest Manage.,* **1990,** *36,* 159-161.
33. Hatfield, L.D.; Staetz, C.A.; McDaniel, S.G. *Performance of Emulsified and Non-Emulsified Permethrin Formulations as Related to Rainfall.* Beltwide Cotton Prod. - Mech. Conf., 1983, pp. 84-86.
34. Baker, J.L.; Shiers, L.E. *Trans. ASAE,* **1989,** *32,* 830-833.
35. Cessna, A.J.; Muir, D.C.G. In *Environmental Chemistry of Herbicides,* Volume II, Raj Grover and Allan J. Cessna, Eds., CRC Press, Inc., Boca Raton, Florida, 1991, pp.199-263.
36. Baker, R.D.; Applegate, H.G.; Texas J. *Science,* **1974,** *25,* 53-59.
37. Leung, J.W., *J. Environ. Sci. Health,* **1994,** *B29,* 341-363.
38. Ford, M.G.; Salt, D.W. In *Pesticides on Plant Surfaces,* H.J. Cottrell, Ed., Society of Chemical Industry, John Wiley and Sons, 1987, pp. 26-81.
39. Hartley, G.S.; Graham-Bryce, I.H. *Physical Principles of Pesticide Behaviour,* Volumes 1 and 2, Academic Press, London, England, 1980.

RECEIVED January 12, 1995

# Chapter 10

# Radiation Protection and Activity Enhancement of Viruses

**Martin Shapiro**

**Insect Biocontrol Laboratory, Plant Sciences Institute, Agricultural Research Service, U.S. Department of Agriculture, Beltsville, MD 20705**

Although several baculoviruses have been registered for use as microbial control agents, none are currently used on a routine, commercial basis in the United States. Two factors influencing the use and performance of these viruses are susceptibility to ultraviolet (UV) radiation and slowness in causing lethal infections. For UV protection, recent emphasis has been placed upon absorption in both the UV-B (280-310 nm) and UV-A portions (320-400 nm) of the solar spectrum. In addition, antioxidants or radical scavengers may also play a critical role in the protection of insect pathogens during solar irradiation. Research will be reviewed on the success of different chemicals as UV screens, with especial emphasis upon dyes and optical brighteners. For the past 25 years, efforts have been made to enhance virus efficacy by selected chemicals and by selection of more virulent biotypes. The most exciting research on activity enhancement has centered upon a viral enhancing factor, and fluorescent brighteners. Research in the area of fluorescent brighteners will be emphasized from both basic and applied aspects.

In recent years there has been a growing interest in the development of microorganisms as alternatives to synthetic, chemical-based insecticides for pest management. Among the reasons for this are an increasing concern for the quality of the environment and the development of pesticide resistance in many major insect species. Insect pathogenic viruses, especially the nuclear polyhedrosis viruses (NPV), frequently cause the collapse of insect populations in nature, and are logical candidates for use in pest management systems. The deployment strategies for these viruses have provided less than expected results. New or modified strategies for pest management with viruses are needed that are based upon more complete information on the interaction of the virus(es) with


Published 1995 American Chemical Society

the target insect(s) and on the use of improved formulations to stabilize and maximize viral activity in the field.

**Sunlight and UV Radiation.** Natural sunlight, especially the ultraviolet (UV) portion of the spectrum (UV-B, UV-A), is responsible for inactivation of insect pathogens (*5,23,36*). In many cases, field-applied pathogens lose at least 50% of their original activity within several days (*23*). The corn earworm, (*Helicoverpa zea*), NPV (=HzSNPV), the cytoplasmic polyhedrosis virus (CPV) from the tobacco budworm, (*Heliothis virescens*), and the entomopoxvirus (EPV) from *Euxoa auxiliaris* are less stable under UV than either the fungus, *Nomurea rileyi*, or the bacterium, *Bacillus thuringiensis*, but are more stable than the protozoan, *Vairimorpha necatrix*, when tested under laboratory conditions (*23*).

**UV Protectants.** During the past two decades, several natural and synthetic compounds have been evaluated as sunlight protectants for entomopathogens such as viruses (*20,26*), bacteria (*30*), protozoa (*67*), and nematodes (*11*). The most successful materials were aromatics (*28*) such as uric acid (*67*), p-aminobenzoic acid (PABA) (*11*), 2-hydroxy-4-methoxy-benzophenone (*31*), 2-phenylbenzimid-azole-5-sulfonic acid (*54*), folic acid (*45*), and Tinopal DCS (*35*). The success of these chemicals was attributed to good absorption in the UV-B portion of the solar spectrum (*25*), although absorption in the UV-A region may also be critical (*16,45*).

**Dyes.** Jaques (*26*), in a pioneering study, evaluated 29 materials or combinations as UV-protectants for the cabbage looper, (*Trichoplusia ni*), NPV, including such stains and dyes as brilliant yellow, buffalo black, methylene blue, and safranin (at 0.1%). In the laboratory, all materials had some protective ability. Under field conditions, brilliant yellow and safranin afforded good protection for NPV on cabbage leaves. Morris (*37*) demonstrated that a sunscreen combination of Uvinul DS49 (benzophenone) and a red dye (Erio acid red B100) provided some protection for *B. thuringiensis* under field conditions. Jones and McKinley (*27*) found that soluble dyes as indigo carmine and Tinopal BRS200 gave some protection to *Spodoptera littoralis* NPV under field conditions in Egypt, but were no more effective than such clays as attapulgite, diatomite, montmorillonite, and colloidal china clay. Shapiro and Robertson (*50*) tested 79 dyes as UV protectants for the gypsy moth, (*Lymantria dispar*), NPV and found that 20 were no more effective than distilled water. Five dyes were effective protectants (*i.e.*, lissamine green, acridine yellow, brilliant yellow, alkali blue, and mercurochrome) and one (*e.g.*, Congo red) provided complete protection. A composite UV absorption profile of the 6 effective dyes was compared with that from a representative sample of 6 ineffective dyes. Both groups of dyes displayed similar absorbency patterns in the UV-B portion of the solar spectrum. In the UV-A portion, however, the total absorbance from 320-400 nm decreased among the ineffective dyes by 16%, while the total absorbance of effective dyes increased by 200% as the spectrum shifted from UV-B to UV-A. Greater than

50% of the absorbance occurred in the UV-B portion among ineffective dyes, while only 25% of absorbance occurred among effective dyes. In other words, effective dyes had a greater capacity to absorb UV-A radiation than did the ineffective dyes.

The study on dyes has been instructive in helping answer two questions: (1) which dyes, or groups of dyes, are effective protectants?; and (2) can effective dyes be separated from ineffective dyes on the basis of their UV absorption spectra? These questions were based upon an assumption that the effectiveness of a given material was directly related to UV absorbance. Congo red was the most effective dye for both the gypsy moth NPV (*46*) and the corn earworm NPV (*24*). While maximum UV absorbance occurs at 321 nm, good absorbance also occurs over the entire UV-A spectrum. Morris (*37*), working with *B. thuringiensis*, concluded that materials should be good absorbers at 330-400 nm to be effective protectants. Absorbance at 400 nm also appeared to be important for a protectant, as Griego and Spence (*13*) reported that mortality of *B. thuringiensis* was caused by irradiation at both UV and visible (400 nm) wavelengths.

Oxygen radical formation (*i.e.,* hydrogen peroxide) occurs during irradiation and is detrimental to germination of *B. thuringiensis* spores (*22*). The addition of a radical scavenger (*e.g.,* peroxidase) increased spore germination, presumably by interacting with peroxide. This study is very important, for it indicates that both UV absorbance and radical scavenging may be important in the protection of entomopathogens.

**Optical brighteners.** Optical brighteners (=fluorescent brighteners) were discovered more than 50 years ago (*9,41*) and are widely used in the detergent, paper, plastics, organic coatings industries (*32*) and as fluorochromes for microorganisms (*4,59*).

These compounds readily absorb UV radiation and transmit light in the blue portion of the visible spectrum. Twenty-three brighteners were tested as UV protectants for the gypsy moth NPV (*47*). A complete spectrum of protection was observed ranging from 0.4% original activity remaining (% OAR) (Synacril White NL) to 100% OAR (Phorwite AR, Phorwite BBU, Phorwite BKL, Phorwite CL, Intrawhite CF, Leucophor BS, Leucophor BSB, Tinopal LPW). Phorwite AR and Tinopal LPW provided the greatest protection at all concentrations (*i.e.,* about 15% OAR at 0.001%, 72-84% OAR at 0.01%; 97-100% OAR at 0.10%; 100% OAR at 1.0%). In all cases, the protective effect was concentration-dependent. The 23 brighteners belonged to several chemical classes (*i.e.,* stilbene, oxazole, pyrazoline, naphthalic acid, lactone, coumarin), and each of these groups contained effective brighteners (*i.e.,* > 70% OAR) groups. The four superior brighteners (*e.g.,* Leucophor BS and BSB, Phorwite AR, and Tinopal LPW) all belong to the stilbene group. These compounds appear to be very promising as radiation protectants not only for insect pathogenic viruses (*47*), but also for such entomogenous nematodes as *Steinernema carpocapsae* (*39*). During the last 30 years progress has been made in the formulation of microbial insecticides (*21*), and microencapsulation technology (*3,20*) will play an increasingly important role.

**Biological activity.** Because insect pathogenic viruses must undergo several cycles of multiplication within susceptible insects, the time required for insects to die may take several days. Larvae continue to feed, defoliate or damage host plants until shortly before death. Therefore, applications in the field may not provide adequate foliage or crop protection. Greater larval mortality and/or faster kill might be achieved by adding chemicals to the microbial preparation (*1,8,73*) or by selection of more virulent biotypes (*42,48,68,71*). During the past decade, we demonstrated that chemicals such as boric acid (*49*), chitinase (*56*), and the dye Congo red (M. Shapiro, unpublished data) reduced the $LC_{50}$s and $LT_{50}$s of gypsy moth NPV suspensions.

**A variable virus population**. Biological activity of insect viruses (= virulence) may be expressed in terms of concentration (*e.g.*, $LC_{50}$), time (*e.g.*, $LT_{50}$), or both. Differences in $LC_{50}$s have been detected among geographical isolates of NPVs by Ossowski (*40*), Smirnoff (*60*), Hamm and Styer (*14*). Magnoler (*34*) found >1000-fold differences among gypsy moth NPV isolates from France, Italy, Japan, United States, and Yugoslavia. Rollinson and Lewis (*43*) found >1000-fold differences among isolates from Japan, United States and Yugoslavia. Shapiro et al. (*55*) observed differences of ~2900-fold among 19 isolates from Asia, Canada, Europe, and the United States. Vasiljevic and Injac (*68*) also demonstrated that NPV isolates from different regions in Yugoslavia varied in activity against larvae from different regions.

The most active NPV isolates generally originated from gypsy moth NPV populations in North America; the least active isolate originated from Asia (Japan) (*55*). Heterogeneity among samples within a given gypsy moth NPV isolate was demonstrated, using geographical isolates from Abington, MA, Hamden, CT, and Dalmatia, Yugoslavia. Samples of each isolate exhibited a skewed distribution of activities at both $LC_{50}$ and $LC_{90}$. Plots of $LC_{50}$ versus $LC_{90}$ identified the samples within each isolate that merit selective propagation. In other words, means were developed to identify the most active samples from each heterogeneous population (*51*). In the case of the *Heliothis* NPV, biological activity among 34 samples varied from 0.7 to 39.0 polyhedral inclusion bodies (PIB) per $mm^2$ of diet surface, and up to 8 activity classes could be obtained. When $LC_{50}$s were graphically displayed as a frequency distribution, the distribution was skewed. Thus, some samples of the HzSNPV population had excellent activity, but the activities of other samples was poor (*48*).

The Abington NPV population comprises many subpopulations with differences in biological activity. In fact, a complete spectrum of activity is obtained. In some cases, a population sample producing fast kill exhibited a high

$LC_{50}$, while some samples were slow in producing a lethal infection (days 7-8) but had low $LC_{50}$ values. In other instances, some samples were slow in producing a lethal infection and had high $LC_{50}$ values. Only samples exhibiting both fast kill and low $LC_{50}$s were considered as inocula for the next passage. Selection for a more active biotype was achieved, in that greater numbers of gypsy moth larvae died by day 8. NPV variability decreased from passage 1 to passage 11 but was not eliminated. Moreover, a greater percentage of the virus population exhibited fast, early kill. The correlation between early kill and a low $LC_{50}$ increased during passage, indicating that the two factors can be linked by *in vivo* selection.

Secondary selection using *in vitro* plaque purification was achieved and the virus biotype (= $\alpha$624) was deposited in the American Type Culture Collection as the "type species" for the Abington strain of the gypsy moth NPV. Using similar methods, Slavicek et al. (*58*) isolated two different plaque variants of the gypsy moth NPV with different biological properties. Previous workers also achieved an increase in activity by *in vivo* serial passage, using changes in the $LC_{50}$ as the sole criterion (*48,61,70,72*). Serial passage probably selects a more active isolate from a heterogeneous population, resulting in a more stable, homogeneous virus population (*48,70*). The use of *in vivo* selection "... is a feasible approach for studying not only activity of virus biotypes and host specificity, but the basic mechanism(s) involved in virulence. At present, *in vivo* selection is an efficient system for measuring genetic changes, in activity and in genetic diversity during directed selection" (*69*).

**Viral enhancing factor.** In 1954, Tanada made a very important discovery on the synergism of viruses. He found that a Hawaiian strain of an armyworm (*Pseudaletia unipuncta*) granulosis virus (GV) synergised the activity of an armyworm NPV (*62*). Moreover, synergism was due to an integral component of the viral inclusion body. Many studies were conducted by Tanada and colleagues over the next 35 years, using both *in vivo* and *in vitro* systems (*19,29,38,63-66,74,75*), and much biochemical and biological information was obtained. Derksen and Granados (*6*) showed that the synergistic factor (SF), now called viral enhancing factor (VEF), altered the paratrophic membrane of the cabbage looper (*Trichoplusia ni*) and enhanced viral activities of the alfalfa looper (*Autographa californica*) and cabbage looper (*T. ni*) NPVs, as well as the cabbage looper GV. Subsequently, a virulence gene product was identified and cloned. This protein (mw = 101 kd) disrupts the larval paratrophic membrane (*pm*) (*12*). This area of research began over 40 years ago by Tanda and collegues, and amplified by Grandos and collegues, is very exciting and may result in the production of more efficacious viruses.

**Optical brighteners.** A better understanding of the role of the host in influencing susceptibility to a given pathogen will enable us to circumvent the host's defense system, thus increasing the activity and/or host range of a given entomopathogen. Within the last several years, we have determined that certain optical brighteners (*e.g.*, selected stilbene brighteners) increased the activity of the gypsy moth NPV (*52*). The brightener (Tinopal LPW), when fed to larvae in combination with the gypsy moth NPV, acts on the larval midgut to (1) increase virus uptake, (2) cause viral replication to occur in a refractory tissue [*e.g.*, the midgut], and (3) cause a cessation in larval feeding within 2 days. Moreover, the addition of stilbene brighteners (Leucophor BS, BSB; Phorwite AR, RKH; Tinopal LPW ) to LdMNPV reduced the average $LC_{50}$s from ~18,000 polyhedral inclusion bodies (PIB) per ml to values between 10 and 44 PIB/ml (*52*). $LT_{50}$s were also greatly reduced by the addition of these brighteners to the gypsy moth NPV. Reduction in $LC_{50}$s and $LT_{50}$s among mature larvae (fourth-fifth instar) were also significant, indicating that the combination of virus and brightener could also be effective in reducing late-instar populations if a late season application of NPV would be desirable (*50*).

The addition of another brightener (Phorwite AR) to the gypsy moth CPV reduced the $LC_{50}$ ~800-fold and the $LT_{50}$ from 13.2 to 8.4 days (at 1 million PIB per ml per cup). Whereas the gypsy moth is insensitive to such viruses as the *Autographa* NPV and *Amsacta* EPV. The addition of Phorwite AR to these virus suspensions resulted in virus replication and virus-caused mortality. With the addition of a selected brightener, it was thus possible to increase the susceptibility of the gypsy moth and to expand the host range of these viruses (*53*). Subsequent cooperative research with John Hamm (ARS-Tifton, GA) demonstrated that a selected stilbene brightener (Tinopal LPW) also enhanced the activities of the fall armyworm NPV (15), the fall armyworm GV against the fall armyworm, (*Spodoptera frugiperda*) and the *Helicoverpa* iridescent virus (IV) against the corn earworm (*Helicoverpa zea*) (*57*). Moreover, this research resulted in the issuance of a U.S. Patent (*57*), and licensing to both American Cyanamid and Sandoz/biosys of this technology.

At this time, it would be beneficial to summarize present knowledge concerning these brighteners. (1) Only stilbenes appear to act as enhancers; (2) Not all stilbene brighteners are active; (3) The brightener-virus combination must be ingested; (4) The virus is not affected by the brightener; (5) The brightener acts in the midgut to affect host susceptibility; (6) The host spectrum of insect pathogenic viruses can be expanded (*53*). At present, the mode of action of these brighteners is not known, but some clues do exist. Several stilbene brighteners are known to interfere with chitin fibrillogenesis (*10,17,18,44*). In insects, the paratrophic membrane (*pm*) lines the midgut and is composed of chitin microfibrils. The *pm* may serve as a barrier for the invasion of microorganisms, including insect viruses (2). Selected brighteners may inhibit or alter the chitinous pm, creating gaps in the lining. In the case of NPV (and CPV) greater uptake of virus into the midgut may also occur in the presence of brightener. Disulfonic acids are known to affect ion transport in mammalian systems and are potent anion transport inhibitors (*7,33*). Since the most active stilbene brighteners (*e.g.*, Tinopal LPW, Blankophor RKH, Leucophor BS,

Leucophor BSB) are stilbene disulfonic acids, it may be inferred that these materials can also act as anion transport inhibitors in insects (*53*).

At this point, we have barely "scratched the surface" of the virus-host-brightener interaction. This research is very exciting from both basic and practical standpoints. From a basic standpoint, these materials may enable us to better understand why a given insect species is susceptible or refractory to a given virus. From a practical standpoint, the use of these brighteners may enable us to manipulate the virus-host interaction for more efficacious insect control (*53*), which would enable insect viruses to become more widely used as effective environmentally acceptable microbial control agents.

## Literature Cited

1. Bell, M. R..; Kanavel, R. F. Potential of bait formulations to increase infectiveness of nuclear polyhedrosis virus against the pink bollworm. *J. Econ. Entomol.* **1975**; 68, 389-391.
2. Brandt,C. R.; Adang, M. J.; Spence, K. D. S. The paratrophic membrane: ultrastructural analysis and function as a mechanical barrier to microbial infection in *Orgyia pseudotsugata*. *J. Invertebr. Pathol.* **1978**; 32, 12-24.
3. Bull, D.L.; Ridgway, R.L.; House, V.S.; Pryor, N.W. Improved formulations of the *Heliothis* nuclear polyhedrosis virus. *J. Econ. Entomol.* **1976**; 69, 731-736.
4. Darken, M.A. Absorption and transport of fluorescent brighteners in biological techniques. *Science.* **1962**; 10, 387-393.
5. David, W. A. L. The effect of ultraviolet radiation of known wavelengths on a granulosis virus of *Pieris brassicae*. *J. Invertebr. Pathol.* **1969**; 14, 336-342.
6. Derksen, A. C. G.; Granados, R. R. Alteration of a lepidopteran paratrophic membrane by baculoviruses and enhancement of viral activity. *Virology.* **1988**; 167, 242-250.
7. Dix, J. A.; Verkman, A. S.; Solomon, A. K. Binding of chloride and a disulfonic stilbene transport inhibitor to red cell band 3. *J. Membrane Biol.* **1986**; 89, 211-223.
8. Doane, C. C.; Wallis, R. C. Enhancement of the action of *Bacillus thuringiensis* var. *thuringiensis* Berliner on *Porthetria dispar* (Linnaeus) in laboratory tests. *J. Insect Pathol.* **1964**; 6, 423-429.
9. Eggert, J.; Wendt, B. Wrapping materials. U.S. Patent 2,171,427; **1939**; 6 pp.
10. Elorza, M. V.; Rico, H.; Sentandreau, R. Calcofluor white alters the assembly of chitin fibrils in *Saccharomyces cerevisiae* and *Candida albicans* cells. *J. Gen. Microbiol.* **1983**; 129, 1577-1582.
11. Gaugler, R.; Boush, G. M. Laboratory tests on ultraviolet protectants of an entomogenous nematode. *Environ. Entomol.* **1979**; 8, 810-813.

12. Granados, R. R.; Corsaro, B. G. *Vth International Colloquium on Invertebrate Pathology and Microbial Control;* Society for Invertebrate Pathology: Adelaide, Australia; **1990**, pp. 174-178.
13. Griego, V. M.; Spence, K. D. Inactivation of *Bacillus thuringiensis* spores by ultra-violet and visible light. *Appl. Environ. Microbiol.* **1978**; 35, 906-910.
14. Hamm, J. J.; Styer, E. L. Comparative pathology of isolates of *Spodoptera frugiperda* nuclear polyhedrosis virus in *S. frugiperda* and *S. exigua*. *J. Gen. Virol.* **1985**; 66, 1249-1261.
15. Hamm, J. J., Shapiro, M. Infectivity of fall armyworm (Lepidoptera: Noctuidae) nuclear polyhedrosis virus enhanced by a fluorescent brightener. *J. Econ. Entomol.* **1992**; 85, 2149-2152.
16. Harms, R. L.; Lane, J. K.; Griego, V. M. *IVth International Colloquium on Invertebrate Pathology and Microbial Control;* Society for Invertebrate Pathology: Wageningen, The Netherlands; **1986**, p.156.
17. Herth, W. Calcofluor white and congo red inhibit microfibril assembly in *Poteriochromonas*: evidence for a gap between polymerization and microfibril formation. *J. Cell Biol.* **1980**; 87, 442-450.
18. Herth, W.; Hausser, J. In *Structure, function, and biosynthesis of plant cell walls*; Dugger, W. M.; Bartnicki-Garcia, S. Eds.; Am. Soc. Plant Physiol., Rockville, MD; **1984**, pp. 89-119.
19. Hukuhara, T.; Zhu, Y. Enhancement of the in vitro infectivity of a nuclear polyhedrosis virus by a factor in the capsule of a granulosis virus. *J. Invertebr. Pathol.* **1989**; 54, 71-78.
20. Ignoffo, C.M., Batzer, O.F. Microencapsulation and ultraviolet protectants to increase sunlight stability of an insect virus. *J. Econ. Entomol.* **1971**. 64, 850-853.
21. Ignoffo, C. M.; Falcon, L. A. Eds. In *Formulation and application of microbial insecticides.* Entomol. Soc. Am., College Park, MD; **1978**, 66 pp.
22. Ignoffo, C. M.; Garcia, C. UV-photoinactivation of cells and spores of *Bacillus thuringiensis* and effects of peroxidase on inactivation. *Environ. Entomol.* **1978**, 7, 270-272.
23. Ignoffo, C.M.; Hostetter, D. L.; Sikorowski, P. P.; Sutter, G.; Brooks, W. M. Inactivation of representative species of entomopathogenic viruses, a bacterium, fungus, and protozoan by an ultraviolet source. *Environ. Entomol.* **1977**; 6,411-415.
24. Ignoffo, C. M.; Shasha, B. S.; Shapiro, M. Sunlight ultraviolet protection of the *Heliothis* nuclear polyhedrosis virus through starch-encapsulation technology. *J. Invertebr. Pathol.* **1991**; 57, 134-136.
25. Jaques,R. P. The inactivation of nuclear polyhedrosis virus of *Trichoplusia ni* by gamma and ultraviolet radiation. *Can. J. Microbiol.* **1968**; 14, 1161-1163.

26. Jaques, R. P. Tests on protectants for foliar deposits of a polyhedrosis virus. *J. Invertebr. Pathol.* **1971**; 17, 9-16.
27. Jones, K. A.; McKinley, D. J. *IVth International Colloquium on Invertebrate Pathology and Microbial Control;* Society for Invertebrate Pathology: Wageningen, The Netherlands; **1986**, p.155.
28. Killick, H. J. *IVth International Colloquium on Invertebrate Pathology and Microbial Control.;* Society for Invertebrate Pathology: Wageningen, The Netherlands; **1986**, pp. 620-623.
29. Kozuma, K.; Hukuhara, T. A synergistic factor of an armyworm granulosis virus contains phosphatidyllcholine. *J. Invertebr. Pathol.* **1992**; 59, 328-329.
30. Krieg, A. Photoprotection against inactivation of *Bacillus thuringiensis* spores by ultraviolet rays. *J. Invertebr. Pathol.* **1975**; 25, 267-268.
31. Krieg, A.; Gröner.; Huber, K.; Metter, M. Über die Wirkung von mittelangewelligen untravioletten Strahlen (UV-B und UV-A) auf insektenpathogene Bakterien und Viren und deren Beeinflussung durch UV-Schutzstoffe. *Nachtichenbl. Dtsch. Pflanzenschutz. Brauschw.* **1980**; 32, 100-105.
32. Lanter, J. Properties and evaluation of fluorescent brightening agents. *J. Soc. Dyes & Colourists.* **1966**; 82, 125-132.
33. Lepke, S.; Fasold, H.; Pring, M.; Passow, H. A study of the relationship between inhibition of anion exchange and binding to the red blood cell membrane of 4,4'-diisthio-cyanostilbene-2,2'-disulfonic acid (DIDS) and its dihydro derivative ($H_2$DIDS). *J. Membrane Biol.* **1976**; 29, 147-177.
34. Magnoler, A. Susceptibility of gypsy moth larvae to *Lymantria* sp. nuclear and cytoplasmic viruses. *Entomophaga.* **1970**; 15, 407-412.
35. Martignoni, M. E., Iwai, P. J.; Laboratory evaluation of new ultraviolet absorbers for protection of Douglas-fir tussock moth (Lepidoptera: Lymantriidae) baculovirus. *J. Econ. Entomol.* **1985**; 78, 982-987.
36. Morris, O. N. the effect of sunlight, ultraviolet, and gamma radiations and temperature on the infectivity of a nuclear polyhedrosis virus. *J. Invertebr. Pathol.* **1971**; 18, 292-294.
37. Morris, O. N. In *Microbial Control of Spruce Budworm and Gypsy Moths;* Canada-United States Spruce Budworm R & D Program; Grimble, D. G.; Lewis, F. B. Eds. USDA/FS, GTR-NE-100; **1985**, pp. 39-46.
38. Nakagaki, M.; Ohba, M.; Tanada, Y. Specificity of receptor sites on insect cells for the synergistic factor of an insect baculovirus. *J. Invertebr. Pathol.* **1987**; 50, 169-175.
39. Nickle, W. R.; Shapiro, M. Use of a stilbene brightener, Tinopal LPW, as a radiation protectant for *Steinernema carpocapsae*. *J. Nematol.* **1992**; 24, 371-373.
40. Ossowski, L. L. J. Variation in virulence of a wattle bagworm virus. *J. Insect Pathol.* **1960**; 2, 35-43.

41. Paine, C.; Radley, J. A.; Rendell, L. P. Fabrics fluorescent to ultraviolet light. U.S. Patent 2,089,413. **1937**; 2 pp.
42. Reichelderfer, C. F.; Benton, C. Y. The effect of 3-methyl-cholanthrene treatment on the virulence of a nuclear polyhedrosis virus of *Spodoptera frugiperda*. *J. Invertebr. Pathol.* **1973**; 22, 38-41.
43. Rollinson, W. D.; Lewis, F. B. Susceptibility of gypsy moth larvae to *Lymantria* spp. nuclear and cytoplasmic polyhedrosis viruses. *Plant. Prot.* **1973**; 24(124-125), 163-168.
44. Selitrennikoff, C. P. Calcofluor white inhibits *Neurospora* chitin synthetase activity. *Exp. Mycol.* **1984**; 8, 269-272.
45. Shapiro, M. Effectiveness of B vitamins as UV screens for the gypsy moth (Lepidoptera: Lymantriidae) nucleoployhedrosis virus. *Environ. Entomol.* **1985**; 14, 705-708.
46. Shapiro, M. Congo red as an ultraviolet protectant for the gypsy moth (Lepidoptera: Lymantriidae) nuclear polyhedrosis virus. *J. Econ. Entomol.* **1989**; 82, 548-550.
47. Shapiro, M. Use of optical brighteners as radiation protectants for the gypsy moth (Lepidoptera: Lymantriidae) nuclear polyhedrosis virus. *J. Econ. Entomol.* **1992**; 85, 1682-1686.
48. Shapiro, M.; Ignoffo, C. M. Nucleopolyhedrosis of *Heliothis*: Activity of isolates from *Heliothis zea*. *J. Invertebr. Pathol.* **1970**; 16, 107-111.
49. Shapiro, M.; Bell, R. A. Enhanced effectiveness of *Lymantria dispar* (Lepidoptera: Lymantriidae) nucleopolyhedrosis virus formulated with boric acid. *Ann. Entomol. Soc. Am.* **1982**; 75, 346-349.
50. Shapiro, M.; Robertson, J. L. Laboratory evaluation of dyes as ultraviolet screens for the gypsy moth (Lepidoptera: Lymantriidae) nuclear polyhedrosis virus. *J. Econ. Entomol.* **1990**; 83, 168-172.
51. Shapiro, M.; Robertson, J. L. Natural variability of three geographic isolates of gypsy moth (Lepidoptera: Lymantriidae) nuclear polyhedrosis virus. *J. Econ. Entomol.* **1991**; 84, 71-75.
52. Shapiro, M.; Robertson, J. L. Enhancement of gypsy moth (Lepidoptera: Lymantriidae) baculovirus activity by optical brighteners. *J. Econ. Entomol.* **1992**; 85, 1120-1124.
53. Shapiro, M.; Dougherty, E. M. In *Pest Management: Biologically Based Technologies;* Lumsden, R. D.; Vaughn, J. L. Eds. Proc. Beltsville Symposium XVIII, Beltsville, Maryland; **1993**, pp. 40-46.
54. Shapiro, M.; Agin, P. P.; Bell, R. A. Ultraviolet protectants of the gypsy moth (Lepidoptera: Lymantriidae) nucleopolyhedrosis virus. *Environ Entomol.* **1983**; 12, 982-985.
55. Shapiro, M.; Robertson, J. L.; Injac, M. G.; Katagiri, K.; Bell, R. A. Comparative infectivities of gypsy moth (Lepidoptera: Lymantriidae) nucleopolyhedrosis virus isolates from North America, Europe,and Asia. *J. Econ. Entomol.* **1984**; 77, 153-156.

56. Shapiro, M.; Preisler, H. K.; Robertson, J. L. Enhancement of baculovirus activity on gypsy moth (Lepidoptera: Lymantriidae) by chitinase. *J. Econ. Entomol.* **1987**; 80,1113-1116.
57. Shapiro, M.; Hamm, J. J.; Dougherty, E. M. Compositions and methods for biocontrol using fluorescent brighteners. U.S. Patent 5,124,149; **1992**, 16 pp.
58. Slavicek, J. M.; Podgwaite, J.; Lanner-Herrera, C. Properties of two *Lymantria dispar* nuclear polyhedrosis virus isolates obtained from the microbial pesticide Gypchek. *J. Invertebr. Pathol.* **1992**; 59, 142-148.
59. Slifkin, M.; Cumbie, R.; Congo red as a fluorochrome for the rapid detection of fungi. *J. Clin. Microbiol.* **1988**; 26, 827-830.
60. Smirnoff, W. A. A virus disease of *Neodiprion swainei* Middleton. J. Insect. Pathol. **1961**; 3, 29-46.
61. Smirnoff, W. A. Adaptation of a nuclear polyhedrosis virus of *Trichiocampus viminalis* (Fallen) to larvae of *Trichiocampus irregularis* (Dyar). *J. Insect Pathol.* **1963**; 5, 104-110.
62. Tanada, Y. Some factors affecting the susceptibility of the armyworm to virus infection. *J. Econ. Entomol.* **1956**; 49, 52-57.
63. Tanada, Y. Synergism between two viruses of the armyworm, *Pseudaletia unipuncta* (Haworth) (Lepidoptera: Noctuidae). *J. Insect Pathol.* **1959**; 1, 215-231.
64. Tanada, Y. A synopsis of studies on the synergistic property of an insect baculovirus: A tribute to Edward A. Steinhaus. *J. Invertebr. Pathol.* **1985**; 45, 125-138.
65. Tanada, Y.; Hukuhara, T. Enhanced infection of a nuclear polyhedrosis virus in larvae of the armyworm, *Pseudaletia unipuncta*, by a factor in the capsule of a granulosis virus. *J. Invertebr. Pathol.* **1971**; 17, 116-126.
66. Tanada, Y.; Himeno, M.; Omi, E. M. Isolation of a factor, from the capsule of a granulosis virus, synergistic for a nuclear-polyhedrosis virus of the armyworm. *J. Invertebr. Pathol.* **1973**; 21, 31-40.
67. Teetor, G. E.; Kramer, J. P. Effect of ultraviolet radiation on the microsporidia *Octosporea muscaedomesticae* with reference to protectants provided by the host *Phormia regina*. *J. Invertebr. Pathol.* **1977**; 30, 348-353.
68. Vasiljevic, L.; Injac, M. A study of gypsy moth viruses originating from different geographical regions. Plant. Prot. **1973**; 24 (124-125), 169-186.
69. Vaughn, J.L.; Ignoffo, C. M. Insect Pathology Workshop, USDA, Agric. Res. Serv., Beltsville, Maryland, **1986**; 43 pp.
70. Veber, J. Virulence of an insect virus increased by repeated passages. Colloq. Int. Pathol. Insectes, Paris, France. **1962**; 3, 403-405.
71. Wood, H. A.; Hughes, P. R.; Johnston, L. B.; Langridge, W. H. R. Increased virulence of *Autographa californica* nuclear polyhedrosis virus by mutagenesis. *J. Invertebr. Pathol.* **1981**; 38, 236-241.

72. Woodward, D. B.; Chapman, H. C. Laboratory studies with the mosquito iridescent virus (MIV). *J. Invertebr. Pathol.* **1968**; 11, 296-301.
73. Yadava, R. L. On the chemical stressors of nuclear polyhedrosis virus of the gypsy moth, *Lymantria dispar*. *Z. Angew. Entomol.* **1971**; 69, 303-311.
74. Yamamoto, T.; Tanada, Y. Phospholipid, an enhancing component in the synergistic factor of a granulosis virus of the armyworm, *Pseudaletia unipuncta*. *J. Invertebr. Pathol.* **1978**; 31, 48-56.
75. Zhu, Y.; Hukuhara, T.; Tamura, K. Location of a synergistic factor in the capsule of a granulosis virus of the armyworm, *Pseudaletia unipuncta*. *J. Invertebr. Pathol.* **1989**; 54, 49-56.

RECEIVED January 31, 1995

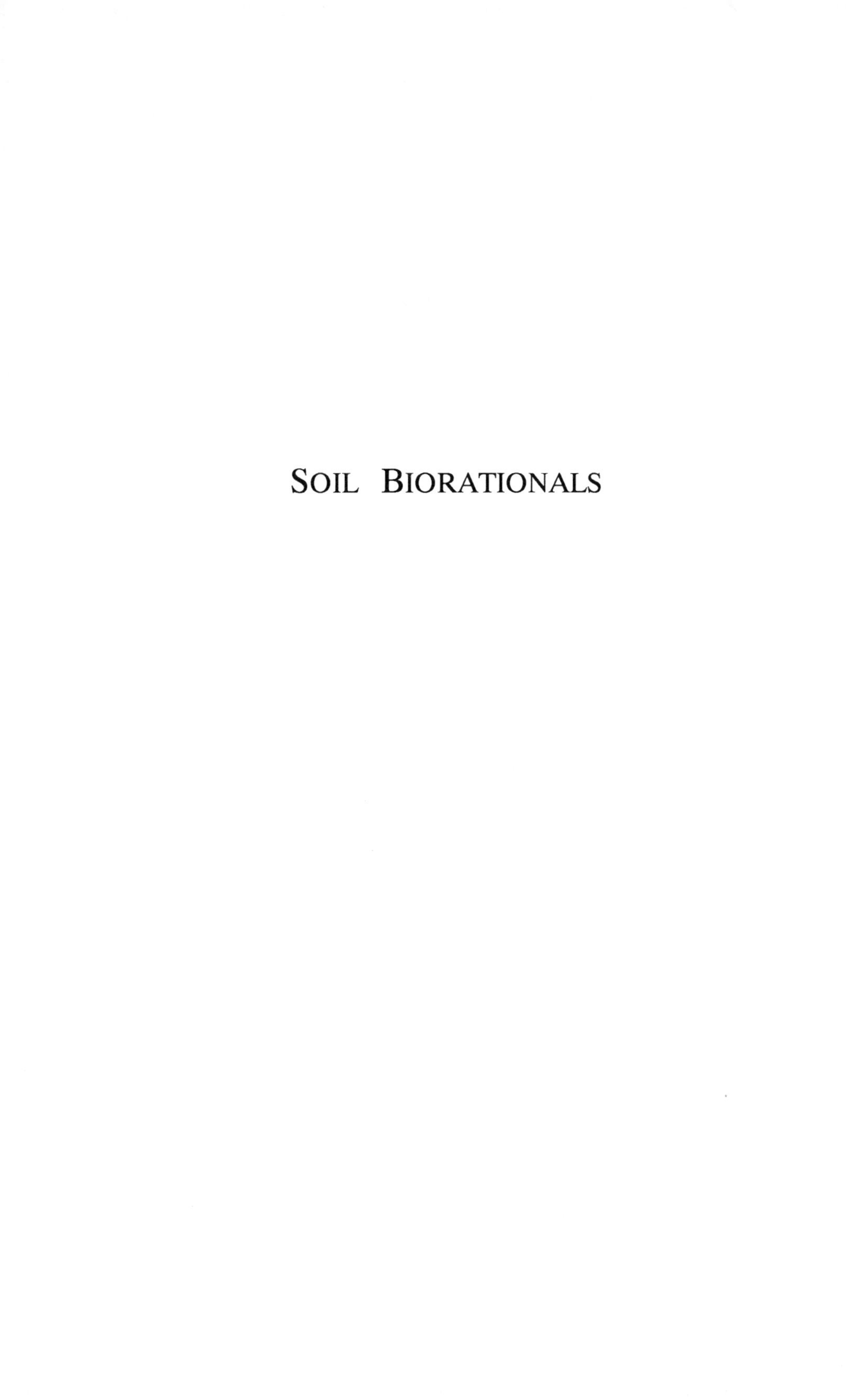

# Soil Biorationals

# Chapter 11

# Formulation and Delivery of Biocontrol Agents for Use Against Soilborne Plant Pathogens

**R. D. Lumsden, J. A. Lewis, and D. R. Fravel**

**Bicontrol of Plant Diseases Laboratory, Agricultural Research Service, U.S. Department of Agriculture, Beltsville, MD 20705**

Because of the obstacles involved in the production of biocontrol agents and their introduction into the environment, limited research has been done on biomass fermentation, formulation, and delivery of microbes for control of soilborne plant pathogens. Considerable work has focused on *Trichoderma* and *Gliocladium* due to ease of isolation, culturing, and fermentation of these common soil fungi. Formulations of biocontrol fungi, bacteria, and streptomycetes have been developed experimentally, although most have not been used commercially. Industrially available products include preparations of *Agrobacterium*, *Streptomyces*, and the first fungus available in the U.S. for control of plant diseases, *Gliocladium virens*. *G. virens* is grown in deep tank liquid fermentation, formulated in alginate prill, and incorporated into soilless potting media for control of the damping-off pathogens *Pythium ultimum* and *Rhizoctonia solani*. This formulation is effective for disease control of vegetable and ornamental seedlings. Isolates of *Trichoderma* are being developed for seed treatment. Cooperative research with private industry has significantly advanced progress in this newly emerging technology. Difficulties and problems involved with commercial production of microbial biocontrol agents, especially fungi, are discussed.

The development of biological control technologies is a newly emerging field in plant pathology compared to the advances made with these systems to reduce damage of economic crops caused by insects and weeds. However, rather than being used as the only method for disease control, it is more likely that biological control will be employed as a component for integrated pest management (IPM) systems and will be used in low-input sustainable agriculture (LISA) production.

As recently as ten years ago, it was estimated that crop loss due to soilborne plant pathogens, most of which are fungi, amounted to at least $4 billion annually (*1*). Current and future restrictive legislation against pesticide use will compound these conservative estimated losses even further. Although the restrictions are being imposed to protect food quality and the environment, chemicals are still the only major resource used worldwide to prevent diseases of food and fiber crops.

The major aspects for consideration of successful biological control technologies include the establishment of production, formulation, and delivery systems for the living microorganisms (fungi, bacteria, and actinomycetes) that result in disease control. In addition, these aspects must be implemented to be compatible with industrial and commercial development methods and field applications. Recent reviews which consider these various aspects are published (*2, 3, 4, 5, 6, 7, 8*).

Each step in the process requires careful attention to disease control agent viability during fermentation and formulation, ease in delivery, effectiveness after application, and adequate survival during storage and transit to enable them to be economically used over several months or years. Recently, it has become evident that a major obstacle in the commercial development of a biocontrol product has been problems associated with "scale-up" operations. Another major concern is the various regulatory statutes that must be satisfied before official registration of the biocontrol product is realized.

Most of the research activities involving biological control have been performed in the laboratory and greenhouse, under less than natural conditions, with easily isolated and cultured fungi and bacteria of the genera *Trichoderma*, *Gliocladium*, *Pseudomonas*, and *Bacillus* (*7*). However, this scenario is rapidly changing as scientists are becoming more actively engaged in "field studies" and investigating novel biocontrol agents such as the fungi *Sporidesmium*, *Talaromyces*, *Laetisaria*, *Cladorrhinum*, and *Stilbella* (*7*). Another attractive approach, but one which is closely regulated, is the use of genetically-improved antagonists (*8*). In this situation, improved strains of the antagonists can be developed to control target pathogens through genetic manipulation to produce more or improved antibiotics, cause physical destruction of the pathogen by lysis or parasitism, or to compete successfully with non-target organisms in the soil and rhizosphere microniches.

This limited review will discuss briefly some aspects which should be considered when producing, formulating, and commercially using microbial biocontrol agents to reduce the impact of disease caused by soilborne plant pathogens. It should be noted that the items discussed pertain generally to the development of any microbial biological agent.

## Fermentation

A major concern in the commercial formulation of a biocontrol product is to achieve adequate growth of the biocontrol agent which is abundant in effective propagules (chlamydospores, conidia, microsclerotia, endospores). In many cases, biomass production is difficult because of the specific nutritional and environmental conditions required for growth of the organism (*9,10,11*). Although both solid and liquid fermentation technologies are employed, the expansion of liquid fermentation technology in developed countries for the production of organic acids, antibiotics and enzymes by fungi and bacteria has resulted in the availability of large-scale deep tank fermentation facilities in industry (*12*). These advances have already resulted in the production of bacterial and fungal biomass for use as biocontrol insecticides and herbicides (*7*). Although considerable advances are continually being made in the production of effective antagonists against plant pathogens, this aspect of the technology is still in the formative stages.

A suitable growth medium should consist of inexpensive, readily-available agricultural by-products with the appropriate nutrient balance. Acceptable materials include molasses, brewer's yeast, corn steep liquor, sulfite waste material and cottonseed and soy flours (*11, 13*). Companies will generally adapt media to utilize readily available substrates. For example, sugar-processing companies would have by-products of corn, sugar-beet and cane sugar available to use. For successful fermentation, not only must appropriate substrates be used, but sufficient biomass containing adequate amounts of effective propagules must be obtained. Research in the Biocontrol of Plant Diseases Laboratory (BPDL) with isolates of *Trichoderma* spp. and *Gliocladium virens* indicated that preparations containing chlamydospores (the survival propagules of the fungus) more effectively prevented damping-off diseases than preparations containing only conidia (*14*). Similarly, selective liquid media result in the production of conidia of the biocontrol fungus *Talaromyces flavus* (*15*). Small-scale fermentation of molasses-brewer's-yeast which resulted in abundant chlamydospore production of isolate GL-21 of *G. virens* has been adapted successfully to large-scale industrial fermentation (*16*) and conidia of *Trichoderma harzianum* have been successfully produced in liquid fermentation systems (*17*).

Operating conditions during fermentation (aeration, pH, temperature) as well as media constituents may affect the quality and quantity of the test organism, especially bacteria. An additional factor to consider in liquid fermentation is the rate of biomass production, which affects the cost of production as well as chance of contamination. It is desirable to obtain the optimum amount of biomass in the shortest time. With isolates of *Trichoderma*, *Talaromyces*, and *Gliocladium*, satisfactory quantities of biomass were obtained in 6-7 days, but this time period is still long compared with that for bacteria (*16*).

Solid or semi-solid fermentation is used successfully for experimental production of fungal biomass. With the advent of deep tank fermentation, however, solid state fermentation facilities have become limited in much of the world. In fact, solid state fermentation is not used extensively in North America (except in mushroom production) because of insufficient consumer demand for the products formed (*18*). There are, however, several companies in both the U.S. and Mexico which specialize in solid state fermentations. It appears evident that the process will be used by industry if effective biocontrol preparations can be produced.

Substrates for the production of inoculum of biocontrol fungi in the genera *Trichoderma*, *Gliocladium*, *Coniothyrium, Chaetomium, Laetisaria,* and *Penicillium* include various grain seeds and meals, bagasse, straw, wheat bran, sawdust, and peat individually or in combination (*19*). The system is especially useful for small scale research laboratory, greenhouse, and field tests which require minimum facilities for implementation. Solid fermentations are also suitable for the production of fungi which either do not sporulate in liquid cultures (*11*) or do not survive the liquid fermentation process (*20*).

The system is also advantageous in countries such as Japan, where fully automated commercial solid fermentation plants are already in existence for other purposes. Also, solid fermentation may be appropriate in underdeveloped countries where agricultural wastes are available, elaborate facilities are limited, and labor is abundant (*21*).

**Formulations**

The problems involved with formulation technology are as important and complex as those associated with growth of biocontrol antagonists (*2*). A formulated product with agricultural application should possess several desirable characteristics: these include adequate market potential, ease in preparation and application, stability during transportation and storage, abundant viable propagules and good shelf-life, sustained efficacy, and acceptable cost (*10, 11, 22*). A major difficulty in formulation is a situation which must be adequately addressed by industry, i.e. scale-up procedures for product development. Problems in large-scale production such as formulation, drying and milling can adversely affect the quality of the product. For example, large-scale slow drying can result in inactivation of the effective propagule as well as an increase in contaminating microorganisms. Uncontrolled or abrasive milling can also inactivate the effective propagules. These factors most often influence quality control and product effectiveness and must be constantly monitored during the entire development process.

After fungal growth, biomass is separated from spent medium by any of various filtration systems (pressure, rotary vacuum drum), centrifugation, or flocculation. The biomass can be incorporated moist into pellets or inert carriers (Celatom, Vermiculite) which are subsequently dried (*23,24*). Before formulation, the biomass may also be dried by systems utilizing pan drying, spray drying or freeze drying, any of which may affect the viability of the propagules in the biomass. Conventional techniques to formulate pesticides may be adapted to biocontrol powders, but care must be taken to avoid drastic treatment during processing (*25*). One of the most intriguing aspects of innovative technology for the formulation of microbials involves the immobilization of wet or dry biomass within cross-linked organic polymers such as alginate, polyacrylamide or carrageenan (*26*). After several years of cooperative research, scientists in the BPDL and W. R. Grace & Co. of Columbia, MD developed an alginate prill product, GlioGard, containing fermentor-produced biomass of the antagonist *G. virens* (Gl-21). GlioGard is used to control damping-off of various ornamental and vegetable transplants caused by the pathogens *Rhizoctonia*, and *Pythium* (*27*). The prill is incorporated into soilless mix before seeding and can significantly reduce the use of current fungicides. The system has potential for use against pathogens of woody ornamental cuttings and turf. GlioGard is the first U.S. Environmental Protection Agency registered fungal biocontrol product effective against soilborne plant pathogens (*28*).

Alginate gel has also been used successfully in the BPDL to prepare formulations of biocontrol bacteria as well as other fungi (*15, 24,29*) for which U.S. patents were issued (Nos. 4,668,512, 4,724,147 and 4,818,530). The antagonists studied included pseudomonads, isolates of *Trichoderma* spp., *G. virens*, *T. flavus*, *Laetisaria arvalis*, and *Stilbella* sp. For all examples, either wet fermentor biomass or the entire fermentor suspension could be used, as well as dry, powdered biomass. With fungi, wet biomass was blended before it was added to the polymer. Use of wet biomass or the whole suspension was advantageous because a drying and milling step was omitted, and some fungi (*e.g. Laetisaria, Arachniotus*), whose biomass cannot be separated easily from spent medium because of gum formation, were easily processed.

The method consists of mixing fermentor biomass and a carrier (bulking agent) with a sodium alginate solution. The carrier may be inert (*e.g.* Pyrax), a food base (*e.g.* powdered wheat bran), or a combination. The mixture is dripped into a calcium salt gellant solution. Each droplet is transformed into a gel bead by the bonding of the sodium alginate with calcium ions. Gel bead size can be manipulated by adjusting the gauge of the droplet-forming tips in the dispensing apparatus. Drying to a percentage moisture content of 5% converts the gel beads or prill into an easily handled product for storage and end-use applications.

Fungal and bacterial biomass can be most conveniently incorporated into dusts, wettable powders, emulsifiable liquids and other types of granules for soil application or seed treatment. In general, conventional techniques used to formulate chemical pesticides may be adapted to biocontrol powders, but care must be taken to avoid drastic treatment during processing (*25*). Experimental powder formulations have been successfully prepared by diluting biomass of isolates of *Trichoderma* and *Gliocladium* with commercially available pyrophyllite clay (Pyrax) as a carrier (*16, 30*).

Seed treatment with biocontrol fungi has recently been improved by adding amendments to stimulate antagonist activity (*31*). Techniques to improve delivery have also included a solid matrix priming system for seed application in which seeds were hydrated to a controlled level with moistened, finely ground carrier (*32*). Formulating dusts or powders from dried biomass of biocontrol agents represents a rapid and efficient approach in delivery, provided the antagonists remain viable. Pyrax dusts containing biomass of *T. viride* and *G. virens* were successfully applied to potato seed pieces for the reduction of disease caused by *R. solani* (*33*). Similar preparations containing *T. flavus* were coated on potato seedpieces for the reduction of Verticillium wilt (*34*).

Fungal formulations are also available as various types of seed treatments to control selected damping-off diseases. One example, called F-Stop, developed in the United States, contains conidia of *T. harzianum,* and has EPA registration (*17*). Formulations of bacterial biomass to control diseases in the spermosphere and rhizosphere include seed treatments and plant dips. Most notable are preparations of the bacterium *Agrobacterium radiobacter* (*e.g.* Galltrol) used chiefly as a woody plant dip. Other seed dressings recently registered by EPA contain the bacteria, *Bacillus subtilis* (*e.g.* Epic, Kodiak) (*35*), *Pseudomonas fluorescens*, (*e.g.* Dagger G) (*36*) and *P. cepacia* (*e.g.* Blue Circle) (*37*). Also, recently registered by EPA is the streptomycete formulation (Mycostop) developed in Finland (*38*). Fungal spores and biomass preparations also have been formulated into pastes (*39*), tablets (*40*), and fluid-drill gels (*41*). The exact nature of such preparations is often proprietary information. Recent research has demonstrated the feasibility of incorporating biocontrol organisms into granules which are starch, flour or gluten based. Although these formulations were initially developed for mycoherbicides, the potential exists for similar incorporation of microbials which control soilborne plant pathogens. Products have been prepared using pearl corn starch, pregelatinized corn starch, corn oil and alpha-amylase (*42*). Several of these preparations effectively prevented damping-off of pepper, eggplant and zinnia seedlings caused by *R. solani* (*Lewis, J. A., Lumsden, R. D., Fravel, D. R., Shasha, B. S. Biol. Cont., in press*). In another formulation, propagules of biocontrol microbials were mixed with semolina wheat flour kneaded thoroughly, extruded through a pasta press and dried. This wheat-gluten/fungus formulation is called "pesta" (*43*). Also a publication describing innovative formulations, their water dispersibility, and modes of delivery has recently appeared (*44*).

In general, products formed from solid or semi-solid state fermentations do not require sophisticated formulation procedures prior to use. For example, grains or other types of organic matter upon which antagonists are grown are simply dried, ground, and added to the area treated. In some instances, these powders are used as seed treatments (*20,45*). However, there are several inherent problems with solid state fermentation which may make the system inappropriate for commercial product development. The preparations are bulky; they may be subject to a greater risk of contamination; and they may require extensive space for processing, incubation, and storage. In addition, they may require drying and milling with the undesirable formation of dusts containing spores; may require costly shipping and transport conditions and special equipment may be necessary for application.

Despite these shortcomings, some solid state fermentation preparations may have potential for commercial development. In the process used by the BPDL for the biocontrol mycoparasite *Sporidesmium sclerotivorum* (effective against lettuce drop and other diseases caused by the pathogen *Sclerotinia minor*), vermiculite moistened with a liquid medium is inoculated with the mycoparasite in a large twin shell blender, aseptically bagged, and incubated until the *S. sclerotivorum* grows and sporulates (*46*). Also, recently we have developed a preparation ('germlings') consisting of sterile wheat bean:water (1:1, w/v) and actively growing hyphae of a wide variety of *Trichoderma* and *Gliocladium* isolates (*14*). In this system, the substrate, inoculated with a conidial suspension of the biocontrol fungus, was allowed to incubate for 3 days. The formulation was immediately mixed into pathogen infested soil whereupon antagonist hyphae in association with the food base (bran), parasitized the pathogen propagules (*47*). In another system from the BPDL, wet or dry fermentor biomass of antagonist isolates was homogenized with dilute acid and mixed with vermiculite (*24*). After drying, the mixture could be stored for at least 12 weeks at 5 and 25° C. Before addition to soil, the dry vermiculite/biomass was remoistened with dilute acid and incubated for 2 days to induce formation of actively growing hyphae of the antagonist. The major advantage of the system is that aseptic conditions do not have to be maintained for formulation, except during production of biomass. Another formulation with potential commercial applications was reported in which conidia of isolates of *Trichoderma* and *Gliocladium* were added to a sterile lignite-stillage mixture (*48*). After incubation and drying, the granules were applied to soil.

A list of commercial formulations, with registration in this country or abroad, is indicated in Table 1. Although only a limited number of products have EPA registration, the list is increasing yearly. For example, in 1993 only six materials were approved for use in the United States. The current list indicates ten EPA-approved products. Additional biocontrol products are under development for both soilborne plant pathogens and foliar pathogens whose causal organisms may be resident in the soil. *Sporidesmium sclerotivorum*, the formulation of which is patented, is being developed in the United States against *Sclerotinia minor* of lettuce (*62*). Formulations of binucleate *Rhizoctonia* spp. and the bacterium, *Bacillus subtilis*, are being prepared in Australia against *R. solani* and *P. ultimum* of field crops and bedding plants (*63, 64*). Commercially, the antagonistic fungus *T. hamatum* and the bacterium *Flavobacterium balusinum* are being used as a compost amendment to reduce diseases of bedding plants caused by *Pythium* spp., and *R. solani* (Earthgro, Inc., Lebanon, CT) (*65*). Research is continuing in Japan to develop a product of nonpathogenic *Fusarium oxysporum* against Fusarium wilt of sweet potato (*66*). Commercial seed coatings with formulations of *Pythium oligandrum* are being investigated in the United Kingdom to control damping-off of cress and sugar beet (*67*).

Various products are also developed for foliar diseases. For example, BINAB-T, containing isolates of *Trichoderma*, has EPA registration for use against *Chondrostereum purpureum* (silver leaf) of several trees and *Cryphonectria parasitica* (blight) of chestnut (*40*). Frostban, produced by Frost Technology, Oakland, CA, is registered to reduce *Erwinia amylovora* (fire blight) of apples and pear trees (*68*). Trigger, a formulation of *Verticillium dahliae*, is produced in the Netherlands by Heidemji Realistre B.V. for use against *Ophiostoma ulmi* (Dutch elm disease) of elms (*69*). A potential product containing hypovirulent *Cryphonectria parasitica* is being developed in France against chestnut blight (*39*).

## Delivery

Having the biocontrol agent in an active state in the right place, at the right time is often the key to successful biocontrol. Activity, placement and timing are usually more important than introducing large populations of the biocontrol agent. Among the many delivery systems available, only a few may be appropriate for a particular situation. In addition to when and where the biocontrol agent is needed, choice of the delivery system is also based on attributes of the host plant, pathogen and biocontrol agent, as well as the control strategy employed. Strategies used for biocontrol of soilborne plant pathogens include i) protection of the infection court (rhizosphere or spermosphere), ii) impeding the progress of the pathogen through soil, iii) inactivation or destruction of overwintering inoculum,

**Table 1. Products with Biocontrol Agents Effective against Soilborne Plant Pathogens**

| *Product/Antagonist* | *Target pathogen/host* | *Source (Country)/Reference* |
|---|---|---|
| Products approved by U.S. E.P.A. | | |
| BINAB-T,W<br>*Trichoderma* spp. | *Verticillium malthousei*/mushrooms | Binab USA, Madison, WI (USA); *(40)* |
| Blue Circle<br>*Pseudomonas cepacia* | *Rhizoctonia solani*, *Pythium* spp./vegetable seed treatment | Stine Microbials, Madison, WI (USA); *(37)* |
| Dagger G<br>*Pseudomonas fluorescens* | *Pythium ultimum*, *Rhizoctonia solani*/cotton seed treatment | Ecogen, Inc., Langhorne, PA (USA); *(36)* |
| Epic<br>*Bacillus subtilis* | *Rhizoctonia solani*, *Fusarium* sp./cotton, legumes | Gustafson, Inc., Dallas, TX (USA); *(35)* |
| F-stop<br>*Trichoderma harzianum* | *Pythium ultimum*, *Rhizoctonia solani*, *Fusarium* sp./row crop seed treatment | TGT, Inc., Cornell Reseach Foundation, Ithaca, NY (USA); *(17)* |
| Galltrol, Norbac 84-C<br>*Agrobacterium radiobacter* | *Agrobacterium tumefaciens*/trees, shrubs/root dip | AgBioChem, Inc., Orinda, CA (USA), New Bioproducts, Inc., Corvallis, OR (USA); *(49)* |
| GlioGard<br>*Gliocladium virens* | *Rhizoctonia solani*, *Pythium ultimum*/bedding plants | W. R. Grace & Co., Columbia, MD (USA); *(28)* |
| Kodiak<br>*Bacillus subtilis* | *Pythium ultimum*, *Rhizoctonia solani*/cotton, legume seed treatment | Gustafson, Inc., Dallas, TX (USA); *(50)* |
| Mycostop<br>*Streptomyces griseovirides* | *Alternaria* sp., *Fusarium* spp./vegetables, ornamentals | Kemira Biotech, Helsinki, (Finland); *(38)* |
| VICTUS<br>*Pseudomonas fluorescens* bv5 | *Pseudomonas tolaasii*/mushrooms | Sylvan, Kattanning, PA (USA);<br>Miller, F. C.; Sylvan, PC[a], 1994 |

*Continued on next page*

Table 1. *Continued*

| Products not in use in the U.S. | | |
|---|---|---|
| ANTI-FUNGUS<br>*Trichoderma* spp. | *Rhizoctonia solani*, *Pythium* spp./bedding plants | Grondontsmettingham De Ceuster (Belgium); *(54)* |
| Bactophyt<br>*Bacillus subtilis* | various fungi/vegetables | NPO Vector, Novosibirsk, Russia; Lisansky, S. and Coombs, R. PC[a] 1993 |
| Biofox C<br>*Fusarium oxysporum* (nonpathogenic) | *Fusarium oxysporum*/ carnation, basil, tomato | S.I.A.P.A., Bologne (Italy); *(58)* |
| Coniothyrin<br>*Coniothyrium minitans* | *Sclerotinia sclerotiorum*/sunflower | (Russia); *(57)* |
| Conqueror<br>*Pseudomonas fluorescens* | *Pseudomonas tolaasii*/mushrooms | Mauri Foods (Australia); Whipps, J. M. Horticulture Research International; PC[a]1994 |
| Fusaclean<br>*Fusarium oxysporum* (nonpathogenic) | *Fusarium oxysporum*/tomato, carnation | Natural Plant Protection, Nogueres (France); *(55)* |
| Intercept<br>*Pseudomonas cepacia* | *Rhizoctonia solani*, *Fusarium* spp., *Pythium* sp./corn, vegetables, cotton | Soil Technologies, Fairfield, IA (USA); Yuen, G., University of Nebraska, PC[a] 1994 |
| Nogall, Diegall<br>*Agrobacterium radiobacter* | *Agrobacterium tumefaciens*/trees | Fruit Growers Chemical Co. (New Zealand), Root Nodule Phy Ltd. (Australia); (53) |
| P.g. Suspension, Rotstop<br>*Phlebia gigantea* | *Heterobasidion annosum*/trees | Ecological Labs., Ltd., Dover, (UK), Kemira Biotech, Helsinki, (Finland); *(51, 52)* |
| Phagus<br>bacteriophage | *Pseudomonas tolaasii*/mushrooms | Natural Plant Protection, Nogueres (France); Davazaglou-Lecant, V. NPP, PC[a] 1994 |

| | | |
|---|---|---|
| Polygandron<br>*Pythium oligandrum* | *Pythium ultimum*/sugarbeet | Vyzkummy ustov (Czech Republic); *(56)* |
| Promot<br>*Trichoderma* spp. | various fungi/fruit, vegetables | J. H. Biotech, Inc., Ventura, CA (USA); Harris, A. R. CSIRO, PC[a] 1994 |
| PSSOL<br>*Pseudomonas solanacearum* (nonpathogenic) | *Pseudomonas solanacearum*/vegetables | Natural Plant Protection, Nogueres (France); Davazaglou-Lecant, V. NPP, PC[a] 1994 |
| Supraavit<br>*Trichoderma harzianum* | various fungi | Bonegaard & Reitzel (Denmark); Funck-Jensen, D. Royal Veterinary and Agricultural University, PC[a] 1994 |
| T-35<br>*Trichoderma harzianum* | *Rhizoctonia solani*, *Fusarium* spp./cucumbers, tomatoes | Makhteshim (Israel); *(60)* |
| Trichodermin-3<br>*Trichoderma lignorum* | *Fusarium* sp./field crops, vegetables | (Russia, Bulgaria); *(57)* |
| Trichodex<br>*Trichoderma harzianum* | *Botrytis cinerea*/fruits, ornamentals | Makhteshim (Israel); *(59)* |
| Trichopel<br>*Trichoderma* spp. | various fungi | Agrimm Biologicals (New Zealand); Lisansky, S. and Coombs, R. PC[a] 1993 |
| TY<br>*Trichoderma* spp. | *Rhizoctonia solani*, *Pythium* sp., *Sclerotium rolfsii*/field cops, vegetables | Mycontrol (Israel); *(61)* |
| Vaminoc<br>mycorrhizae | *Botrytis* spp., *Pythium* sp./cucumbers | AGC Microbiol (UK); Whipps, J. M., Horticulture Research International, PC[a] 1994 |

[a]PC=Personal Communication

and iv) induced systemic resistance or cross protection. In general, biocontrol agents that work through competition or antibiosis are used for protecting the infection court and impeding the progress of the pathogen, while mycoparasites are often used for destruction of overwintering inoculum.

Much attention has focused on protection of the rhizosphere or spermosphere as a biocontrol strategy. One of the appeals of this strategy is that there is no need to distribute the biocontrol agent in the soil bulk. Root and seed exudates provide nutrients which promote establishment and proliferation of biocontrol agents and may extend the time period during which the roots are protected. Root and seed exudates may also stimulate production of antibiotics, siderophores, agglutination factors and other compounds involved in biocontrol (*70*).

Biocontrol agents protect roots or seeds by mechanisms which include competition for carbon, nitrogen, iron and other nutrients, by antibiosis, physical exclusion, mycoparasitism/lysis, or a combination of these mechanisms. For example, siderophores chelate ferric iron, thus enabling the bacteria to compete for iron under iron-limiting conditions. Siderophores make important contributions to biocontrol in some cases, but are not related to biocontrol in other systems (*71-74*). Competition is also evident in the interception of signals from the root so that the pathogen does not recognize the presence of the root (*75, 76*). Mycorrhizal fungi can protect roots through physical exclusion of the pathogen (*77*).

Several systems have been used to deliver biocontrol agents used to protect the infection court. Control of *A. tumefaciens* by *A. radiobacter* is accomplished by root dip into an aqueous suspension of the biocontrol agent at transplant (*49, 53*). The pathogen *Heterobasidion annosum* gains entry to healthy pine trees from root grafts with infected trees or where it has colonized the stump of a harvested tree. Spores of *Phlebia gigantea* can be applied in aqueous suspension or in chain saw oil to the freshly cut stumps of pine (*51*).

Many forms of seed treatment have been used to deliver biocontrol agents to protect the rhizosphere or spermosphere. For the simplest seed treatments, either liquids or dusts containing the biocontrol agent are applied to seed (*33, 34, 78-84*). Fluid-drilling gels have been used to deliver *T. harzianum* and *Laetisaria arvalis* for control of *R. solani* and *S. rolfsii* on apple (*41*). A liquid seed treatment containing an aqueous binder (Pelgel or Polyox N-10), a finely ground solid particulate and homogenized cultures of *T. harzianum* was used to protect cucumber seeds from *Pythium*-induced damping-off (*85, 86*). A modified commercial process was used to incorporate *P. oligandrum* in seed pelleting for control of *P. ultimum* and *Mycocentrospora acerina*-induced damping-off on sugar beet, cress and carrot (*87*). Seed priming, in which seeds are mixed with an organic carrier and the moisture content is brought to a level just below that required for seed germination, has been used to deliver *T. harzianum* to control *Pythium*-induced damping-off on cucumber (*88*) and *P. fluorescens* on corn to prevent damping-off induced by *P. ultimum* (*89*).

When the soil has been disinfested or when plants are seeded into soilless potting medium, the strategy of impeding the progress of the pathogen has been successfully used to protect plants. The advantage of inoculating disinfested soil or potting mix is that the biocontrol agent can become established when competitive interactions in soil are at a minimum, creating a suppressive soil and making subsequent colonization of the soil by pathogens difficult. Biocontrol agents used for this purpose have been delivered in a variety of ways including polymer prills, seed treatments, drenches, products of solid fermentation and soil. Except for the soil and drench treatments, microbes used to impede the progress of the pathogen are applied to the soil surface and then raked or rotovated into the soil prior to planting. For greenhouse crops and transplants, these materials can be mixed with the potting mix or soil.

Incorporation of bacterial cells, spores of biocontrol fungi, or fermenter biomass into polymer prills with various carriers has been used to deliver biocontrol agents (3, 15). GlioGard (Table 1), a preparation of *G. virens*, is incorporated into potting mix before seeding bedding plants or vegetable transplants (*27, 28*). *G. virens* protects the plants from damping-off caused by *Pythium* and *Rhizoctonia*. Similarly, "pesta," a pasta-like formulation, has been used to deliver fungi for biocontrol of plant pathogens as well as mycoherbicides (*43*). Anti-Fungus, a *Trichoderma* product which is applied to methyl bromide-treated soil to prevent subsequent invasion by *Pythium*, *Rhizoctonia*, and *Verticillium* spp. (Table 1), is available in several formulations. An aqueous drench containing conidia of *T. harzianum* controlled wilt of chrysanthemum by preventing reinvasion of *F. oxysporum* (*90, 91*). Liquid fermentation was used to produce biomass of *G. virens, T. hamatum, T. harzianum, T. viride, and Talaromyces flavus* for biocontrol (*16*). Production of biocontrol agents in bulk organic matter or vermiculite carrier for incorporation into soil has been used for a number of biocontrol fungi including *T. harzianum* for control of *S. rolfsii* (*92*). *T. harzianum* applied as a conidial suspension, seed coating or wheat bran preparation successfully competed for carbon and nitrogen to prevent germination of chlamydospores of *F. oxysporum* (*93*). Soil has also been used to deliver biocontrol agents. A quantity of suppressive soil was placed around papaya transplants, preventing invasion of *Phytophthora palmivora* to the root zone of the transplant (*94*). When the plant was large enough that the roots grew out from this protective soil, it had matured to a stage when it was no longer susceptible to the pathogen.

Because of the monocyclic nature of the life cycles of most soilborne plant pathogens, destruction of overwintering inoculum has a proportionally greater affect on disease reduction than for polycyclic foliar diseases. Yet, the strategy of inactivating or destroying overwintering inoculum has received little attention, due to the perceived difficulty of locating pathogen propagules throughout bulk soil. Pathogen propagules occur in aggregate distribution in association with crop residues, providing at least two advantages for delivery of biocontrol agents. First, for a biocontrol agent that can grow through soil, the aggregate distribution facilitates spread of the biocontrol agent. Secondly, for many pathogens, if the biocontrol agent is applied to the above-ground portions of the plant, then it will be in close proximity to the pathogen propagules after the residue is incorporated into the soil. For example, sclerotia of *Sclerotinia minor* are formed primarily on the above-ground portion of lettuce. The biocontrol agent *S. sclerotivorum* can be applied to lettuce residue and then the residue incorporated into the soil. Some sclerotia will be infected from the initial delivery. New spores of the biocontrol fungus are formed on the infected sclerotia and the biocontrol fungus grows though soil to infect other sclerotia. Using this delivery system, rates as low as 0.2 kg/ha of a preparation of *S. sclerotivorum* can result in economically meaningful biocontrol (*62, 95*).

## Conclusions

It is becoming increasingly apparent throughout the world that efficient and productive agriculture is practiced at great cost to the environment. There are increasing reports of loss of pesticides from the market due to resistance and revocation of registration. Pesticides contaminate ground water and food stuffs. It is inevitable that biological control strategies, probably as a component with other control measures, will fill the vacuum. Although many of the basic concepts necessary for the implementation of biological control against soilborne plant pathogens are in place, apparent obstacles in formation of biomass, formulation of a product, time and site of application, and registration difficulties exist because this is a new technology.

The causes of the problems which are inhibiting the rapid advancement of biological control fall into several sectors which encompass basic research approaches to application of a final product. Initially, research scientists are reluctant to follow a complete assessment of a biological control agent from discovery to field trials. Scientists are aware of this deficiency in research because of the increasing pressure to "publish or perish," the time necessary to develop a prototype formulation, and the years needed to evaluate the formulation in field or horticultural tests.

A considerable degree of reluctance for biological implementation also lies with industry. Of course, the main consideration for commercial development is the profit incentive. Does it pay for a company, especially a large one, to devote personnel, time, and expense to develop a product which may have specific and selective application? The majority of biocontrol products represent niche markets too small for development by large companies. This results in the potential for smaller companies to become involved in biocontrol technology and is exemplified in the type of company which formulate the products such as those that appear in Table 1. This leads to the question of "seed" money or capital expenditures for small companies or even money as an incentive to large companies. Is it in the interest of the federal government, for a variety of reasons, to invest in biocontrol technology? Is it imperative that a certain percentage of government dollars is devoted to the technology which improves the quality of our food stuffs and reduces environmental pollution?

An important consideration during the entire research and development process should be careful monitoring for quality control of the product (*96*). Simple, but well defined quality control assays should be implemented to determine viability, stability, and efficacy. These parameters should also be monitored during marketing to assure a high quality, efficacious product.

Another major problem which industry faces is "scale-up." This includes not only strain selection, delivery system, or compatibility with the minimal use of other pesticides, but, perhaps more importantly, the logistics involved in large-scale biomass production and formulation. It is apparent that industry must overcome severe problems which are associated with these processes.

Biological control systems consist of interactions among antagonists-pathogens-other soil microflora-hosts, compounding the problems of variability. Variability of efficacy is especially acute in field applications. These difficulties are less obvious in horticultural greenhouses where environmental extremes are less. These problems are not insurmountable with the application of advanced technology and with commitment and persistence.

**Literature Cited**

1. Papavizas, G. C. In *Proceedings of the 1984 British crop protection conference: Pests and disease*; Pressman, E. C., Ed.; British Crop Protection Council: Thornton Heath, Surrey, U.K., 1985, 371-378.
2. Lumsden, R. D.; Lewis, J. A. In *Biotechnology of Fungi for Improving Plant Growth;* Whipps, J. M.; Lumsden, R. D.; Eds., Cambridge University Press, Cambridge, U. K., 1989, 171-190.
3. Lewis, J. A. In *The Rhizosphere and Plant Gowth;* Keister, D. L.; Cregan, P. B.; Eds.; Kluwer Academic Publishers: Dortdrecht, The Netherlands, 1991, 279-287.
4. *Pest Management: Biologically Based Technologies;* Lumsden, R. D.; Vaughn, J. L., Eds.; American Chemical Society: Washington, D. C., 1993.
5. Bothast, R. J.; Schisler, D. A.; Jackson, M. A.; Vancauenberge, J. E.; Slininger, P. J. In *Pesticide formulations*; Berger, P. D.; Devisetty, B. N.; Hall, F., Eds.; ASTM: Philadelphia, PA, 1993, 45-46.
6. *Biological Control for Plant Diseases: Progress and Challenges for the Future*; Tjamos, E. C.; Papavizas, G. C.; Cook, R. J., Eds.; Plenum Press: New York, NY, 1992.
7. *New Directions in Biological Control: Alternatives for Suppressing Agricultural Pests and Diseases*; Baker, R. R.; Dunn, P. E., Eds.; Alan R. Liss, Inc.: New York, NY, 1990.
8. Cook, R. J. *Annu. Rev. Phytopathol.* **1993**, *31*, 53-80.
9. Bowers, R. C. In *Biological Control of Weeds with Plant Pathogens*; Charudattan, R.; Walker, H. L., Eds.; John Wiley and Sons: New York, NY, 1982, 157-173.
10. Churchill, B. W. In *Biological Control of Weeds with Plant Pathogens*; Charudattan, R. Walker, H. L., Eds.; John Wiley and Sons: New York, NY, 1982; 139-156.
11. Lisansky, S. G. In *Biological Pest Control*; Hussay, N. W.; Scopes, N., Eds.; Blandford Press: Poole, U. K., 1985, 210-218.
12. *Basic Technology*; Bu'lock, J.; Kristiansen, B., Eds.; Academic Press: New York, NY, 1987.
13. Stack, J. P.; Kenerley, C. M.; Pettit, R. E. In *Biocontrol of Plant Diseases*; Mukergi, K. C.; Garg, K. L., Eds.; CRC Press: Boca Raton, FL, 1988, Vol. 2; 43-54.

14. Lewis, J. A.; Papavizas, G. C. *Phytopathology* **1984**, *74*, 1240-1244.
15. Fravel, D. R.; Marois, J. J.; Lumsden, R. D.; Connick, Jr., W. J. *Phytopathology* **1985**, *75*, 774-777.
16. Papavizas, G. C.; Dunn, M. T.; Lewis, J. A.; Beagle-Ristaino, J. *Phytopathology* **1984**, *74*, 1171-1175.
17. Harman, G. E.; Jin, X; Stosz, J. E.; Peruzzotti, G.; Leopold, A. C.; Taylor, A. G. *Biol. Cont.*, **1991**, *1*, 23-28.
18. Cannel, E.; Moo-Young, M. *Process Biochem.* **1980**, *9*, 24-28.
19. Papavizas, G. C. *Annu. Rev. Phytopathol.* **1985**, *23*, 23-54.
20. Lemaire, J. M.; Alabouvette, C.; Davet, P. Tramier, R. *Symbiosis* **1986**, *2*, 287-301.
21. Carrizales, V.; Jaffe, W. *Interscience* **1986**, *11*, 9-15.
22. Powell, K. A.; Faull, J. L. In *Biotechnology of Fungi for Improving Plant Growth*; Whipps, J.; Lumsden, R. D., Eds.; Cambridge University Press: Cambridge, U.K., 1989, 259-276.
23. Lewis, J. A.; Papavizas, G. C. *Plant Pathol.* **1985**, *34*, 571-577.
24. Lewis, J. A.; Papavizas, G. C.; Lumsden, R. D. *Biocont. Sci. Technol.* **1991**, *1*, 59-69.
25. Polon, J. A. In *Pesticide Formulations*; van Valkenburg, W., Ed.; Marcel Dekker: New York, NY, 1973, 144-234.
26. Woodward, J. *Immobilized Cells and Enzymes*; IRL Press: Washington, D. C., 1985.
27. Lumsden, R. D.; Locke, J. C. *Phytopathology* **1989**, *79*, 361-366.
28. Knauss, J. F. *Florida Foliage,* **1992**, *18*, 6-7.
29. Papavizas, G. C.; Fravel, D. R.; Lewis, J. A. *Phytopathology* **1987**, *77*, 131-136.
30. Papavizas, G. C.; Lewis, J. A. *Plant Pathol.* **1989**, *38*, 277-286.
31. Nelson, E. B. *Plant Dis.*, **1988**, *72*, 140-142.
32. Harman, G. E.; Taylor, A. G. *Phytopathology* **1988**, *78*, 520-525.
33. Beagle-Ristaino, J.; Papavizas, G. C. *Phytopathology* **1985**, *75*, 560-564.
34. Davis, J. R.; Fravel, D. R.; Marois, J. J.; Sorensen, L. H. *Biol. Cult. Tests* **1986**, *1*, 18.
35. Brannon, P. M.; Backman, P. A. *Proc. Beltwide Cotton Prod. Res. Confer.* **1993**, 194-196.
36. Currier, T. C.; Skwara, J. E.; McIntyre, J. L. *Proc. Beltwide Cotton Prod. Res. Confer.* **1988**, 18-19.
37. McLoughlin, T. J.; Quinn, J. P.; Bettermann, A.; Bookland, R. *Appl. Environ. Microbiol.* **1992**, *58*, 1760-1763.
38. Tavonen, R.; Avikainen, H. *Agric. Sci.* **1987**, *59*, 199-208.
39. Dubos, B. In *Innovative Approaches to Plant Disease Control*; Chet, I., Ed.; John Wiley and Sons: New York, NY, 1987, pp 107-135.
40. Ricard, J. L. *Biocont. News Inf.* **1981**, *2*, 95-98.
41. Conway, K. E. *Plant Dis.* **1986**, *70*, 835-839.
42. Dunkle, R. L.; Shasha, B. S. *Environ. Entomol.* **1988**, *17*, 120-126.
43. Connick, W. J., Jr.; Boyette, C. D.; McAlpine, J. R. *Weed Sci.* **1991**, *41*, 678-681.

44. *Pesticide Formulations*; Berger, P. D.; Devisetty, B. N.; Hall, F. R., Eds; American Chemical Society for Testing and Materials: Philadelphia, PA, 1993.
45. Chet, I.; Sivan, A.; Elad, Y. *U.S. Patent No. 4,713,342;* **1987**.
46. Adams, P. B.; Ayers, W. A. *Phytopathology* **1982**, *72*, 485-488.
47. Lewis, J. A.; Papavizas, G. C. *Plant Pathol.* **1987**, *36*, 438-446.
48. Jones, R. W.; Pettit, R. E.; Taber, R. A. *Phytopathology* **1984**, *74*, 1167-1170.
49. Kerr, A. *Plant Dis.* **1980**, *64*, 25-30.
50. Backman, P. A.; Brannon, P. M.; Mahaffee, W. F. In *Improving Plant Productivity with Rhizosphere Bacteria*; Ryder, M. H.; Stephens, P. M.; Bowen, C. D., Eds.; CSIRO: Canberra, Australia, 1994, 3-9.
51. Rishbeth, J. In *Biology and Control of Soil-borne Plant Pathogens*; Bruehl, G. W. Ed.; American Phytopathological Society: St. Paul, MN, 1975, 158-162.
52. Korhonen, K.; Lipponen, K.; Bendz, M.; Johansen, M.; Ryen, I.; Venn, K.; Seiskari, P.; Niemi, M. *IUFRO Working Party Root and Butt Rots* **1993**, 54.
53. Kerr, A. *Agric. Sci.* **1989**, *2*, 41-48.
54. Pauwels, F. *Acta Horticult.* **1989**, *255*, 31-35.
55. Alabouvette, C.; Lemanceau, P.; Steinberg, C. *Pestic. Sci.* **1993**, *37*, 365-373.
56. Vésely, D. *Folia Microbiol.* **1987**, *32*, 502-506.
57. Lipa, J. J. In *Biological Pest Control: The Glasshouse Experience*; Hussey, N. W.; Scopes, N., Eds.; Blandford Press: Poole, U.K., 1985, 23-29.
58. Aloi, C.; Bergonzoni, P.; Arteconi, M.; Mallegni, C.; Gullino, M. L. *Atti Giorn. Fitopatol.* **1992**, *1*, 73-78.
59. Cohen, A. *Phytoparasitica* **1993**, *21*, 165-166.
60. Sivan, A.; Chet, I. *Phytopathol. Z.* **1986**, *116*, 39-47.
61. Chet, I.; Elad, Y. *Colloq. INRA* **1983**, *18*, 35-40.
62. Adams, P. B.; Fravel, D. R. *Phytopathology* **1990**,
63. Harris, A. R.; Schisler, D. A.; Ryder, M. H. *Soil Biol. Biochem.* **1993**, *25*, 909-914.
64. Harris, A. R.; Schisler, D. A.; Correll, R. L.; Ryder, M. H. *Soil Biol. Biochem.* **1994**, *26*, 1249-1255.
65. Kwok, O. C. H.; Fahy, P. C.; Hoitink, H. A. J.; Kuter, G. *Phytopathology* **1987**, *77*, 1206-1212.
66. Ogawa, K.; Komada, H. *Ann. Rev. Phytopathol. Soc. Japan* **1986**, *52*, 15-20.
67. McQuilken, M. P.; Whipps, J. M.; Cooke, R. C. *Plant Pathol.* **1990**, *39*, 452-462.
68. Wilson, M.; Lindow, S. E. *Phytopathology* **1993**, *83*, 117-123.
69. Scheffer, R. J. *Phytopathology* **1990**, *130*, 265-276.
70. Tari, P. H.; Anderson, J. *Appl. Environ. Microbiol.* **1988,** *54*, 2037-2041.
71. Loper, J. E. *Phytopathology* **1988,** *78*, 166-172.
72. Park, C. -S.; Paulitz, T. C.; Baker, R. *Phytopathology* **1988.** 78, 190-194.

73. Kloepper, J. W.; Leong, J.; Teintze, M.; Schroth, M. N. *Curr. Microbiol.* **1980**, *4*, 317-320.
74. Loper, J. E. In *New Directions for Biological Control: Alternatives for Suppressing Agricultural Pests and Diseases;* Baker, R. R.; Dunn, P. E., Eds.; Alan R. Liss: New York, NY, 1990, 735-747.
75. Nelson, E. B. In *The Rhizosphere and Plant Growth;* Keister, D. L.; Cregan , P. B., Eds.; Kluwer Academic Publishers: Dortdrecht, The Netherlands, 1991, 197-209.
76. Paulitz, T. C. *Phytopathology* **1991**, *81*, 1282-1287.
77. Marx, D. *Annu. Rev. Phytopathol.* **1972**, *10*, 429-454.
78. Burr, T. J.; Schroth, M. N.; Suslow, T. *Phytopathology* **1978.** 68, 1377-1383.
79. Chao, W. L; Nelson, E. B.; Harman, G. E.; Hoch, H. C. *Phytopathology* **1986**, *76*, 60-65.
80. Chang, I. P.; Kommedahl, T. *Phytopathology* **1968**, *58*, 1395-1401.
81. Hubbard, J. P.; Harman, G. E.; Eckenrode, C. J. *Can. J. Microbiol.* **1982**, *28*, 431-437.
82. Huber, D. M.; El-Nasshar, H,; Moore, L. W.; Mathre, D. E.; Wagner, J. E. *Biol. Fert. Soils* **1989**, *8*, 166-171.
83. Thomashow, L. S.; Weller, D. M.; Bonsall, R. F.; Pierson, L. S. III.; *Appl. Environ. Microbiol.* **1990**, *56*, 908-912.
84. Howell, C. R. *Phytopathology* **1991**, *77*, 992-994.
85. Taylor, A. G.; Min, T. -G.; Harman, G. E.; Jin, X. *Biol. Cont.* **1991,** *1*, 16-22.
86. Harman, G. E.; Taylor, A. G. In *Biological Control of Soil-Borne Plant Pathogens*; Hornby, D., Ed. CAB Ineternational: Oxford, UK., 1990, 415-426.
87. Lutchmeah, R. S.; Cooke, R. C. *Plant Pathol.* **1985**, *34*, 528-531.
88. Harman, G. E.; Taylor, A. G. *Phytopathology* **1988**, *78*, 520-525.
89. Callan, N. W.; Mathre, D. E.; Miller, J. B. *Plant Dis.* **1990**, *74*, 368-372.
90. Locke, J. C.; Marois, J. J. Papavizas, G. C. *Plant Dis.* **1985**, *69*, 167-169.
91. Marois, J. J.; Mitchell, D. J. *Phytopathology* **1981**, *71*, 1251-1256.
92. Wells, H. D.; Bell, D. K.; Jaworski, C. A. *Phytopathology* **1972**, *62*, 442-447.
93. Sivan, A.; Chet, I. *Phytopathology* **1989**, *79*, 198-203.
94. Ko, W. H. *Phytopathology* **1971**, *61*, 780-782.
95. Fravel, D. R.; Adams, P. B.; Potts, W. E. *Biocont. Sci. Technol.* **1992**, *2*, 341-348.
96. Mintz, A. S.; Walter, J. F. In *Pest Management: Biologically Based Technologies*; Lumsden, R. D.; Vaughn, J. L., Eds; American Chemical Society: Washington, D. C., 1993, 398-403.


RECEIVED January 31, 1995

# Chapter 12

# *Metarhizium anisopliae* for Soil Pest Control

**M. R. Schwarz**

**Miles, Inc., Miles Research Park, Stilwell, KS 66085**

The case history of BIO 1020, a mycelial granule formulation of *Metarhizium anisopliae*, illustrates the scientific and economic hurdles precluding commercial acceptance of entomopathogenic fungi for soil insect control. This formulation provides notable field activity against several economic soil pests. Application rates as low as 0.7 g/l soil result in conidial densities sufficient for high infectivity. Since *M. anisopliae* conidia have limited movement in soil, application strategies have been developed to aid in soil dissemination. Overall performance, however, has been inconsistent. The effects of soil microflora, temperature, moisture, pH, and density can be detrimental to granule sporulation, and conidial survival and infectivity. Storage stability of the granules is temperature dependent, with optimum viability maintained when granules are refrigerated. Some virulent *M. anisopliae* isolates can not be formulated as granules, and production, delivery, and storage costs are estimated to be relatively high. Successful formulation and delivery of this product must conform to an integrated set of specific criteria for field performance, but also to storage stability, production potential, registration, and profitability parameters.

The fungal species *Metarhizium anisopliae* is recognized as having potential for commercial exploitation. Products using natural strains of this fungus can be easily shipped from one part of the world to another with a minimum of regulatory constraints because it has a worldwide distribution, mainly in soil (9). A product of this fungus has widespread marketing possibilities since *M. anisopliae* attacks economically important insect pests from at least nine different insect orders (Table 1). The most predominant hosts are the beetles in the families *Curculionidae*, *Elateridae*, and *Scarabaeidae* (9). The fungus will infect most insect developmental life stages. It can be targeted for specific insect species because it demonstrates strain-specific host ranges. The

0097–6156/95/0595–0183$12.00/0

fungus is environmentally safe and poses no serious non-target effects. In addition, *M. anisopliae* lends itself readily to commercial production because it grows and sporulates on natural and selective nutrient substrates under controlled environmental conditions (1). Because of its potential for use in practical pest control, development of *M. anisopliae* insecticidal products began in 1985, and led to the experimental formulation, BIO 1020 (6).

Table 1. Some of the economically important insect orders containing hosts of the entomopathogenic fungus *Metarhizium anisopliae*

| |
|---|
| Coleoptera (beetles) |
| Diptera (flies, mosquitoes) |
| Hymenoptera (ants, wasps, bees) |
| Isoptera (termites) |
| Homoptera (cicadas) |
| Orthoptera (grasshoppers, cockroaches) |
| Hemiptera (true bugs) |
| Dermaptera (earwigs) |
| Lepidoptera (butterflies, moths) |

BIO 1020, a mycelial granule formulation, is one of the first commercially viable formulations of an entomopathogenic fungus. BIO 1020 is registered for use in some parts of Europe, and is under development in the United States and Japan. The purpose of this paper is to use the case history of BIO 1020 development to illustrate the general set of criteria, from proprietary position, production potential, storage stability, and biological activity, to field performance and market potential, necessary for commercial acceptance of entomopathogenic fungi for soil insect control.

A new, patented formulation process has been developed specifically for *M. anisopliae*. In this procedure, granules are produced by first growing fungal biomass in liquid fermentation under closely controlled conditions. The fungal biomass is then separated from the nutrient broth through centrifugation, and pelleted by passage through a rotating screen. The fungal pellets are then dried by slowly withdrawing water with a fluidized-bed dryer. Gentle and controlled removal of water induces the cells to enter a "resting state". The resultant granules are finally vacuum-sealed in plastic to retain viability and purity (1). Not all *M. anisopliae* strains, and other entomopathogenic fungi can be formulated using this process (Andersch, personal communication).

Unfortunately, the production process is fairly expensive. Production research and consultation with ten outside companies specializing in the manufacture of microbial products show that BIO 1020 will likely cost more than $50 per kilogram to the consumer. In addition, consumer cost might be substantially higher since production cost is inversely correlated with production volume, and BIO 1020 is anticipated to be a small volume product (Andersch, personal communication).

Nevertheless, the mycelial granule formulation of BIO 1020 has several physical characteristics which make it well suited for commercial pest control. The product is a dry, dust-free granule consisting solely of fungal mycelium. The "cross-linking" of the fungal hyphae in each granule enables it to resist high levels of mechanical stress. The formulation contains no extra carbon source which might stimulate the growth of soil microflora antagonistic to *M. anisopliae* after application. The granules are a uniformly round 0.5-0.8 mm diameter, and have a bulk density of 100 g/200 cc. The uniformity of the formulation gives it excellent flow characteristics and enables it to be easily applied through traditional granule-spreading equipment (1).

Finally, the vacuum-packaged granules show acceptable storage stability (Table 2), especially under refrigeration. Viability is maintained for 70 weeks at 4 C, although at 20 C, 100% viability is maintained only for 20 weeks (1). Therefore, the formulation partially conforms to commercial standards that specify no appreciable loss of viability for at least 12 months. However, the fungus would probably require special shipping and handling procedures to guarantee optimum viability. The granules remain in a biologically quiescent state during storage but can be reactivated by rehydration.

Table 2. Effect of storage temperature and duration of storage on the viability (LT100) of *M. anisopliae* granules (BIO 1020)

| Temperature | Duration of Storage |
|---|---|
| 20 C | 20 weeks |
| 15 C | 30 weeks |
| 10 C | 40 weeks |
| 4 C | 70 weeks |

(Andersch, personal communication)

The granules are "reactivated" after application to the soil, when the granules rehydrate and form conidiophores and subsequently conidia after 4-7 days relative to temperature. Each granule has the potential to produce more than one million conidia. Once reactivated, the granules must remain moist in order to complete conidial formation. Therefore, the granules are most suited for soil application where consistent moisture through natural or artificial means can be managed during the first several days following application.

Extensive testing of BIO 1020 granules shows that application rates at 1 g/liter soil under greenhouse or nursery conditions commonly produce conidial concentrations of between $1 \times 10^7$ to $1 \times 10^8$ conidia/g soil (Figs. 1 & 2). *M. anisopliae* conidial concentrations of at least $1 \times 10^6$ conidia/g soil are considered necessary to achieve commercially acceptable target pest mortality (4). The conidia can remain viable in the soil at infective levels for at least 10 months. The half-life of BIO 1020 conidia in soils under greenhouse or field nursery conditions is approximately 2 1/2 to 3 months.

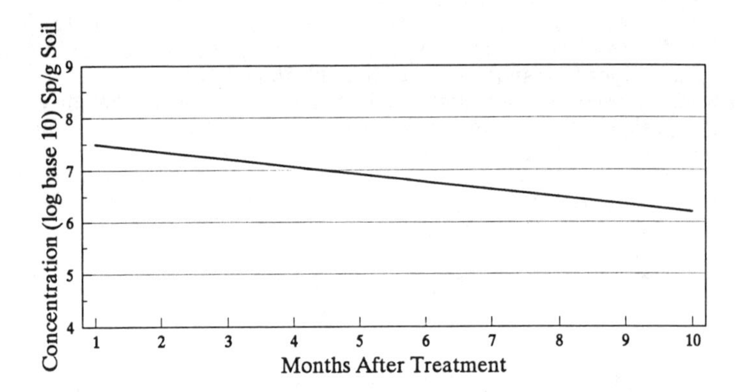


Figure 1. Residual concentration of Metarhizium anisopliae conidia, BIO 1020 strain, in greenhouse soils sampled in 1990. (Stenzel, personal communication)

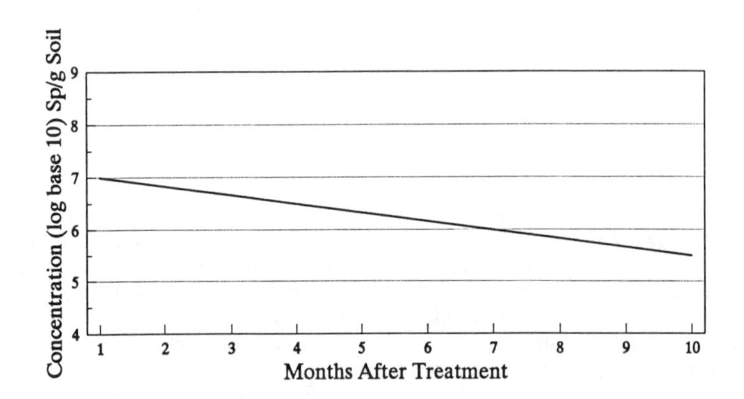


Figure 2. Residual concentration of Metarhizium anisopliae conidia, BIO 1020 strain, in nursery soils sampled in 1990. (Stenzel, personal communication)

After conidiation, the mycelial portion of the granules appear to be largely broken down by soil microbial action. Highly concentrated "pockets" of conidia remain at their site of hyphal formation. Depending on soil type, conidia do not move more than a centimeter or so from their point of origination unless the soil is disturbed. For example, at least 95% of conidia remain in the upper 3 cm of soil following broadcast applications of granules to the surface of soils found in small planted "roof gardens", turf, or nursery plant containers (Table 3). Since conidial movement is generally limited, application and post-application management strategies are being developed to optimize the likelihood that the target pest will come in contact with an infective dose of conidia.

The BIO 1020 formulation was first tested extensively in Europe against a wide variety of economically important insects. BIO 1020 was found to be highly active against black vine weevil (*Otiorhynchus sulcatus*). This pest was selected for further intensive study because it represented an economically important and growing pest problem in greenhouses and nurseries. Efficacy studies showed that the material could be mixed in the potting soil at the optimum rate of 1 g product/liter soil and provide efficacy of 65 to 80% control of *O. sulcatus* (6). In addition, it was found that control could be improved by mixing the desired amount of BIO 1020 granules in a sub-sample of soil ("premix"), allowing 3-7 days for sporulation, and then mixing the sub-sample with the final volume of soil to achieve the desired rate of 1 g/l soil (Fig. 3). This "premix" ensured closer regulation of environmental conditions necessary for optimum granular sporulation, achieved more uniform distribution of the conidia in the soil, and was less sensitive to low soil temperature effect on sporulation and infectivity of granules in soil (3) (Table 4). Premix preparations of BIO 1020 generally achieved 5-10% better control of black vine weevil compared to granules applied directly to the final volume of soil (Fig. 3).

Table 3. Efficacy of BIO 1020 against *Tenebrio molitor* and spore titer at different soil depths after broadcast application

| | Roof Garden | | Turf | | Container (7.5l) | |
|---|---|---|---|---|---|---|
| cm depth | % spores | % effic. | % spores | % effic. | % spores | % effic. |
| 0-3 | 96.1 | 100 | 95.2 | 100 | 99.2 | 82 |
| 4-7 | 1.5 | 38 | 4.8 | 40 | | |
| 8-10 | 1.5 | 38 | | | 0.4 | 0 |
| 11-14 | 0.9 | 0 | | | | |
| 15-20 | | | | | 0.4 | 0 |

(Stenzel, personal communication)

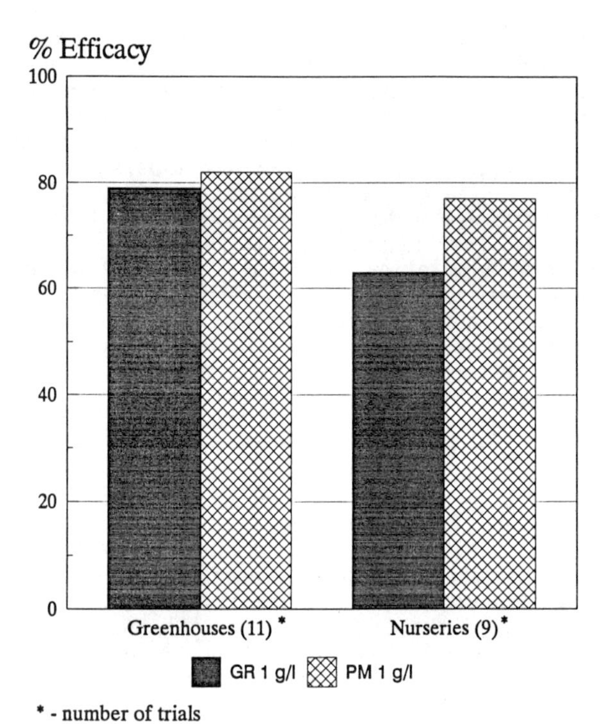


Figure 3. Average percent control of Metarhizium anisopliae, BIO 1020 strain, applied as formulated granules (GR) or as "premixed" soil (PM) against black vine weevil in commercial greenhouse and nursery conditions. (Reproduced with permission from reference 6. Copyright 1992.)

Table 4. Influence of temperature on activity of BIO 1020 granules and premixture applied at 1.0 g/l soil against *Tenebrio molitor*

| | Percent control of *T. molitor* | | | |
|---|---|---|---|---|
| | C Temperature | | | |
| | 4 | 10 | 15 | 18 |
| Granules | 0 | 3 | 70 | 100 |
| Premixture | 0 | 100 | 100 | 100 |

(Stenzel, personal communication)

The BIO 1020 formulation was also tested in the United States against black vine weevil. Since considerable losses to black vine weevil were found in perennial potted plants in outdoor nurseries, an application strategy was devised whereby "premix" was added over the top layer of soil already present in the pots. This "top dressing" would make possible treatment of second year potted perennials without having to repot each plant. BIO 1020 generally showed acceptable levels of control when the premix was applied as a "top dressing" (Table 5). In addition, other strains of the fungus, such as the MADA strain, first isolated from West Indian sugarcane root borer, *Diaprepes abbreviatus*, larvae found in a Florida citrus grove (obtained from Dr. C. W. McCoy, University of Florida), showed similar or superior activity to BIO 1020 when they were formulated according to BIO 1020 specifications.

Table 5. Number of Surviving Larvae of Black Vine Weevil (*Otiorhynchus sulcatus*) Larvae per Pot at 90 Days After Treatment

| Treatment | Rate (g Prod/L Soil) | % Black Vine Weevil Control |
|---|---|---|
| Untreated | | (27.0)[1] |
| BIO 1020 | 0.7 | 73 |
| BIO 1020 | 1.3 | 86 |
| MADA | 0.7 | 99 |
| MADA | 1.3 | 100 |

(M. Browning and S. Alm, U of R.I., personal communication)

[1] Number in parens indicates the number of surviving larvae

After the BIO 1020/MADA formulations were found to be effective at 1 g/l soil, a more critical look at market potential revealed the estimated cost per liter of treated soil at $0.05 to be high, but not unacceptable to growers. The volume of treated soil needed in greenhouses and nurseries could result in a grower's expenditure of several thousand dollars per acre under glass. In addition, the total market size for insecticidal products used against black vine weevil in nurseries and greenhouses was estimated to be only $1 million nationwide. Since there are effective products already available, the total market penetration by an expensive biological product could at best be about 50%, or $500,000 total return per year. Therefore, the return on investment was considered marginal in light of the registration costs, estimated to be approximately $250,000-500,000, and the special scale-up for production, packaging, and refrigerated storage and distribution of this biological product. For further development to be economically feasible, it is deemed necessary that several other larger volume markets, such as citrus root weevil or termite control, be identified in order to recoup research and development costs, and realize additional monetary returns.

The principle component of the citrus root weevil (CRW) complex is *D. abbreviatus*. It threatens all 92 million citrus trees grown on 791,000 acres of citrus in Florida, and represents a much larger market potential (approximately $6.5 million) than black vine weevil. *D. abbreviatus* causes economic losses by larval feeding on the root and crown areas of the citrus tree, eventually girdling and killing trees. Therefore, a comprehensive research program was initiated in 1992 to identify any environmental/cultural factors present in citrus groves which might adversely affect the performance of *M. anisopliae* formulations and isolates, and to develop IPM strategies for *M. anisopliae* products against CRW.

The conidiation kinetics of MADA and BIO 1020 granules were compared in seven soil types (including four from citrus groves) under different soil moistures, pH's, and microbial competition (2). Soil temperatures beneath citrus trees remain relatively stable throughout the spring and summer and were not considered an important variable. The MADA strain produced large amounts of conidia, measured as colony forming units (CFUs), between 1.9 X $10^3$ CFU/g soil in candler soil to 3.0 X $10^4$ CFU in potting soil, when applied at 1 granule per gram of soil (Table 6). MADA was not adversely affected by the normal microflora found in non-autoclaved citrus grove soils since these soils showed equivalent CFUs when they were autoclaved to remove the microbes. Soil pH had no adverse affect on conidiation from 5.0-8.0 (Table 7). The MADA strain sporulated in soils containing 2.5% to 30% moisture, with 5-10% being optimum (Table 8). Sporulation was slightly reduced in soil containing 30% versus 20% moisture. There were no CFUs produced in dry soils, with only 0-1.25% soil moisture. These investigations showed soil moisture was the single most critical factor for adequate sporulation of *M. anisopliae* granules. MADA demonstrated higher conidiation potential compared to BIO 1020 under the conditions of these tests (Table 8) and was subsequently selected for further testing in the field.

MADA was evaluated in a commercial citrus grove in the summer when weevils were most abundant. The fungus was applied at 1, 5, or 25 g active material/$m^2$ to small plot areas located under the drip-line of commercial citrus trees. The conidiation potential of each treatment was evaluated weekly by collecting soil samples from the upper 3 cm of soil in each plot and determining soil titer as MADA CFUs. MADA was distinguished from indigenous *M. anisopliae* strains using selective media and isozyme characterization (7). The soil titer of *M. anisopliae* was substantially increased by the addition of MADA at all rates (Table 9). Both the 5 g, and 25 g rate gave conidial titers which were up to log 3 higher (9.5 X $10^6$ and 1.7 X $10^7$, respectively) than pretreatment soils or non-treated check soils. The conidial populations remained at this higher level for at least 10 weeks post-treatment. These titers were within the range (1 X $10^6$ or higher/g soil) to cause economically acceptable larval mortality.

Table 6. Mean Number of CFU's Recovered from Autoclaved or Non-autoclaved Soil Inoculated with MADA Mycelial Granules [1]

| Soil Type | Autoclaved | Non-Autoclaved |
|---|---|---|
| Potting Soil | $3.0 \times 10^4 \pm 6.0 \times 10^3$ a | $4.5 \times 10^4 \pm 8.8 \times 10^3$ a |
| Copeland | $1.9 \times 10^4 \pm 4.7 \times 10^3$ a | $1.4 \times 10^4 \pm 2.7 \times 10^3$ a |
| Ona | $1.5 \times 10^4 \pm 4.7 \times 10^3$ a | $1.5 \times 10^4 \pm 5.3 \times 10^3$ a |
| Builders Sand | $1.6 \times 10^4 \pm 3.1 \times 10^3$ a | $5.9 \times 10^3 \pm 8.6 \times 10^2$ b |
| Homestead Muck | $6.7 \times 10^3 \pm 2.1 \times 10^3$ a | $1.1 \times 10^4 \pm 2.2 \times 10^3$ a |
| Myakka | $5.3 \times 10^3 \pm 2.4 \times 10^3$ a | $4.7 \times 10^3 \pm 1.2 \times 10^3$ a |
| Candler | $1.9 \times 10^3 \pm 7.4 \times 10^2$ a | $1.5 \times 10^3 \pm 5.3 \times 10^2$ a |

(Harrison, McCoy, Schwarz 1993)

[1] Means followed by the same letters between sterilization columns are not significantly different ($P \geq 0.05$, Duncan's Multiple Range Test)

Table 7. Mean Number of CFU's Recovered from Candler Soil Inoculated with MADA[a] at Different Soil pH's[b]

| Soil pH | Mean CFU's |
|---|---|
| 5.0 | $2.1 \times 10^6$ a |
| 5.5 | $1.3 \times 10^6$ a |
| 6.0 | $1.3 \times 10^6$ a |
| 6.5 | $1.9 \times 10^6$ a |
| 7.0 | $1.4 \times 10^6$ a |
| 7.5 | $0.6 \times 10^6$ a |
| 8.0 | $2.2 \times 10^6$ a |

(Harrison, McCoy, Schwarz 1993)

[a] 1 granule per gram of dry soil.
[b] Means followed by the same letters are not significantly different ($P \geq 0.05$, Duncan's Multiple Range Test).

Table 8. Mean Number of CFU's BIO 1020/MADA Recovered from Candler Soil Incubated for Three Weeks with 1 Granule/g Soil[1]

| Percent Soil Moisture | Isolate BIO 1020 CFU's/g of Soil | | Isolate MADA CFU's/g of Soil | |
|---|---|---|---|---|
| 0.00 | 0 | c | 0 | c |
| 1.25 | 0 | c | 0 | c |
| 2.50 | 3 ± 3 | bc | $3.0 \times 10^4 \pm 2.3 \times 10^3$ | b |
| 5.00 | 9 ± 3 | a | $4.0 \times 10^4 \pm 3.4 \times 10^3$ | a |
| 10.00 | 5 ± 1 | b | $4.0 \times 10^4 \pm 3.3 \times 10^3$ | a |
| 20.00 | 2 ± 2 | bc | $2.4 \times 10^4 \pm 1.6 \times 10^3$ | b |
| 30.00 | 0 | c | $9.7 \times 10^2 \pm 94$ | c |

(Harrison, McCoy, Schwarz 1993)

[1] Means followed by the same letters within columns are not significantly different ($P \geq 0.05$, Duncan's Multiple Range Test).

Table 9. Estimated Population Density of *M. anisopliae* in Upper Two Inches of Soil

| Treatment | Rate g/m$^2$ | Mean Number CFU's/cm$^3$ of soil; wks: pre(-) and post (+) treatment[a] | | | | | |
|---|---|---|---|---|---|---|---|
| | | -2 | -1 | +1 | +2 | +4 | +10 |
| M. anisopliae | 1 | $3.6 \times 10^3$ | $4.7 \times 10^3$ | $1.0 \times 10^6$ | $1.1 \times 10^6$ | $1.9 \times 10^5$ | $2.0 \times 10^5$ |
| M. anisopliae | 5 | $1.1 \times 10^3$ | $2.3 \times 10^4$ | $9.5 \times 10^6$ | $5.0 \times 10^6$ | $8.8 \times 10^6$ | $7.2 \times 10^5$ |
| M. anisopliae | 25 | $3.8 \times 10^2$ | $2.2 \times 10^4$ | $1.7 \times 10^7$ | $1.2 \times 10^7$ | $2.3 \times 10^7$ | $6.0 \times 10^6$ |
| Check | -- | $1.1 \times 10^3$ | $5.2 \times 10^3$ | $3.7 \times 10^4$ | $1.5 \times 10^4$ | $4.4 \times 10^4$ | $7.3 \times 10^3$ |

(McCoy, Harrison, Ferguson, and Schwarz 1993)

[a] Mean values based on four counts/replicate, 10 replicates per treatment.

The soil samples were also placed into vertical column bioassay units, neonate larvae were placed on the soil at the tops of the columns, and larval mortalities and mycoses determined after 10 days for the number of larvae successfully emerging from the soil at the bottom of the soil columns. MADA applications at 1, 5, and 25 g/m$^2$ provided 43.5, 61.2, and 73.3% larval mortality, respectively, at 1 week after application (Table 10). However, no treatment provided control at 2 weeks or later after application. Heavy rainfall between 1 and 2 weeks post-treatment resulted in saturated soils that had no effect on conidial density in soil but appeared to inhibit conidial infection of larvae. Saturated soils were thought to inhibit infection of larvae by conidia. A second series of field tests showed the same adverse affect by soil moisture on infectivity of conidia to *D. abbreviatus* larvae. Further field tests showed that infective conidial titers could be maintained through the application of mycelial granules, but saturated soil at any time during the control period might prevent further larval mortality. How long this inhibition period lasts in the soil is unknown.

Table 10. Percent mycosis to Neonatal Larvae of *Diaprepes abbreviatus* by Field Soil

| Treatment | Rate g/m$^2$ | Mean % larval mycosis; wks: pre (-) and post (+) treatment[a] | | | | | |
|---|---|---|---|---|---|---|---|
| | | -2 | -1 | +1 | +2 | +4 | +10 |
| M. anisopliae | 1 | 3.6 a | 12.7 a | 43.5 a | 0.0 a | 1.4 a | 0.0 a |
| M. anisopliae | 5 | 0.0 a | 5.2 a | 61.2 a | 6.5 a | 4.0 a | 0.0 a |
| M. anisopliae | 25 | 10.2 a | 12.0 a | 73.3 a | 8.6 a | 0.0 a | 3.7 a |
| Check | -- | 2.2 a | 1.2 b | 28.9 b | 3.3 a | 0.0 a | 1.6 a |

(McCoy, Harrison, Furguson and Schwarz 1993)

[a] Means followed by the same letters within a column are not significantly different (P = 0.01, Duncan's multiple range test).

A simple assessment of market feasibility using the effective rates for citrus and the projected cost to the consumer indicated that the product was not competitive with traditional pest control strategies. The citrus field trials demonstrated that a minimum of 5 $g/m^2$ of product was necessary for acceptable control, and the projected cost of MADA granules, at least $50/kg, coincided with a minimum cost to the grower of $1012/treated acre of citrus. This cost is at least five times the cost of conventional pest control methods. It is obvious at this juncture that a market of sufficient size to justify development costs, lower rates of product to keep costs competitive, and new techniques to improve the efficiency of the microbial component are not present in the combination necessary for successful development of a *M. anisopliae* product.

Development work for *M. anisopliae* products against several other key pests is underway against subterranean termites, *Reticulitermes flavipes*. Since the control of termites is a multi-million dollar market, there is hope that this market might provide the right combination of size and cost structure which would support a relatively high-cost product, and sufficient performance to justify development of *M. anisopliae*. Simultaneously, new insecticides, such as imidacloprid representing the new class of chloronicotinyl insecticides, and several pyrethroid compounds, are being developed to join the organophosphate compounds already registered for use to control termites. When imidacloprid, cyfluthrin, and other compounds were tested for efficacy alone at sublethal doses, termite workers reduced or stopped their normal feeding behavior and/or noticeably altered their intricate social behavior. In sterile soils, the termites then either recovered, or probably starved following exposure. In non-sterile soils, indigenous entomogenous fungi, such as *M. anisopliae* and *Beauveria bassiana*, caused higher mortality than would normally be observed (8). For example, *M. anisopliae* at the rate of only 1 X $10^3$ conidia/g soil gave 97% mortality when the termites had fed on imidacloprid-treated baits, compared to 13% mortality when the termites had fed on non-treated baits (Table 11). This improved control was observed to various degrees when other chloronicotinyl, organophosphate, and pyrethroid insecticides were integrated with the entomogenous fungi and bacteria. The same general response was noted when the termite colonies were deprived of food, or were in other ways placed under stress.

Table 11. Interaction in termite activity in bait application of Imidacloprid and *Metarhizium anisopliae*

| No. of Spores per Gram of Soil | % Mortality after 10d | |
|---|---|---|
| | Plain Bait | Bait treated with 0.001% Imidacloprid |
| $10^7$ | 95 | 100 |
| $10^6$ | 53 | 97 |
| $10^5$ | 45 | 85 |
| $10^4$ | 23 | 95 |
| $10^3$ | 13 | 97 |
| 0 | 0 | 27 |

(B. Monke, personal communication)

The projected cost for a *M. anisopliae* product in the termite market is more favorable. Based on a projected MADA rate of 1 g/l, it would cost about $250 in fungal materials to treat 5000 liters of soil, enough to treat a 6 by 12-inch trench around a house containing roughly 2000 square feet. This does not include costs for application or for other traditional insecticides. Field tests using chemical and biological combinations are currently underway and a market decision is expected in the near future.

In conclusion, most of the important parameters necessary for a commercially feasible biological control product have been met by BIO 1020/MADA formulations of *M. anisopliae*. The product(s) is effective against economic pests such as black vine weevil, citrus root weevil, and perhaps termite.

The formulation technology will be secured in a broad-ranging patent that covers many types of applications and entomogenous fungi. It demonstrates acceptable shelf life, suitable for refrigerated storage and distribution through normal distributor networks with minor adaptations. Finally, the formulation appears to be compatible with existing technologies and management strategies.

However, the two most important parameters necessary for a commercially feasible biological control product have not been met. The relatively high cost of production has crippled the product potential in large markets by rendering it too expensive for broadcast applications at effective rates (i.e. citrus root weevil). In addition, for uses which show more favorable cost/effectiveness, such as black vine weevil, the markets are too small to realize any clear return on investment.

Finally, the termite work suggests a more general conclusion, that is, looking at microbial control agents simply as a one-on-one system, to control the target pest, may seldom provide a favorable economic picture for the biological component alone. All evolving strategies for successful pest control, therefore, must look at integrating chemical, and non-chemical methods.

## ACKNOWLEDGEMENTS

The author thanks J. M. DeAngel for manuscript preparation and Drs. S. Alm, W. Andersch, R. Harrison, C. McCoy, B. Monke, K. Stenzel, and A. Terranova for their contributions to this manuscript.

## LITERATURE CITED

1. Andersch, W. 1992. Production of Fungi as Crop Protection Agents. Pflanzenschutz-Nachrichten Bayer 45:128-141.
2. Harrison, R. D., McCoy, C. W., and Schwarz, M. R. 1993. Effect of Different Abiotic Factors on Conidiation of *Metarhizium anisopliae* Mycelial Granules in Soil. Proceedings of the Society for Invertebrate Pathology, Ashville (Abstract).
3. Hartwig, J., and Oehmig, S. 1992. BIO 1020 - Behavior in the soil, and Important Factors Affecting its Action. Pflanzenschutz-Nachrichten Bayer 45:159-176.
4. McCoy, C. W. 1990. Entomogenous Fungi as Microbial Pesticides. In: New Directions in Biological Control: Alternatives for Suppressing Agricultural Pests and Diseases (eds R. R. Baker and P. E. Dunn) pp. 139-159.

5. McCoy, C. W., Harrison, R.D., Ferguson, J. S., and Schwarz, M. R. 1993. Integration of Entomopathogenic Fungi into an IPM Strategy for Citrus Root Weevils. Proceedings of the Society for Invertebrate Pathology, Ashville (Abstract).
6. Stenzel, K. 1992. Mode of Action and Spectrum of Activity of BIO 1020. Pflanzenschutz-Nachrichten Bayer 45:143-158.
7. Terranova, A. C., McCoy, C. W., Harrison, R.D., Ferguson, J. S., and Schwarz, M. R. 1993. Protocol for Monitoring the Fate of Artificially Applied Entomopathogenic Fungi in Field Soil Via Isozyme Characterization. Proceedings of the Society for Invertebrate Pathology, Ashville (Abstract).
8. Zeck, W. M., and Monke, B. J. 1992. Synergy in Soil and Bait Applications of Imidacloprid with Entomophageous Fungi in Termite Control. Abstract. 1992 Annual Meeting of the Entomological Society of America.
9. Zimmermann, G. 1992. *Metarhizium anisopliae* - An Entomopathogenic Fungus. Pflanzenschutz-Nachrichten Bayer 45:113-128.

RECEIVED January 31, 1995

# Chapter 13

# Formulation of Entomopathogenic Nematodes

**R. Georgis, D. B. Dunlop, and P. S. Grewal**

**biosys, 1057 East Meadow Circle, Palo Alto, CA 94303**

Entomopathogenic nematodes in the genera *Steinernema* and *Heterorhabditis* are commercially available for the control of soil-inhabiting insects. Stable formulations have been achieved by immobilizing and/or partially desiccating infective stage (IJ) nematodes. These formulations allowed introduction of nematode products with acceptable shelf-life into various market segments. A breakthrough in nematode formulation was accomplished with the development of a unique water dispersible granular formulation that allows nematodes to enter into a hydrobiotic state extending nematode survival and pathogenicity for up to 6 months at 4-25°C and up to 8 wks at 30°C. This formulation is easy and quick to apply and is well suited for a wide variety of agricultural and horticultur applications.

Entomopathogenic nematodes (Steinernematidae and Heterorhabditidae) are attractive alternatives to chemical pesticides. Desirable attributes such as ease of mass-production, efficacy comparable to most insecticides in favorable habitats, and safety to non-target organisms have invoked commercial interest in these parasites. Significant progress achieved in the last 5 years in liquid culture and application and formulation technology has strengthened the position of nematode-based products in the marketplace (2,3). Marketing and/or research agreements between nematode producers and agrochemical companies and distributors have given nematode products a world wide recognition (Table I).

In the USA, Japan, Canada and Western Europe, successful market introduction was only achieved after steinernematid-based products were proven comparable with chemical insecticides based on cost and ease of application. Since the technology of liquid fermentation and formulation stability of

0097–6156/95/0595–0197$12.00/0

heterorhabditids lags behind, their cost is 1-3 times higher than steinernematids and chemical insecticides (3).

The last 5 years also had its share of disappointments. The development efforts to establish efficacious data against corn rootworms (*Diabrotica* spp.), root maggots (*Delia* spp.) and wireworms (Elateridae) were unsuccessful (3).

Table I. Major Products and Formulations of *Steinernema* and *Heterorhabditis*-Based Products

| Formulation | Nematode Species | Product | Company/Distributor |
|---|---|---|---|
| Alginate gel | *S. carpocapsae* | Exhibit | Ciba-Geigy, USA, W. Europe |
| | | Sanoplant | Dr. R. Maag, Switzerland |
| | | Boden-Nützlinge | Celaflor, Germany |
| | | BioSafe | Pan Britanica Ind., United Kingdom |
| | | | Rhone-Poulenc, Italy |
| Clay | *H. bacteriophora* | Otinem | Ecogen, USA |
| | *H. megidis* | Nemasys-H | A.G.C., UK |
| | | Larvanem | Koppert, Holland |
| | *S. feltiae* | Nemasys | A.G.C., UK |
| | | Entonem | Koppert, Holland |
| | *S. scapterisci* | ProAct | BioControl, USA |
| Flowable gel | *S. carpocapsae* | BioVector | biosys, USA |
| | | BioSafe | SDS Biotech, Japan |
| | *S. feltiae* | Stealth | Ciba-Geigy, Canada |
| | | Exhibit | Ciba-Geigy, USA, W. Europe |
| | | Magnet | Amycel, USA |
| Water dispersible granule | *S. carpocapsae* | Vector T&L | Lesco, USA |
| | | BioFlea Halt | Farnam, USA |
| | | Interrupt | Farnam, USA |
| | | Defend | Pittman-Moore, USA |
| | | BioSafe | biosys, USA |
| | | Vector PCO | V.W. & Rogers, USA |
| | *S. riobravis* | Vector MC | Lesco, USA |

All life stages of these nematodes occur in the insect host, except the infective third-stage juvenile (IJ). The infective juveniles locate and penetrate into the insects, and release mutualistic bacteria *Xenorhabdus* or *Photorhabdus* into the hemocoel. The bacteria multiply killing the insects within 24 - 48 hours. The nematodes feed on the bacteria and host tissues, reproduce, and produce IJs after two or three generations. The IJs leave the cadavers hosts host after the depletion of their nutrients, seeking new insects. Depending on the nematode species and the insect host, the life cycle is completed in 6 - 14 days at 20-28°C. The duration and the life cycle of the nematodes in the liquid culture is similar to that in insects. At present, effective production of steinernematids has been achieved in 15,000-80,000 liter fermenters with a yield capacity as high as 150,000 IJs per $cm^3$.

## Formulation

Formulation of nematodes serves two main purposes. One it extends shelf-life of the product, and second, it enables easy transport, delivery and application.

Limited shelf-life of infective stage nematodes is a major obstacle in expanding their commercial potential. Factors affecting storage stability of nematodes are least understood. Holding large volumes of nematode suspensions in tanks has met with numerous problems including contamination. Theoretically, shelf-life of the non-feeding infective juveniles would be a function of stored energy and rate of its utilization. Lipid is a major energy reserve for non-feeding infective stages (8). We found that initial lipid level of nematodes have a direct impact on shelf life. The rate of utilization of stored energy depends upon many factors such as temperature, environmental stress, and activity. Behavior of nematodes during storage in water suspensions differs among species. *S. carpocapsae* and *S. scapterisci* are less active during storage, but *S. glaseri, S. feltiae,* and *H. bacteriophora* are highly active. This behavior has a direct influence on energy burn-rate, and therefore, impacts nematode shelf-life (9). However, we have also observed that nematode batches with lower lipid content were generally more pathogenic,suggesting that trade-offs may exist.

A large number of formulations have been developed for nematodes. In most of these formulations, the nematode movement is restricted to preserve stored energy (Table II). Generally, handling, shipping and application of large quantities of product is suitable in the high and medium value crops such as mushrooms, berries, artichokes, citrus, mint, and turfgrass. However, more stable formulations are needed for the nematodes to become commercially competitive with chemical insecticides in most of these markets and in low-value traditional agricultural markets such as cotton and corn. In this regard, recently a breakthrough in nematode formulation was achieved with the development of water dispersible granular formulation that allows the nematodes to enter partially into an anhydrobiotic state extending nematode survival and pathogenicity for up to 6 months at 4-25°C and up to 8 wks at 30°C (Table II). This formulation is scaleable and is easy to apply without any time consuming preparation steps (Table III). The formulation is suited for a wide variety of consumer, agricultural and horticultural applications.

The successful market acceptance of the nematode-based formulations will depend greatly on their consistent performance under field conditions. Therefore, it is important to maintain nematode quality throughout all stages of product development. The first step in standardization is aimed at obtaining reliable and consistent nematode production. Inoculum batches from *in vivo* cultures are produced from stocks of nematode strains that are stored by cryopreservation (8) to minimize variation in nematode pathogenicity between various production lots (3). Subsequent steps are focused on maintaining the viability and pathogenicity of the nematodes immediately after the nematodes are harvested from the fermenter until the product is applied by the end-user. To assure this process, $LT_{50}$ (the time needed to kill 50% of test insects) performance standards and optimum lipid contents of infective stage nematodes have been determined and are used to measure product stability. To assure stable products, nematodes are stored in large aerated tanks and are formulated within 1-3 months of the completion of production (Table IV).

Table II. Description and Storage of Major *Steinernema* and *Heterorhabditis*-Based Products

| Formulation and Description | Product Storage[1] 20-25°C | 4-10°C |
|---|---|---|
| Alginate gel[2] 20X10$^6$ nematodes (in 0.5 liter container) or 250X10$^6$ (in 4 liter container) trapped into a gel matrix and coated on a mesh screen | 3-5 mo | 6 mo |
| Clay[3] 60X10$^6$ nematodes spread on 80 g clay | 0 | 3 mo |
| Flowable gel[3] Up to 1X10$^9$ nematodes suspended in a gel matrix enclosed in a special film (18cm$^2$). Usually 4 to 6 films are enclosed in 20 cm$^3$ container | 4-6 wk | 3 mo |
| Water dispersible granule[3] 100X10$^6$ nematodes (formulated in 350 g) or 250X10$^6$ nematodes (formulated in 680 g) of granule material (each approximately 5 mm diameter) enclosed in 700 ml and 1200 ml container, respectively | 6 mo | 6 mo |

[1] Based on product labels.
[2] Form of appliction is aqueous spray after dissolution of the alginate gel with sodium citrate.
[3] Form of application is aqueous spray after placement of the formulation in the spray water.

Table III. Characteristics of *Steinernema carpocapsae* - Water Dispersible Granular Formulation

| Character | Data |
|---|---|
| Stability[a] | up to 6 months 4-25°C<br>up to 8 wks at 30°C<br>6 days at 36°C<br>2 days at 38°C |
| Ease of Use | Dissolve quickly in water<br>Compatible with most agrochemicals<br>Compatible with most commercial sprayers |
| Product Size/Coverage | Adequate for use in various market segments<br>350 gm (100X10$^6$ nematodes) treats up to 500 m$^2$<br>Size comparable to chemical products<br>Minimal product/packaging disposal requirements |
| Cost<br>3.39/50m$^2$<br>$1.36/100m$^2$ for | Competitive. For example BioSafe (Solaris Ortho) costs $2.99-<br>compared to $1.80-2.50/50m$^2$ for Dursban (DowElanco); and<br>Exhibit (Ciba) compared to $0.86 and $0.65-1.30/100m$^2$ for<br>Triumph (Ciba) and Oftanol (Miles), respectively. |

[a] Nematode viability over 90% with stable pathogenicity

Another aspect of product assurance is the timing of production according to the market need. Most of the production is accomplished from January to March for products needed from May to August. However, for August to December markets, nematodes are produced from March to June. Certainly, such considerations are dependent on the nematode species, formulation type, storage requirements, market forecast and the distribution channels.

## Application Technology

Steinernematids and heterorhabditids have been proven efficacious against various soil inhabiting insects (Table V). These nematodes differ in virulence to specific hosts, tolerance to adverse environmental conditions, ability to seek out hosts, and behavior in the soil (5,6). Based on these characteristics, efforts made in recent years have led to matching optimal strain or species in a particular habitat against a particular insect species.

Many factors affect our ability to place quantities of nematodes on or in close proximity to the target host in order to produce optimal results at the lowest possible cost. To compensate for the impact of abiotic and biotic factors on nematode efficacy and persistence, the inundative application of a high concentration of a specific nematode species (approximately $2.5X10^9$ - $7.5X10^9$ infectives/ha) is needed (Table VI). Certainly careful considerations to optimal strain, irrigation requirements, timing of application, and method of application are needed to achieve predictable control (4).

Table IV. Storage and Shipping Requirements for Steinernematid-Based Products

| Steps in product development and distribution | Storage Parameters | |
|---|---|---|
| | Condition | Period |
| Post-harvest | 4-10°C | 1-3 months |
| Formulation | 4-8°C | 3-6 months |
| Shipping | Non-refrigeration | 2-15 days |
| Distributor and/or end user<br>(Product Storage) | 4-10°C<br>20-25°C | 3-6 months<br>See Table II |

As with chemical insecticides, spraying nematodes as a curative treatment directly onto the soil surface is the most commonly used application method. This method is quick, simple, and provides good coverage. In some situations, nematodes have demonstrated control potential when applied at planting time. In certain soil environments where the target stage does not appear until 4-6 wk after planting (e.g., corn rootworm, *Diabrotica* spp.), insect damage was not prevented by nematodes applied at planting. However, efficacy improved when nematodes were applied 4 wk after planting (10).

When the duration of nematode persistence is less than the time period during which a pest can damage a crop, multiple nematode applications at intervals of two or more weeks may be necessary to achieve the desired level of control. This is especially important for insects such as mole crickets (*Scapteriscus* spp.) and fungus gnats (Sciaridae) that have multiple overlapping generations. The number of applications needed is the subject of current research.

Table V. Major Pests for Commercially Available *Steinernema* and *Heterorhabditis* species in Western Europe, Japan, and North America

| Segment | Nematode Species[a] | Common Name | Scientific Name |
|---|---|---|---|
| Artichoke | Sc | Artichoke plume moth | *Platyptilia cardiuidactyla* |
| Citrus | Sc, Sr | Blue green weevil | *Pachnaeus litus* |
| | | Sugarcane rootstalk borer | *Diaprepes abbreviatus* |
| Cranberry and berries | Sc,Hb,Hm | Black vine weevil | *Otiorhynchus sulcatus* |
| | Sc | Cranberry girdler | *Chrysoteuchia topiaria* |
| | Sc | Crown borers | Sesiidae |
| | Sc,Hb,Hm | Strawberry root weevil | *O. ovatus* |
| Fruit Trees | Sc | Stem borers | Sesiidae |
| Greenhouse and Nursery Plants | Sc,Hb,Hm | Black vine weevil | *O. sulcatus* |
| | Sf | Sciarid flies | Sciaridae |
| | Sc | Stem borers | Sesiidae |
| | Sc, Hb, Hm | Strawberry root weevil | *O. ovatus* |
| Mint | Sc | Cutworms | Noctuidae |
| | | Mint flea beetle | *Longitarsus waterhousei* |
| | | Mint root borer | *Fumibotys fumalis* |
| | | Black vine weevil | *O. sulcatus* |
| Mushroom | Sf | Sciarid fly | *Lycoriella* spp. |
| Sugar beet | Sc | Sugar beet weevil | *Cleonus mendikus* |
| Turf | Sc | Armyworm | *Pseudaletia unipuncta* |
| | | Billbugs | *Sphenophorus* spp. |
| | | Black cutworm | *A. ipsilon* |
| | | Bluegrass webworm | *Parapediasia teterrella* |
| | | European crane fly | *Tipula paludosa* |
| | | Japanese lawn cutworm | *Spodoptera depravata* |
| | Sr, Ss | Mole crickets | *Scapteriscus* spp. |
| | Sg, Hb | White grubs | Scarabaeidae |
| Vegetable and field crops | Sc | Cutworms | Noctuidae |
| | | Cucumber beetles | Chrysomelidae |
| | | Flea beetles | Chrysomelidae |
| Pet/Vet | Sc | Cat flea | *Ctenocephalides felis* |

[a]: Sc = *S. carpcapsae*, Sr = *S. riobravis*, Sg = *S. glaseri*, Ss = *S. scapterisci*, Sf = *S. feltiae*, Hb = *H. bacteriophora*, Hm = *H. megidis*

Nematodes should be applied to moist soil. Post-application irrigation and continuous moderate soil moisture are essential for nematode movement, persistence, and pathogenicity (2,4), and for nematodes to achieve insect control to a level comparable to standard insecticides (3). Although nematodes are recommended to be applied during early morning or evening to avoid the effects of ultraviolet radiation and temperature extremes, in many situations nematodes can be applied at any time of the day as long as post-application irrigation is employed within 30 minutes (3,4).

Low temperature limits the pathogenicity of steinernematids and heterorhabditids, either by its influence on the activity of the nematode, the bacterial symbiont, or both. In the field, soil temperatures below 12-14°C resulted in unsuccessful insect control (3).

As with most soil pesticides, a spray volume of 750 - 1890 liter/ha is usually required for most nematode species to reach the depth occupied by the target insect. Nematodes can be applied to the target zone with almost any commercially available spray equipment (Table VI). These include small pressurized sprayers, mist blowers, electrostatic sprayers, as well as aerial application via helicopters. In addition, nematodes are commonly applied using drip and sprinkler irrigation systems. Pressures of up to 1068 kPa have no detrimental effect on nematodes. The nematodes can pass easily through sprayer screens with openings as small as 100 microns in diameter.

Table VI. Primary Equipment Systems Used to Apply Nematode-Based Products

| System | Segment | Treatment Site | Dosage |
|---|---|---|---|
| Overhead Irrigation | Berries | Broadcast | $2.5X10^9$-$7.5X10^9$/ha |
| Helicopter | | Broadcast | $7.5X10^9$/ha |
| Drip Irrigation | | Furrow | $2.5X10^9$-$5.0X10^9$/ha |
| Microjet Irrigation | Citrus | Spotted | $2.0X10^6$/tree |
| Ground Sprayers | Turfgrass | Broadcast | $2.5X10^9$/ha |
| Ground Sprayers | Mint | Furrow | $2.5X10^9$/ha |
| Hose-end Sprayer | Mushrooms | Broadcast | $0.5X10^6$-$1.0X10^6/m^2$ |
| Sprinkler Irrigation | Greenhouse and | Broadcast | $7.5X10^9$/ha |
| Ground and Hand Sprayers | Nursery Plants | Pot | $2X10^4$-$5X10^4$/4 liter pot |

Numerous field data generated over the last decade showed that nematodes can protect crops from insect damage, thus, are not slow biological control agents. Insects such as white grubs (Scarabaeidae) and root weevils (Curculionidae) are controlled successfully within 2-4 wk post-application, whereas, 3-7 d is generally sufficient for the control of lepidopterans.

The compatibility of steinernematids and heterorhabditids with various chemical pesticides is a major concern when considering their inclusion in integrated pest management systems. Nematodes can be mixed safely with commercial preparations of *Bacillus thuringiensis*, pyrethroids, and various pesticides and fertilizers. Some pesticides can adversely affect nematodes, however they still can be used together if nematodes are applied before the pesticide or vice versa, thus allowing time for the pesticide to become absorbed or degraded to a level non-toxic to the nematode. To date, several successful attempts have been made to further increase nematode efficacy when employed in conjunction with chemical and microbial agents. (2,7).

## Conclusion

The quality of commercial nematode products is critical if insect-parasitic nematodes are to realize their full potential as biological insecticides. The stability and ease of use of dispersible granular formulation and the excellent quality of nematodes grown in liquid culture are significant steps towards this goal. All commercial formulations including water dispersible granular have been developed to maintain product stability during storage and transportation, and they are applied as a spray in water against the target pest. Granular, capsules and bait pellets (1,2) that can be applied by aircraft and standard granular applicators that protect and (or) release nematodes in the soil are also desirable and worth further investigation.

Steinernematid and heterorhabditid species and strains differ in virulence, tolerance to adverse environments and behavior in soil. Therefore, future markets will require the introduction of various species and stable formulation to optimize and to expand the market potential of nematode-based products. Additionally, it is important to emphasize that the increased use of nematode products will require a change in the attitudes and behavior of the technical advisers and the end users. For example, managers will need to focus more attention on monitoring of pest populations and timing applications. The key to successful marketing and acceptance of nematode products will depend largely on how well producers and distributors response to consumer needs, anticipate and react to the changing environment, and develop quality products to solve their problem.

Companies involved in commercialization of these nematodes are focusing on increasing their market shares as well as attempting to introduce nematode products against insects of large volume - low value markets such as corn and cotton. To achieve this goal, research is directed towards optimizing the production process, improving the application technology (e.g., timing and method of application), and utilizing more virulent nematode species/strains (through genetic manipulation or natural isolation) that may strengthen the position of the nematode-based products in the marketplace (2,7). Genetic engineering may be the solution for the development of an optimum formulation by inducing the nematodes to enter into a full anhydrobiotic state.

## LITERATURE CITED

1. **Connick, W.J., Jr., Nickle, W.R. and Vinyard, B.J.** J. Nematol. 1993, 25, 198-203.
2. **Gaugler, R. and Kaya, K.K.** Entomopathogenic nematodes in biological control. CRC Press. Boca Raton, Florida, 1990.
3. **Georgis, R.** Biocont. Sci. and Tech. 1991, 2,, 83-99.
4. **Georgis, R. and Gaugler, R.** J. Econ. Entomol. 1991, 84, 713-720.
5. **Grewal, P.S., Lewis, E.E., Gaugler, R. and Campbell, J.F.** Parasitology. 1994, 108, 207-215.
6. **Grewal, P.S., Selvan, S. and Gaugler, R.** J. Therm. Biol. 1994, (in press).
7. **Kaya, H.K. and Gaugler, R.** Annu. Rev. Entomol. 1993, 38, 181-206.
8. **Popiel, I. and Vasquez, E.M.** J. Nematol. 1991, 23, 432-437.
9. **Selvan, S., Gaugler, R. and Campbell, J.F.** J. Econ. Entomol. 1993, 86, 353-360.
10. **Wright, R.J., Witkowski, J.F., Echtenkamp, G. and Georgis, R.** J. Econ. Entomol. 1993, 86, 1348-1354.

RECEIVED January 31, 1995

# Foliar Biorationals

# Chapter 14

# Pheromone Formulations for Insect Control in Agriculture

**Janice Gillespie[1], Scott Herbig[2], and Ron Beyerinck[2]**

**[1]Consep, Inc., 213 S.W. Columbia Street, Bend, OR 97702–1013**
**[2]Bend Research, Inc., 64550 Research Road, Bend, OR 97701–8599**

Pheromones are useful for insect control in Integrated Pest Management (IPM) programs. Most of the semiochemicals that have been commercialized are sex pheromones for pests of cotton, specialty row crops, and orchard crops. Two major classes of formulations have been registered: 1) hand-applied dispensers, and 2) sprayable formulations. The hand-applied dispensers typically provide long durations of efficacy but can be inconvenient to apply to large acreages. Sprayable formulations can be applied quickly and to large areas using conventional ground or aerial equipment, but these formulations typically have shorter durations of efficacy (weeks versus months). Efficacy is the primary goal in the design of all pheromone products. Optimizing formulations to minimize cost and making their use economically compelling, rather than just economically competitive with conventional insecticides, is the key to acceptance and use of pheromone-based products.

Many products containing insect pheromones have been introduced in recent years for control of insect pests in agricultural, household and, to a lesser extent, forestry applications. The active ingredients (AI) of most of these products are the sex pheromones of the Lepidoptera (moths, skippers, and butterflies). Sex pheromones are biological chemicals emitted by an individual insect to attract a receiving, conspecific individual of the opposite sex for mating. Products based on alarm pheromones of the Homoptera (aphids and their relatives) and aggregation pheromones of Coleoptera (beetles)—as well as naturally occurring chemicals that function as insect attractants—also have been developed and/or commercialized in these markets.

0097–6156/95/0595–0208$12.00/0

Major uses of pheromone-based products developed since the early 1970s include insect survey, detection, and monitoring in Integrated Pest Management (IPM) programs, insect eradication, and insect control.

## Pheromone Products for Survey, Detection, Eradication, and Monitoring

State and federal agencies are using pheromone-baited traps for the survey, detection, and eradication of exotic pests. Applications include the detection of Japanese beetles (*Popillia japonica* Newman) in certain western states, and the identification of locations requiring insecticide treatment in areawide eradication programs. For example, an eradication program in the southeastern United States targets boll weevil (*Anthonomus grandis* Boheman) (*1*) and an eradication program in the southwestern United States targets pink bollworm (*Pectinophora gossypiella* Saunders).

Products for monitoring the presence, growth, and development of damaging insect populations are available for major insect pests in row, vegetable, and tree fruit and nut crops. Often the monitoring products are used to establish the time of emergence of the overwintered pests (Biofix, the time that the first moth of the season is captured in pheromone-baited traps) and to initiate heat-unit-based population models useful for accurately targeting insecticide applications.

Products in the household pest-control market assist homeowners in identifying home and garden pests, in removing nuisance pests such as yellowjackets, and in planning appropriate control procedures for garden pests. These products can also be used by the structural pest-control industry to locate infestations in the home environment.

## Products for Control of Pests

Certain descriptive terms are useful to classify the principal mode of action of commercial pheromone products for control of insect and mite pests. These are 1) mating disruptants, 2) attracticides, 3) bioirritants, and 4) deterrents.

Mating disrupants, which act by permeating the crop environment with sex pheromone to prevent reproduction and reduce infestations, have been and continue to be a primary target of commercial R&D efforts (*2,3*). Mating-disruptants are designed to flood the crop environment with a synthetic replica of the sex pheromone, thereby preventing the sexes from locating each other and mating. Reduced reproduction rates prevent or delay growth of the population to the economic threshold and eliminate or significantly reduce the need to use conventional insecticides.

Some attracticides also control insect populations through disruption of mating (*4*). In these products, a small amount of a conventional insecticide is added to the pheromone product (as a tank mix) to kill moths seeking a mate (*5*). The addition of insecticide enhances the mating-disruption system by removing "sexually active" individuals from the population. Pheromones that attract pests to a substrate impregnated with insecticide (e.g., Plato Industries' Boll Weevil Attract And Control Tube) are an example of an attracticide product that does not involve mating disruption (*6*). This type of product is designed to kill all pests attracted when they feed or otherwise contact the substrate impregnated with conventional insecticide.

The term "bioirritant" describes certain tank-mix applications that contain low levels of sex pheromones (below effective disruption levels) and conventional insecticides. The mode of action of bioirritants is to increase contact between moths excited in the presence of the pheromone and foliar residues of the insecticide.

A deterrent mode of action may best describe certain pheromones—e.g., trans-β-farnesene, an alarm pheromone of aphids—wherein conspecific individuals respond to the pheromone by evacuating the substrate. Products based on these pheromones may prevent colonization of a host crop and subsequent disease transmission and crop loss (*7*).

## Problems in Early Commercialization Efforts (1970s and 1980s)

Many problems became apparent in early efforts to commercialize pheromone-based products during the 1970s and 1980s, some of which are listed below:

- special application equipment required,
- short field life of products,
- limited knowledge of IPM,
- introduction of synthetic pyrethroids,
- poor efficacy related to low application rates, and
- limited R&D for new uses.

The first commercial pheromone products were registered to control the pink bollworm in cotton. Brooks et al. in 1979 (*8*) and Kydonieus et al. in 1981 (*9*) described two controlled-release pheromone products for pink bollworm and the specialized equipment required to apply them. These early formulations were efficacious for 1 to 3 weeks, depending on temperatures and development of crop canopy to hold the hollow-fiber and laminated-flake products, respectively. Products with longer durations were needed for pests of tree fruit and nut crops, where the overwintered generation may emerge for up to 90 days after the first moth emerges. Licensed Pest Control Advisors (PCAs) responsible for crop protection often had a sketchy understanding of integrated pest control and limited knowledge of practical strategies for implementation of integrated programs in the field. Although they were expensive, synthetic pyrethroids competed with nontoxic control programs in the mid-1970s since pyrethroids promised insecticidal solutions that offered lower mammalian toxicity for hard-to-control pests. The efficacy of pyrethroid products initially delayed the impetus for biorational alternatives to all conventional insecticides.

Economic considerations also affected development of pheromone-based products. Because only small quantities of pheromones were needed, their cost was high—as is true with other low-volume specialty chemicals. Due to the high cost of pheromones, their application rates were kept low to keep the costs of pheromone-based pest-control products competitive with those of conventional toxic insecticides. In some cases, these low application rates resulted in insufficient efficacy. Finally, exploratory R&D was severely limited by regulatory constraints on test-plot size (limited to 10 acres before an Experimental Use Permit is required) and crop-purchase requirements (when regulatory approval is not obtained).

In the mid-1980s, Flint et al. and Rice et al. reported on field trials of manually applied polyethylene tube dispensers designed to last several insect generations (*10,11*). These dispensers were applied at high rates of active ingredient (e.g., 75 g AI/ha). Sprayable bead and granule products that can be applied using conventional ground and aerial dispensing equipment also became available in the mid-1980s. These products are repeatedly applied at high rates (20 to 25 g AI/ha) at appropriate intervals to provide control through mating disruption. Many product forms are available today, and other efficacious forms will become available in coming years.

The manually applied dispensers typically provide long durations of efficacy but can be inconvenient to apply to large acreages. Sprayable formulations can be applied quickly and to large areas using conventional ground and aerial equipment but typically show shorter durations of efficacy (weeks versus months) and are affected more by weather conditions (e.g., rain, UV radiation).

## Regulatory Authority

The pheromone-based products described above for detection and/or monitoring are exempt from registration requirements mandated by the Federal Insecticide, Fungicide and Rodenticide Act (FIFRA), but not from regulation by the Environmental Protection Agency (EPA). Pheromone-based products used for control of insect pests are now classified as biorational pesticides and are registered as such by the EPA. Registration testing requirements are significantly reduced for biorational pesticides, as compared with conventional insecticides, and the review process required before registration is significantly shorter (typically 12 months). Rules enacted between December 1993 and July 1994 control evaluation of these products and have provided significant regulatory relief to the biorational pesticide industry. These rules include 1) generic exemptions from the requirement of a tolerance for inert ingredients in manually applied dispensers and for pheromones formulated in manually applied dispensers, and 2) a provision allowing evaluation of these products on up to 250 acres without an Experimental Use Permit.

## Current Directions in Product Development

Formulation optimization to maximize field efficacy and minimize cost are imperative to increase product use in IPM programs, especially because implementation of these programs often necessitates more-thorough field monitoring by the PCA. Also key to increased on-farm use (to meet President Clinton's policy goal that 70% of U.S. farm acreage will be under IPM by 2000) is improved pest-scouting techniques to facilitate the shift from calendar-based spray programs to IPM programs. A legitimate goal of the biorational pesticide industry is to make the use of pheromone-based products economically compelling rather than just economically competitive with conventional insecticides.

## Literature Cited


1. Ridgway, R.L., in *Proc. Beltwide Cotton Prod. Res. Conf.*, National Cotton Council, Raleigh, NC, **1978**, 108-109.
2. Beroza, M., *Agric. Chem.*, **1960**, *15*, 37.
3. Babson, A.L., *Sci.*, **1963**, *142*, 447.
4. Conlee, J., and Staten, R.T., U.S. Patent No. 4,671,010.
5. Butler, G.W., Henneberry, T.J., and Barker, R.J., *USDA ARS ARM W-35*, **1983**, 1-13.
6. McKibben, G.H., Smith, J.W., and McGovern, W.L., *J. Entomol. Sci.*, **1990**, *25*, 581-586.
7. Briggs, G.G., Dawson, G.W., Gibson, R.W., and Pickett, J.A., in *Proceedings 5th International Congress of Pesticide Chemistry*, Miyamoto, J. and Kearney, P.C. Eds.; Pergamon, Oxford, 1983, pp. 117-122.
8. Brooks, T.W., Doane, C.C., and Staten, R.T., in *Chemical Ecology: Odour Communication in Animals*, Ritter, F.J., Ed.; Elsevier, New York, NY, 1979, pp. 375-388.
9. Kydonieus, A.F., Gillespie, J.M., Barry, M.W., Welch, J., Henneberry, T.J., and Leonhardt, B.A., in *American Chemical Society Symposium Series 190. Insect Pheromone Technology: Chemistry and Applications*, Leonhardt, B.A., and Beroza, M., Eds.; 1981.
10. Flint, H.M., Merkle, J.R., and Yamamoto, A., *J. Econ. Entomol.*, **1985**, *78*, 1431-1436.
11. Rice, R.E., and Kirsch, P., in *Behavior-Modifying Chemicals for Insect Management—Applications of Pheromones and Other Attractants*, Ridgway, R.L., Silverstein, R.M., and Inscoe, M.N., Eds.; Marcel-Dekker, New York, NY, 1990, pp. 193-211.


RECEIVED January 12, 1995

# Chapter 15

# Commercial Development of Entomopathogenic Fungi

## Formulation and Delivery

David W. Miller

Research and Development, EcoScience Corporation,
377 Plantation Street, Worcester, MA 01605

To be effective in the control of insect pests, insect pathogens must come into contact with the target insect, either by ingestion or external contact leading to cuticle penetration. This distinguishes pathogens from chemical pesticides, where the method of exposure is less critical and can be reliant on indirect means such as translocation leading to systemic action or activity via the vapor phase. This requirement for direct contact places unique and strong demands on the *formulation* and application methods for microbial pesticides. In fact, the operational concept of *delivery* is much more useful than that of application. The microbial active ingredient must be placed or brought into contact with the target, *delivered* to it, as it were. The principle commercially relevant genera of entomopathogenic fungi, *Metarhizium* and *Beauveria*, are contact pesticides. The infective fungal conidia penetrate the insect cuticle to initiate the infection process, leading to insect death.

The development of pest control products based on fungal conidia has been hampered by problems with shelf-life, formulation, delivery and cost effectiveness. (1)

These problems largely remain in spite of the over 100 years these agents have been in use. EcoScience has begun to develop tools which will form the technical core for a variety of fungal-based insect control products. To achieve this, the microbial product development pathway has been conceptualized as a series of discrete steps, specifically tailored to the active ingredients with which we deal. This perspective has already yielded registered and nearly-registered microbial insecticide products.

### The Microbial Product Development Pathway

The pathway illustrated in Figure 1 captures the steps in the development of microbial products in a way which gives insight to their unique aspects and

0097–6156/95/0595–0213$12.00/0

requirements. Of particular note is the fact that these are living organisms which must, of course, remain alive through production and into and through the product distribution pipeline. A shelf-life of perhaps several years under frequently unfavorable conditions is required.

Focusing on those pathway elements which must impact shelf life; that is, stabilization, formulation, and packaging, makes clear the unique challenge of developing microbial products. We have found that these technical challenges are initially most successfully and productively handled as individual steps. At the same time though, they must be seen to be of a piece in providing a finished product of acceptable commercial characteristics. To identify and employ the conditions which keep an organism viable and healthy, that is, stabilized, for a long period of time requires an understanding of the physiological demands of the organism and providing for them through appropriate means. We have purposefully placed the stabilization step ahead of formulation and packaging because this maintains the maximum process flexibility. Stabilization, therefore, is best viewed as a process option. For example, organisms which are most effectively stabilized under dry conditions are best dried before any formulation components are added. Once dry, a considerable number of formulation options will then exist (*or* remain).

Formulation in the figure refers to the modifications made to the active ingredient, already stabilized as discussed above, and subsequently formulated into a form where it can be most easily delivered for its intended pest control purpose. Formats might include dusts, wettable powders, granules or liquids. Again, addressing the microorganism's requirements for long-term stability can either narrow the formulation options or increase them.

Packaging is an historically unappreciated component of microbial pesticide product development. In part, this reflects the lack of appreciation for the role of the physiology of the active ingredient in the product development effort. Now, as will be seen below, this is being addressed, and packaging provides the means to maintain conditions which provide for these physiological requirements, yielding products with considerable shelf life. Packaging is our friend and, with stabilization and formulation, provides a set of tools which lead to commercially acceptable products.

How the technical elements captured in the product development pathway come together will be illustrated with two examples from the product portfolio of EcoScience.

**Bio-Path® Cockroach Infection Chamber**

Cockroaches are a human pest of legendary proportions familiar to most everyone. (2) The german cockroach, *Blatella germanica*, is the predominant pest species. Large amounts of chemical insecticides are used in the attempt to control roach populations with the attendant concerns of resistance on the part of both the insect and the consumer. It is chiefly a forager and will investigate narrow spaces for food and shelter, preferring dark, humid spaces. The fact that insect pathogenic fungi are contact insecticides would seem to make them ideal agents for a microbial pesticide directed at cockroaches. We have pursued this opportunity and developed a product now fully registered for sale by the U.S. Environmental Protection Agency, the Bio-Path® Cockroach Infection Chamber for the control of german cockroaches.

The chamber is a plastic device similar in size and appearance to a chemical insect bait station but very different in its mode of action. The chamber (see Figure 2) contains 4 entrances equally placed around its perimeter. Located on the inside of the top section is a layer of the fungus *Metarhizium anisopliae* with the fungal conidia facing down into the chamber. Ramps lead up from the entrances so that entering roaches are brought into contact with the fungus layer; that is, the chamber serves as the *formulation* to *deliver* the active ingredient to the target. Roaches are infested with the fungus and leave soon after entering, taking fungal inoculum with them. A roach directly contaminated with the fungus will die within two weeks.

Roaches so infested play a key role in the further spread of the fungus to their colleagues. Through the Horizontal Transfer™ effect, an infested roach, coming into contact with its nestmates, spreads the fungus to them. This leads to an amplification of control beyond just those insects which entered the chamber. This phenomenon is, of course, another method of *delivery* and illustrates how microbial pesticides offer not only environmental benefits but, as a new class of actives, provide unique features.

The Bio-Path product has been through many tests in both the laboratory and the field. In the field, it is providing performance comparable to chemical bait stations (see Figure 3). Currently, the product is being handled by professional pest control operators and has just been introduced to the retail marketplace. It is being produced through solid state fermentation and assembled in a modern facility capable of producing several million chambers per year.

A consumer-oriented product needs a significant shelf-life; assume, two years at room temperature. This has been a major commercial limitation for most microbial pesticides, including fungi. To achieve this goal, our efforts have been directed at determining the requirements of the fungus – those both environmental and physiological. The fungus in the Bio-Path chamber consists of the conidia attached to the vegetative mycelia. These remain attached to the solid agar substrate and constitute an actively metabolizing culture, although the implications of this were not initially obvious to us, especially as regards the spores. Under the assumption that packaging could provide a key to maintaining the appropriate environmental conditions to stabilize the fungus, we examined several standard, commercial packaging materials for their impact and influence on long-term fungal viability.

Packaging materials can be either opaque or of varying degrees of transparency to gases. Aluminum foil packaging of, in general, 0.007 in. thickness is a complete

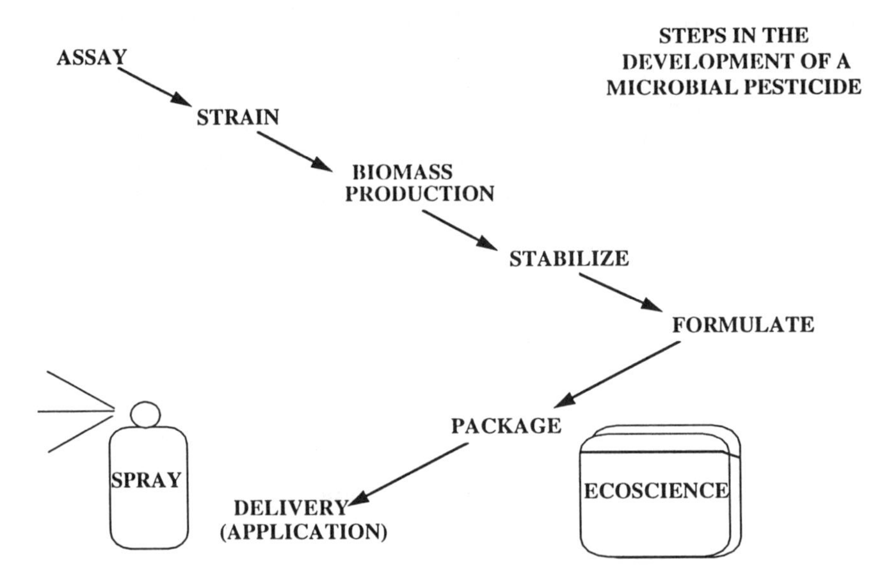


Figure 1. The microbial product development pathway.

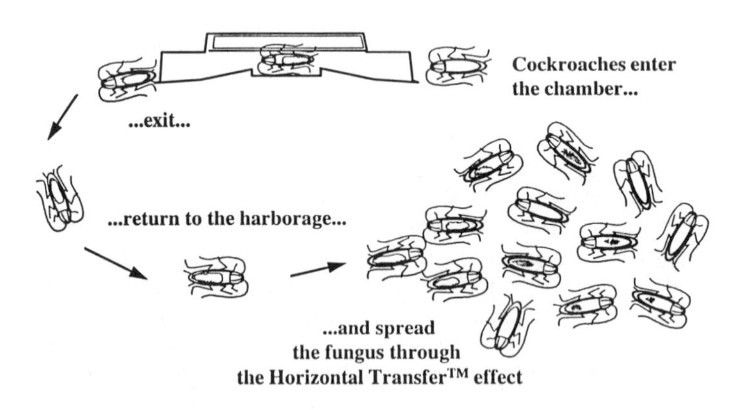


Figure 2. Schematic for Bio-Path chamber.

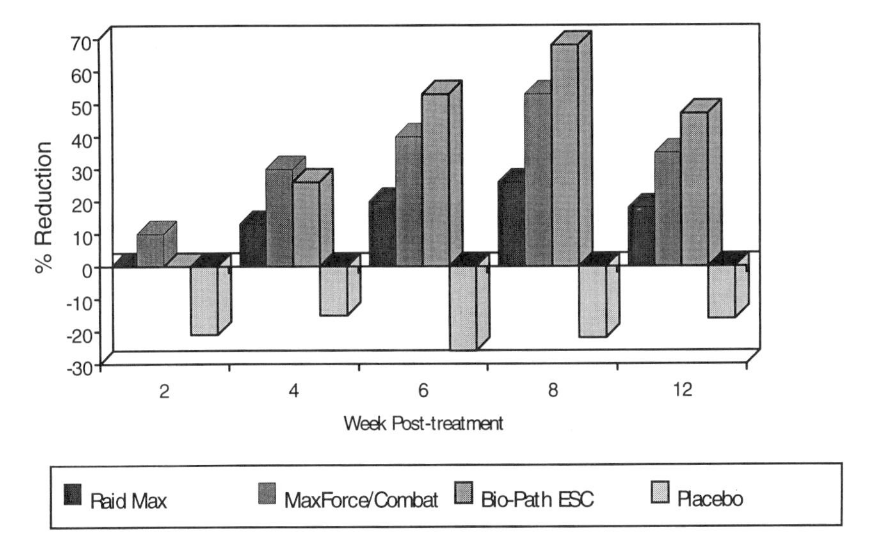


Figure 3. This figure illustrates the % reduction achieved in a German cockroach population through the use of several commercially available treatments. Each treatment represents about fifteen apartments. In each case, an average number of roaches is determined for the treatment group through a one week pre-count of each of the apartments. The specific treatments are then introduced, and the apartments are monitored at the times indicated. The % reduction indicates the current average roach population for each treatment relative to its pre-count.
The treatments are: Combat (active ingredient: hydramethylnon), Raid Max (active ingredient: sulfluramid), Bio-Path, and placebo (empty Bio-Path chambers).

barrier to gases of importance to living organisms: oxygen, carbon dioxide and water vapor. On the other hand, polyethylene film allows the relatively free passage of oxygen and carbon dioxide but is a significant barrier to water vapor, with the degree of these properties depending on film thickness and density.

We sealed chambers containing fully-developed *Metarhizium* cultures into pouches of either 0.007 in. aluminum foil or low density polyethylene (LDPE) of 0.008 in thickness. These were then incubated for several months at 25$^0$C or 37$^0$C before analyzing fungal conidia viability. The interior of the pouches was also monitored for atmospheric composition. As can be seen in Table 1, the viability of the fungus suffered dramatically in the foil pouches. This was correlated with a total loss of oxygen and a great increase in carbon dioxide. The fungus in LDPE, however, survived well with an elevated carbon dioxide level but a good oxygen level. These results and others similar indicate the importance of oxygen to (and removal of $CO_2$ from) the fungus in storage. They also indicate the actively respiring state of the fungus culture and the importance of the packaging choice in maintaining fungal viability. A similar experiment indicated that in the presence of oxygen, high relative humidity was essential for long-term fungal viability providing another reason to select LDPE packaging.

A two-year test of the completed packaging concept demonstrated high viability and efficacy of *Metarhizium* in the Bio-Path chamber for the full period.

*To summarize this formulation and delivery concept*: The fungus *Metarhizium anisopliae* has been formulated into a chamber designed to maximize delivery of the pathogen to the cockroaches which enter it. Roaches leave the chamber and deliver the fungus to yet additional roaches. The product, manufactured through solid state fermentation, is packaged with a system which, by addressing the physiological requirements of the fungus, provides a commercially relevant shelf-life of two years. The patented chamber delivery concept has been shown to be of utility against a variety of insect pests.

**Fungal Conidia Sprays**

The insect pathogenic fungi, *Metarhizium* and *Beauveria*, have a very broad host range measured as either breadth of recorded natural species infections or by laboratory bioassay. (3) Nonetheless, most efforts to employ them as insecticide active ingredients have been based on the use of a particular isolate against a particular insect pest, usually an isolate found associated with the pest in some natural infection. There are two problems with this: One, technology developed for this unique situation doesn't really carry to other systems, requiring continual re-invention and, as a result, two, the potential for a broad spectrum, environmentally friendly, natural insect control agent is not realized.

Our approach has been to develop a set of technologies which create a fungal product that can be thought of *and* used as a broad spectrum insecticide, having the performance and handling characteristics of classical chemical products while

Table 1. Stability of fungal conidia under different storage conditions.

| Temperature | % $O_2$ | % $CO_2$ | % Viability* |
|---|---|---|---|
| **Conditions within Foil Pouch at Three Months** | | | |
| 22$^0$C-A | 0.3 | 23.4 | 15.00 |
| 22$^0$C-B | 0 | 22.5 | 11.15 |
| 30$^0$C-A | 0 | 23.7 | 0.00 |
| 30$^0$C-B | 0 | 22.0 | 0.00 |
| **Conditions within LDPE Pouch at Three Months** | | | |
| 22 $^0$C-A | 8.5 | 3.0 | 92.1 |
| 22$^0$C-B | 8.8 | 2.5 | 90.9 |
| 30$^0$C-A | 8.7 | 3.0 | 86.0 |
| 30$^0$C-B | 4.5 | 4.2 | 87.8 |

* % viability is determined as the number of germinating conidia versus total conidia after 13 hours on potato dextrose agar media at 28$^0$C.

maintaining the environmentally favorable characteristics of a product based on a naturally occurring microorganism.

In the development of such a product, the key issues to address are: Manufacturing, shelf-life under unfavorable conditions, ease of application and efficacy. This abbreviated discussion will touch on those issues related to *formulation* and *delivery*.

As the above discussion on the Bio-Path product revealed, a respiring culture in a high humidity environment requires atmospheric conditions comparable to a natural environment. However, a more standard format for a broad spectrum, wide applicability insect control product is as a dry formulation. Such a product could be more easily distributed and stored. When needed, it could, for example, be re-suspended and delivered as a spray for insect control. There are some intriguing possibilities here: A dry product is light and compact, a dry product provides a variety of formulation options, and, for a microorganism, decreasing the water content holds out the possibility of improved stability, particularly at higher temperature.

As has been in published under the Patent Cooperation Treaty in number W93/24013, this goal has been achieved for *Metarhizium*. In particular, it was found that storage under dry conditions yields a product with exceptional stability characteristics. It was also found that removal of the oxygen further improved stability, presumably protecting the biological machinery from oxidative effects more pronounced at elevated temperatures.

Formulations composed of dry fungal conidia, amended with bulking agents and appropriate surfactants and then correctly packaged, provide exceptional products, having good shelf stability, ease of use and efficacy. We have such products now in extensive field trials, where substantially similar formulations, used as aqueous suspensions and sharing a common fungal species, are showing good activity against insects as diverse as termites, aphids, and whiteflies. Current storage data indicates at least one year of good fungal viability at 37°C.

The concept we have tried to establish in this presentation is the importance of addressing the requirements of both the fungus and the insect target in developing products with desirable (and commercially required) product characteristics. Not surprisingly, our principal areas of focus then are formulation and delivery, taking advantage of and addressing the unique characteristics of fungi.

## Acknowledgments

David W. Miller, Ph.D. acknowledges the major contributions of the EcoScience research staff to the work described in this paper. Thank you to Ms. Robin Gay for administrative assistance.

## Literature Cited

(1) Fungi in Biological Control Systems, Burge, M.N., Ed., Manchester University Press: Manchester, UK. 1988
(2) Understanding and Controlling the German Cockroach, Rust, M.K., Owens, J.M., Reierson, D.A., Eds. Oxford University Press: New York, NY. 1995.
(3) Goettel, M.S., et al. In Safety of Microbial Insecticides; Laird, M., et al, Eds. CRC Press, Inc.: Boca Raton, Florida. 1990. pp. 209-231.

RECEIVED March 13, 1995

Chapter 16

# Development of Novel Delivery Strategies for Use with Genetically Enhanced Baculovirus Pesticides

H. Alan Wood and Patrick R. Hughes

Boyce Thompson Institute for Plant Research, Tower Road, Ithaca, NY 14853

Phytophagous insect pests account for billions of dollars of losses in agricultural and forest production each year. Synthetic chemical insecticides have limited these losses, but are becoming more costly and more restricted in availability. Because of the increasing development and registration costs, the number of new synthetic pesticides brought to market each year has steadily declined over the past 10 years. At the same time, insect populations have developed resistance to many products, resulting in the need to increase application doses or to abandon pesticides. In response to the documented and potential health/environmental risks associated with synthetic chemical pesticides, in 1988 the U.S. Congress mandated that pesticides registered prior to 1984 must undergo a re-registration process. Because of the costs of satisfying current registration requirements, registration of many pesticides for use in small markets has been abandoned. Subsequently, particularly with medium to small market crops, the need for effective insect control is increasing, but the availability of acceptable chemicals is decreasing.

The need to develop "safer pesticides" has become a priority of both the current administration and the Environmental Protection Agency (EPA) (1). The current EPA policy is to facilitate the testing and registration of pesticides which have "reduced risks". In addition the Clinton administration has announced plans to propose legislation which will reduce the use of chemical pesticides, increase the availability of alternatives and promote sustainable agriculture.

Use of biological control agents, most notably baculoviruses, as alternatives to chemical pesticides is among the most promising approaches to reaching these goals (2, 3). Hundreds of baculoviruses have been described. They all infect invertebrates and about 90% infect lepidopterous insects, many of which are major agricultural pests (4). In many instances, baculoviruses appear to play an integral role in the natural regulation of insect populations. Extensive health safety and environmental testing with baculoviruses has been conducted over the past 25 years (5). This testing has indicated a total lack of environmental and health concerns with the use of baculovirus pesticides.

Based on their safety and potential to replace chemical pesticides, five baculoviruses have been registered as pesticides - *Helicoverpa zea* nuclear polyhedrosis virus (HzSNPV) in 1975, *Orgyia pseudotsugata* (Op) MNPV in 1976, *Lymantria dispar* (Ld) MNPV in 1978, *Neodiprion sertifer* (Ns) SNPV in 1983, and *Spodoptera exigua* (Se) MNPV in 1993. However, in the U.S. the only privately produced and commercially available viral pesticide is the SeMNPV (Spod-X™)

0097–6156/95/0595–0221$12.00/0

manufactured by Crop Genetics International. The LdMNPV and OpMNPV are only produced for and used by the U.S. and Canadian forest services. Accordingly, very few viral pesticides are currently available as alternatives to chemical pesticides.

A major deterrent to the commercial development of viral pesticides has been the relatively slow pesticidal action of the viruses. After infection, it may take from 5 to 15 days to kill the insects, during which time they can continue to cause significant damage. Based on this, baculoviruses as well as many other microbial pesticides generally have been considered to have limited commercial efficacy.

Genetic engineering offers a potential solution to this shortcoming (6, 7). A foreign pesticidal gene can be inserted into the viral genome (8), and expression of the pesticidal gene product during replication will allow the virus to kill insects faster or cause quick cessation of feeding (9). Foreign genes which have been inserted into baculoviruses for this purpose include the *Buthus eupeus* insect toxin-1 (10), the *Manduca sexta* diuretic hormone (11), the *Bacillus thuringiensis* ssp. *kurstaki* HD-73 delta-endotoxin (12, 13), the *Heliothis virescens* juvenile hormone esterase (14), the *Pyemotes tritici* TxP-I toxin (15), *Androctonus australis* neurotoxin (16, 17), *Dol m* V gene (18) and T-urf 13 (19) genes. The most efficacious gene inserts have been the neurotoxins and the T-urf 13 gene which is responsible for cytoplasmic male sterility of maize. Several major pesticide companies are currently involved in the commercial development of these and other genetically enhanced viral pesticides. They include American Cyanamid, Dupont and FMC.

**Potential Problems.**

Although the insertion and expression of foreign genes in baculoviruses can significantly improve their commercial value, the release of recombinant organisms into the environment is cause for concern and caution. Although naturally occurring baculoviruses appear to have no deleterious environmental/health properties, a careful assessment of the properties of recombinant baculovirus pesticides needs to be performed before they can be used as alternatives to synthetic chemical pesticides.

Among the environmental issues raised by the field use of a genetically enhanced baculovirus are the precise determinations of virus host range. One of the major benefits of viral pesticides is their apparent host specificity and inability to infect beneficial insects. Although host range expansion could be commercially attractive, precise host range determinations need to be performed with recombinant viruses to ensure that, if host range expansion has occurred, it has not been expanded to include beneficial or desired host species.

From an ecological standpoint, there is a concern that release of any recombinant organism may displace other organisms from the ecological niches which they occupy. Accordingly, displacement of natural virus populations, particularly in non-agricultural settings, could result in unanticipated ecological perturbations over large areas. Although commercially attractive, the release of recombinant viral pesticides with properties which give them a selective advantage over naturally occurring viruses (i.e. increased environmental persistence or improved efficiency of infection) should be viewed with caution and carefully evaluated.

Another concern is that recombinant viral pesticides have the potential for reassortment and/or gene transfer. Accordingly, it will be important to consider the frequency and possible consequences of a foreign gene transfer from the released virus into naturally occurring virus populations.

It should also be appreciated that, even after the most thorough laboratory evaluation of the biological properties and potential environmental interaction of a recombinant organism, there may be unanticipated and unwanted properties associated with the organism. Following release of organisms such as baculoviruses, their removal from the environment would be highly problematic. Baculoviruses occluded within polyhedra or granules can survive in the soil for years (20). Accordingly, the construction of recombinant baculoviruses which have a limited survival capacity in nature is very attractive from an ecological standpoint.

The genetic enhancement of baculovirus pesticides has been an extension of the baculovirus expression vector technology developed in the mid 80's (21, 22). This technology has been extremely useful for the expression of foreign proteins of basic research and commercial interest.

**Baculovirus Expression Vector Technology.**

Most genetic engineering of baculoviruses has been the construction of baculovirus expression vectors with the *Autographa californica* nuclear polyhedrosis virus (AcMNPV). The baculovirus expression vector technology is based on the following biological properties. The naturally occurring virus persists in nature as virions occluded within a crystalline viral protein matrix called a polyhedron. An AcMNPV polyhedron typically measures 0.8 to 2 µm in diameter and contains approximately 50 virions. When an insect larvae ingests a polyhedron, the alkaline nature of the midgut region results in dissolution of the crystal, releasing the virions for infection of the midgut cells.

Early in the replication cycle the progeny virus particles bud through the nuclear membrane into the cytoplasm, lose this membrane and then bud through the plasma membrane. The budded virions are responsible for secondary infections, resulting in eventual systemic infection of the larvae.

Late in the virus replication cycle, the virus particles become membrane-bound within the nucleus. At the same time, polyhedrin protein is produced in high concentrations and crystallizes around the membrane-bound virus particles within the nucleus.

The baculovirus expression vector system was originally based on the fact that the polyhedrin matrix protein is a late, nonessential protein which is required only for the occluded or late phase of replication. The polyhedrin gene is under the control of one of the strongest eukaryotic transcriptional promoters known. At cell death approximately 50% of the total protein may be polyhedrin protein. It is easy to locate the polyhedrin gene, delete the coding region and insert the coding region for a foreign gene which is then under the transcriptional control of the polyhedrin gene promoter.

When the polyhedrin gene is replaced with a foreign gene, the early production of budded virus particles proceeds in a normal fashion. Late in the infection cycle, large amounts of the foreign protein are produced instead of polyhedrin matrix protein. Consequently, the progeny virus particles do not become occluded, and, when the larva dies and its tissues disintegrate, the unprotected virus particles are inactivated by the proteolytic enzymes in the decaying larval tissues.

Accordingly, there is a stabilization problem associated with a polyhedrin-minus, genetically enhanced baculovirus pesticide. One solution to this problem has been to leave the polyhedrin gene intact and to replace the viral p10 gene with the foreign gene sequences (16, 23). In this way the foreign gene is expressed under the transcriptional promoter of the nonessential p10 gene. The development of occluded, recombinant baculovirus pesticides is rather straightforward and can use previous technologies developed for the production, purification and application of naturally occurring viral pesticides.

However, the environmentally persistent nature of occluded baculoviruses is cause for special consideration when dealing with recombinant viruses. Firstly, if an occluded, recombinant viral pesticide is found to have unwanted environmental properties following release, it might be impossible to eliminate such a virus from the environment. Secondly, a genetically enhanced viral pesticide may compete with natural viruses in the environment. Because of the dearth of information concerning the ecology of baculoviruses, it would not be possible to predict the probability of a recombinant virus displacing a natural virus under field conditions. Therefore it is generally considered that several in depth ecological studies will be needed to evaluate the issues related to the commercial application of occluded, recombinant viral pesticides.

**Co-occlusion Virus Technology.**

A second solution to inactivation of nonoccluded virions is the co-occlusion process (24, 25). The basic concept of the co-occlusion technology is that if a host cell is infected simultaneously with a wild-type virus particle (containing a polyhedrin gene) and a recombinant virus (lacking a polyhedrin gene), the polyhedrin protein produced by the wild-type virus will occlude both the wild-type and recombinant virus particles. Through co-infection, one achieves co-occlusion.

In laboratory studies Hamblin et al. (25) modeled the persistence of a polyhedrin-minus AcMNPV in a virus population as polyhedra containing polyhedrin-plus and polyhedrin-minus virus particles were passaged from insect to insect. The model predicted that persistence of the polyhedrin-minus virus required that insect larvae ingest polyhedra containing equal amounts of the two types of virions at dosages 100 times the dosage required to infect 100% of the larvae. These conditions could not be maintained in nature, and the model predicted that the polyhedrin-minus virus would be eliminated from the progeny virus polyhedra within a few passages.

Based on this laboratory data, in 1989 the model was tested in the first field test of a genetically engineered virus in the United States (26). During the first year of the study, AcMNPV polyhedra containing 48% polyhedrin-minus (no foreign gene insert) and 52% wild-type virus particles were applied three times to cabbage plants artificially infested with *Trichoplusia ni* larvae, a susceptible host. The large progeny polyhedra population produced in the seeded larvae during year one was found to contain 42% recombinant virus particles. During the second and third years, additional progeny polyhedra populations were monitored and found to contain 9% and 6% recombinant virions, respectively. Accordingly, the laboratory model correctly predicted the loss of the recombinant virus during passage.

Based on these results, a second field release was conducted in 1993. The test was designed to compare the dynamics of the co-occlusion technology in a forest ecosystem with the row crop study performed in the cabbage field. The recombinant virus was an isolate of the gypsy moth baculovirus, *Lymantria dispar* (Ld)MNPV (27). The coding region of the LdMNPV polyhedrin gene was replaced with the bacterial lacZ gene which codes for ß-galactosidase. The polyhedrin-minus, ß-gal-plus virus was occluded within polyhedra following co-infection with the wild-type LdMNPV. The expression of ß-galactosidase by the recombinant virus is being used to track the recombinant virus in time and space. The 1993 release of the co-occluded, recombinant and wild-type LdMNPV established a virus population in a forest plot which will be monitored for the next two years to determine the spread and persistence of the recombinant virus.

The co-occlusion technology clearly allows for the release of a recombinant viral pesticide which will not compete with natural virus populations. As it cycles from insect to insect, it will be lost from the progeny polyhedra. The 1989 field test did illustrate however a potential problem with the co-occlusion process. A large amount of the progeny virus population produced during year one remained infectious in the soil throughout the study (Wood et al. in press). Over 1,600 biologically active polyhedra per gram dry weight of soil were found in years two and three. With continued applications of co-occluded virus preparations, the concentration of polyhedra containing recombinant virus particles would continue to increase and would persist in the soil with unknown consequences (20, 28, 29).

It should also be appreciated that with the co-occlusion technology typically only half of the virus applied in the field would have the genetically enhanced genotype. In addition, although the co-occlusion technology was easily applied to the AcMNPV system, production of co-occluded virus preparations with other host:virus systems can be problematic (Wood, unpublished data).

**Pre-occluded Virus Technology.**

In appreciation of the stability of occluded, recombinant viruses in the soil, a suicide strategy was developed to eliminate environmental persistence. The suicide strategy is based on the pre-occluded virus technology (30). Following the removal of the polyhedrin gene (with or without insertion of a foreign gene), the early events of budded virus production proceed in a normal fashion. Late in the replication cycle, progeny virus particles, which are normally destined to become occluded, still become membrane bound and accumulate in high concentrations in the nucleus. These virions are called pre-occluded virus particles. Since occluded virions were known to be highly infectious when released and fed to larvae *per os* (orally) (31), it was reasoned that the pre-occluded virus particles might also be highly infectious *per os*.

Following infections of insect cell cultures with a polyhedrin-minus isolate of AcMNPV, it was shown that infected nuclei contained high concentrations of pre-occluded virus particles. Using neonate droplet feeding bioassays (32), it was found that the pre-occluded virions were highly infectious *per os*. In addition, it was shown that tissue culture samples infected with a polyhedrin-minus virus (producing only pre-occluded virions) always contained higher levels of infectivity than samples infected with wild-type virus (containing polyhedra and pre-occluded virions). It is considered that the reduced metabolic requirements, when large amounts of polyhedrin are not produced, may allow for the production of more virus particles.

It was subsequently discovered that infectious pre-occluded virus preparations could be produced in insect larvae. Following death of infected larvae, the occlusion process is required to protect the virions from inactivation. In order to avoid this inactivation process, infected larvae were freeze dried immediately prior to death and release of the proteolytic enzymes. Not only was the infectivity of the pre-occluded virions preserved by freeze drying, but the larval tissues also contained higher levels of infectivity per weight of tissue than are produced with wild-type virus infections. Accordingly, the pre-occluded virus technology can be applied to the *in vitro* and *in vivo* production of genetically enhanced viral pesticides.

In 1993 the pre-occluded virus technology was field tested by AgriVirion Inc. (unpublished data). The test compared the field efficacy of pre-occluded virus preparations with polyhedra samples produced in equivalent numbers of *T. ni* larvae. The pre-occluded virus samples and polyhedra were sprayed onto cabbage plants infested with *T. ni* larvae. As expected, the pre-occluded virus samples were highly infectious under field conditions. When compared to comparable samples of polyhedra, the pre-occluded samples infected as many or slightly more larvae than the polyhedra samples. Soil samples were taken one month after death of the larvae and bioassayed (Wood et al. in press). The soil from the polyhedra plots contained infectious virus, but, as expected, soil samples from the pre-occluded virus plots were not infectious.

Accordingly, the pre-occluded virus technology presents a delivery strategy for recombinant viruses with zero environmental persistence. Following infection and death of a target pest, the pre-occluded virus particles are completely inactivated. Following death of larvae injected with budded virions of a polyhedrin-minus recombinant, Bishop et al. (33) also observed that the virions were quickly inactivated. Besides possessing this highly desirable environmental property, the production cost (*in vivo* or *in vitro*) per unit activity with the pre-occluded virus technology is approximately half that of production of polyhedrin-plus (wild-type) viral pesticides. The pre-occluded virus technology has been used with the AcMNPV and LdMNPV systems, and, unlike the co-occlusion technology, should be readily applicable to all occluded baculovirus systems.

**Conclusions.**

Although baculoviruses have been proposed as alternatives to synthetic chemical pesticides for over 20 years, these and other biological control agents have not been able to compete with chemical pesticides based on the cost/benefit factors. During the past decade there has been an increase in costs and reduction in benefits associated with the development, commercialization, and use of synthetic pesticides. This has lead to the successful development and marketing of numerous Bt toxin products, the success and public acceptance of which have pointed the way for the development of other biological control agents such as baculoviruses.

With the advent of biotechnology, the commercial efficacy of viral pesticides can now be improved significantly. Also, technologies have been developed recently that will allow for significant reductions in production costs (Hughes, unpublished). Concurrent with these advances, public perceptions and government regulatory policies about pesticides clearly are favoring the development of genetically enhanced baculoviruses and other microbial pesticides. Together, it is anticipated that these advances and conditions will foster further investment and development of technologies leading to additional improvements both in formulation and application technologies required for successful commercialization of viral pesticides and other biological control agents.

These changes are bringing the widespread application of viruses for insect control closer to realization. With this increased interest and likelihood of success has come increased concern and attention particularly with respect to possible ecological problems associated with recombinant viruses. To address this concern, two delivery strategies, co-occlusion and pre-occluded virus, have been developed to limit or eliminate persistence of the modified viruses in the environment. These approaches reduce the possibility of unwanted interactions and obviate the question of how to remove the viruses from the environment should they have some unforeseen deleterious property. The current academic and commercial interest in viral pesticides will undoubtedly lead to the discovery of several new such technologies for the development and use of viral pesticides.

Based on current commercial interest, it is anticipated that in the next five years several new baculovirus pesticide products will be commercially available for forest and agricultural pest management. With changing public attitudes accompanied by more science-based regulatory policies, the commercial development of genetically enhanced viral pesticides will follow in the near future.

**Acknowledgments**

Preparation of this manuscript was partially supported by National Science Foundation Grant BCS-9208905 and U.S. Forest Service Fed. Agr. No. 23-570.

**References**

1. Office of Science and Technology Policy February 27, 1992. *Federal Register* **1992,** *57,* 6753.
2. Cunningham, J.C. *Microbial and Viral Pesticides.* Marcel Dekker: NY, **1982,** pp. 335-386
3. Leisy, D. J.; van Beek, N.A.M. *Chemical Industry,* **1992,** *7,* pp. 233-276.
4. Martignoni, M.E.; Iwai, P.J. *USDA Forest Service PNW-195,* Washington, DC: **1986,** 50 pp.
5. Groner, A. *The Biology of Baculoviruses Vol 1,* CRC Press: Boca Raton, Florida, **1986,** pp. 177-202.
6. Wood, H.A.; Granados, R.R. *Ann. Rev. Microbiol.,* **1991,** *45,* pp. 69-87.
7. Wood, H.A.; Hughes, P.R. *Advanced Engineered Pesticides,* Marcel Dekker, Inc.: New York. **1993,** pp. 261-281.
8. Luckow, V.A. *Recombinant DNA Technology and Applications,* McGraw-Hill: New York, **1990,** pp. 97-112.
9. Possee, R.D. *Advanced Engineered Pesticides,* Marcel Dekker, Inc.: New York, **1993,** pp. 99-112.
10. Carbonell, L.F.; Hodge, M.R.; Tomalski, M.D.; Miller, M.K. *Gene* **1988,** *73,* pp. 409-418
11. Maeda,S. *Biochemical Biophysical Research Communication,* **1989,** *165,* pp. 1177-1183.
12. Merryweather, A.T.; Weyer, U.; Harris, M.P.G.; Hirst, M.; Booth,T.; Possee, R. D. *Journal of General Virolology,* **1990,** *71,* pp. 1535-1544.
13. Martens, J.W.M.; Honee, G.; Zuidema, D.; van Lent, J.W.M.; Visser, B.; Vlak, J.M. *Applied and Environmental Microbiology,* **1990,** *56,* pp. 2764- 2770.
14. Hammock, B.D.; B.C. Bonning, B.D.; Possee, R.D.; Hanzlik, T.N.; Maeda, S. *Nature,* **1990,** *344,* pp. 458-461.
15. Tomalski, M. D.; Miller, L.K. *Nature,* **1991,** *352,* pp. 82-85.
16. Stewart, L.M.D.; Hirst, M.; Ferber, M.L.; Merryweather, A.T.; Cayley, P.J.; Possee, R.D. *Nature (Lond.),* **1991,** *352,* pp. 85-88.
17. Maeda, S.; Volrath, S.L.; Hanzlik, T.N.; Harper, S.A.; Majima, K., Maddox, D.W.; Hammock, B.D.; Fowler, E. *Virolology,* **1991,** *184,* pp. 777-780.
18. Tomalski, M.; King, T.P.; Miller, L.K. *Archives of Insect Biochemistry & Physiology* , **1993,** *22,* pp. 303-313.
19. Korth, K.; Levings III, C.S. *Proceedings of the National Academy of Science,* **1993,** *90,* pp. 3388-3392.
20. Thompson, C. G.; Scott, D.W.; Wickham, B.E. *Environmental Entomology,* **1981,** *10,* pp. 254-255.
21. Smith, G. E.; Summers, M.D.; Fraser, M.J. *Molecular and Cellular Biology,* **1983,** *3,* pp. 2156-2165.
22. Pennock, G.D; Shoemaker, C.; Miller, L.K. *Molecular and Cellular Biology,* **1984,** *4,* pp. 399-406.
23. McCutchen, B.F.; Choudary, P.V.; Crenshaw, R.; Maddox, D.; Kamita, S.G.; Palekar, N.; Volrath, S.; Fowler, E.; Hammock, B.D.; Maeda, S. *Biotechnology,* **1991,** pp. 848-852.
24. Miller, D.W. *Biotechnology for Crop Protection;* Am. Chem. Soc.: Washington, DC., **1988,** pp. 405-421.

25. Hamblin, M.; vanBeek, N.A.M.; Hughes, P.R; Wood, H.A. *Applied and Environmental Microbiology,* **1990,** *56,* pp. 3057-3062.
26. Wood, H.A.; Hughes, P.R.; Shelton, A. *Journal of Environmental Entomology,* in press.
27. Yu, Z.; Podgwaite, J.D.; Wood, H.A. *Journal of General Virology,* **1992,** *73,* pp. 1509-1514.
28. Thomas, E.D.; Reichelderfer, C.F.; Heimpel, A.H. *Journal of Invertebrate Pathology,* **1972,** *20,* pp. 157-164.
29. Jaques, R. P. *Baculoviruses for Insect Pest Control: Safety Considerations;* Am. Soc. Microbiol.: Washington, DC. **1975,** pp. 90-99.
30. Wood, H.A.; Trotter, K.M.; Davis, T.R.; Hughes, P.R. *Journal of Invertebrate Patholology,* **1993,** *62,* pp. 64-67.
31. van Beek, N.A.M.; Wood, H.A.; Hughes, P.R. *J. Invertebr. Pathol.,* **1988,** *51,* p. 58.
32. Hughes, P.R.; van Beek, N.A.M.; Wood, H.A. *Journal of Invertebrate Patholology,* **1986,** *48,* pp. 187-193.
33. Bishop, D.H.L.; Entwistle, P.F.; Cameron, I.R.; Allen, C.J.; Possee, R.D. *The Release of Genetically-engineered Micro-organisms;* Academic Press: New York, **1988,** pp. 143-179.

RECEIVED January 31, 1995

# Chapter 17

# Starch Encapsulation of Microbial Pesticides

**M. R. McGuire and B. S. Shasha**

**Plant Polymer Research, Agricultural Research Service, U.S. Department of Agriculture, 1815 North University Street, Peoria, IL 61604**

Microbial insecticides, when formulated within starch or flour matrices are more efficacious and have longer residual activity than commercial formulations. Three starch formulations have been developed: a sprayable formulation and two granular baits. The sprayable formulation is composed of a premixed combination of sucrose and commercially available pregelatinized cornstarch or pregelatinized corn flour that can be tank-mixed at solids rates of 2-6%. Bioassays of cotton or cabbage leaf tissue treated with the sprayable formulations demonstrated increased residual activity of *Bacillus thuringiensis* (Bt) after simulated (greenhouse) or actual (field) rainfall. The two types of granular formulations are: (1) a conventional granule which remains discrete through wet and dry periods, and (2) an adherent granule which, upon contact with water, will partially swell and remain stuck to leaf tissue after drying. Field and laboratory tests have demonstrated that Bt will remain active longer when encapsulated in starch or flour under rainy conditions than non-encapsulated Bt. This technology has also been used for a wide variety of microbial pesticides including bacteria, viruses, fungi, protozoa, and nematodes.

Due to increased perceptions of the hazards associated with chemical pesticides, there is a need to develop alternative non-chemical pest control tools. The recent establishment of guidelines aimed at reducing pesticide application and increasing the use of integrated pest management techniques further fuels this need. One option is to use microbial insecticides typified by the bacterium, *Bacillus thuringiensis* (Bt). Bt is very effective at controlling certain insect species. Strains have been comercially available for use against insects in the orders Lepidoptera, Coleoptera, and Diptera. At this time many additional strains are under commercial development to control a wider variety of insects. Despite this success, the use of Bt has encountered certain problems with acceptance by growers and has lagged behind market expectations over the past few years. Acceptance has been hindered, in part, by the short residual activity of Bt and by reduced efficacy of some of the commercial formulations available to the grower. Residual activity is affected by several environmental factors including degradation by sunlight (*1*) and washoff by rainfall (*2*). In addition, some insects find Bt unpalatable and may not consume a lethal dose of the active agent (*3*).

The purpose of this paper is to summarize our work with novel Bt formulations and to asses the feasibility of providing new formulations for Bt and other microbial insecticides. During the last several years, we have examined formulating ingredients that may increase residual activity or enhance feeding on Bt laced substrates. Starch and flour derived from corn can be modified into film-forming materials that will entrap or encapsulate agents. These materials are inexpensive, in surplus and milling companies are eager to work towards creating new markets for their products. Besides Bt, other microbial control agents have been formulated using similar techniques.

## Preparation of Soluble Starch and Flour

When dry starch granules are exposed to water at 0-40° C, they undergo limited, reversible swelling. Excess water and high temperature (80-100° C) causes hydration accompanied by irreversible swelling, a process known as gelatinization. Upon cooling and especially at high starch concentration, a three-dimensional gel network results, a process known as retrogradation. During this process, the reassociated molecules become insoluble in water. A water-soluble starch powder, called pregelatinized starch, can be obtained if the water is removed before the retrogradation occurs. The pregelatinized starch, when added in high concentrations to water, will entrap other agents present in the water upon retrogradation. Because the resulting product is insoluble, the matrix holds the active agent inside and prevents leaching. Desirable characteristics of the active agent can be enhanced by the addition of sunscreens or other additives. If flour is substituted for the starch, a similar reaction occurs, except the protein content of the flour does not become soluble upon heating. The benefits of flour over starch include lower cost and the protein may act as a feeding stimulant (*3*) and/or a sunlight screen.

## Granular Formulations

In a process described by Dunkle and Shasha (*4*), a ratio of 1 part pregelatinized starch was added to 2 parts water containing Bt and corn oil to form a single large mass. After several hours at 4° C, the mass was chopped in a Waring blendor with the addition of pearl (unmodified) cornstarch to obtain granules of a desired size. We have since learned that equal amounts of starch and water can be used to form the initial mass. Besides lowering the overall water content, this ratio hastens retrogradation and the resulting mass can be chopped into granules without the addition of pearl starch. Granules made with this process have been extensively studied as formulations for control of the European corn borer, *Ostrinia nubilalis* Hübner, in field corn. Preliminary tests demonstrated that feeding stimulants could be added to the formulation that would enhance the acceptance of granules containing

Bt to corn borer larvae (*5*). These authors also demonstrated the utility of using relatively inert starch granules to test the usefulness of additives that may enhance the efficacy of the formulation. In essence, Bartelt *et al.* (*5*), showed that starch by itself was not readily accepted by corn borer larvae. However, if a combination of feeding stimulants, composed of a lipid, a protein and a sugar is added, larvae would feed on the granules in preference to plant tissue. Wheat germ, for example, contains all three ingredients. If one or more of the components of wheat germ was omitted, larvae did not respond as well to the granule. Coax (CCT Corporation, Litchfield Park, AZ), a commercial product, was also highly preferred by corn borer larvae in these tests. In greenhouse tests, using corn as a test plant and corn borer larvae as the test insect, Bartelt *et al.* (*5*) demonstrated that if Coax was present in the granule, Bt content could be reduced by 75% without loss of insecticidal activity. Subsequent work reported by McGuire *et al.* (*6*) documented a similar response by corn borer larvae under field conditions (Table I and Field Evaluation, below). However, these field

**Table I. Efficacy of *Bacillus thuringiensis* in granular formulations for control of *Ostrinia nubilalis* in whorl-stage field corn**

| Formulation | Dose[a] | Avg. % Control[b] | Sites Tested |
|---|---|---|---|
| Starch | 400 | 75 | 2 |
| Starch | 1600 | 90 | 2 |
| Starch + Congo Red | 400 | 77 | 2 |
| Starch + Congo Red | 1600 | 88 | 2 |
| Starch + 1% Coax | 400 | 85 | 2 |
| Starch + 1% Coax | 1600 | 89 | 2 |
| Starch + 10% Coax | 400 | 88 | 2 |
| Starch + 10% Coax | 1600 | 97 | 2 |
| Flour + $CaCl_2$ | 400 | 76 | 3 |
| Flour + $CaCl_2$ | 1600 | 63 | 4 |
| Flour + $CaCl_2$ + 10% Coax | 400 | 80 | 3 |
| Flour + $CaCl_2$ + 10% Coax | 1600 | 66 | 4 |
| Flour + Molasses | 1600 | 75 | 1 |
| Dipel 10G | 1600 | 77 | 7 |
| Furadan 15G | | 57 | 1 |

[a] International Units of *Bacillus thuringiensis* per mg formulation. All formulations applied at 11.2 kg/ha.
[b] Tunneling in treated plants / tunneling in untreated control plants x 100.

tests were strictly efficacy tests and no measure of residual activity was taken. To determine residual activity of starch formulations, Bt preparations were exposed to direct sunlight (*7*). Starch alone was not sufficient to protect activity of the Bt but the addition of sunscreens could effectively maintain the potency over a 12 day period (*7*). To measure residual activity under more realistic conditions, granules were placed in the whorl of corn plants in the field and then collected over time. Assays were done in the laboratory to measure relative activity of the granules and results were reported as percent original activity remaining (*8*). From this study, it was apparent that sunlight alone was not an important factor in degradation of Bt activity in the whorl of a corn plant. Sunlight intensity, measured with a LiCor Spectroradiometer (LiCor, Inc. Lincoln, NE), demonstrated the lack of ultraviolet light in the whorl or within a leaf axil of a corn plant. These sites are where granules come to rest after application and where corn borer larvae feed. Instead, rainfall was the predominant mechanism that acted to remove Bt from commercial granules and the feeding site and render it ineffective. After 70 mm rain, a commercial granule lost 40% of its insecticidal activity while starch formulations lost <10% activity. Over a 12 day period, with >100 mm rainfall, all starch formulations tested maintained significantly higher activity than the commercial formulation (*8*).

While the effectiveness of these granules has been verified, the processing demands to make large quantities of the granules reduces the feasibility of this technology being used extensively. Therefore, we began developing methods to produce the granules with smaller amounts of water. If water simply is added in less than equal amounts to the pregelatinized starch or flour, the result is a mixture of a large mass and a lot of dust; clearly an unacceptable product. However, we found we could reduce the water by mixing it with substances that would not gel the starch or at least delay the water from starting the gelling reaction. This allowed time for complete dispersion of the water throughout the starch and resulted, upon mixing, in discrete granules being formed. One example of a substance that does not gel the starch is alcohol. Therefore, by adding 25 ml of a 30% isopropanol solution to 25 g pregelatinized starch, we have reduced the amount of water by 30% and eliminated the need for grinding and sieving. By altering the percentage of alcohol, granule size can be controlled (*9*). We have since found that high concentrations of a salt, such as $CaCl_2$, or sugar solution, such as molasses, will allow a bigger reduction of water while maintaining adequate granule production (*10*). Testing of these formulations led to the discovery that granules made with this process would adhere to surfaces if the surfaces were wet upon application. Following drying, the granules would remain stuck to the surface and resist washoff by simulated rain (*9*).

**Field Evaluation**. Both types of granules have been extensively tested in field corn over the last several years. These tests were aimed at determining efficacy of Bt against the European corn borer in field corn with different formulations. For these tests, newly hatched European corn borer larvae were applied to the whorl of each plant using the procedures of Guthrie (*11*). Approximately 7 days later, granules were applied over the row using a high clearance vehicle equipped with metering applicators calibrated to deliver 11.2 kg/ha. Approximately 6 weeks later, after the larvae had pupated, damage was assessed by splitting each corn stalk from base to tassel and measuring vertical tunnelling. For a more detailed description of these general procedures see McGuire *et al.* (*6*). Results of these tests (Table I) strongly indicated that each of the formulations tested with the exception of some of the granules made with $CaCl_2$ performed as well as or better than a commercial formulation. All starch formulations were made with the first process (*i.e.*, large amounts of water) while the flour granules were made with less water. Granules formulated at the same Bt concentration as a commercial product (1600 IU/mg) provided control equal to the commercial formulation. Granules with 400 IU/mg Bt containing Coax, however, also provided control similar to formulations with more Bt. Because information pertaining to the cost of manufacturing Bt is unavailable, it is unclear if it is economical to replace part of the Bt with a feeding stimulant such as Coax. However, these data, coupled with the extreme versatility of the granule formation process which allows incorporation of sunlight screens, feeding stimulants, attractants, etc. suggest that this process could have utility for a wide range of approaches to pest control.

## Sprayable Formulations

When pregelatinized starch or flour is added to water in proportions of less than 10% solids (*i.e.*, <10 g/100 ml), a gel will not form. Instead, a suspension is created that can be sprayed through common on-farm spray equipment. As the spray deposit dries, the concentration of starch increases and an insoluble film is formed on the leaf surface, entrapping other components of the spray solution. However, this film will peel off within a few days. If the starch or flour is premixed with an equal amount of powdered sucrose, the resulting powder goes into suspension easier and the film does not peel. Preliminary greenhouse tests (*12*) suggested that a 6% solids suspension of Mirasperse (A.E. Staley, Inc. Decatur, IL) and sucrose remained on cotton leaf surfaces for up to 15 days even if water was applied to the leaf surfaces every 2 days. Under similar conditions, other formulations peeled or washed off. Bioassays with this formulation also revealed a possible feeding stimulant effect for the European corn borer. Field tests were conducted with the starch/sucrose formulation on cabbage. Suspensions of 4% solids and Bt were applied in approximately 200 L of water per ha. Leaf tissue was collected at 0, 3, and 5 days after application and assayed against diamondback moth larvae. Residual activity was significantly better with the starch/sucrose formulation than with the commercial standard (Table II). By 3 days after application, Dipel 2X (Abbott Laboratories, North Chicago, IL) started to lose activity while the starch formulations maintained activity. By 5 days after application, the differences were more striking. However, if a field is sprayed with 200 L water per ha, 8 kg material is necessary to reach that 4% solids level. The large amount of material may limit the usefulness of this sprayable formulation. However, this formulation has potential for use under low volume applications (*e.g.*, cotton, range, or forestry) and to protect environmentally sensitive pesticides used on high value crops.

**Table II. Residual activity of *Bacillus thuringiensis* on cabbage as measured by percent mortality of diamondback moth larvae**

| Formulation | Days Post Application[a] | | | |
|---|---|---|---|---|
| | 0 | 3 | 5 | Combined |
| Starch | 100 | 100a | 88a | 96a |
| Starch + Congo Red | 100 | 100a | 84a | 94a |
| Starch + 10% Coax | 100 | 100a | 92a | 97a |
| Dipel 2X | 100 | 79b | 19b | 65b |

[a] Formulations (all at 4% solids composed of a premix of pregelatinized starch and sucrose) were applied on Day 0. Leaf disks were collected and fed to laboratory-reared larvae. After 3 days, mortality was assessed. Means within a column followed by the same letter not significantly different ($P<0.05$, protected least significant difference).

**Table III. Microbial pesticides formulated in pregelatinized starch**

| Pathogen Group | Species | Reference |
|---|---|---|
| Bacteria | *Bacillus thuringiensis* | *2-10* |
| Viruses | *Autographa californica* NPV | (unpublished data) |
| | Celery looper NPV | (unpublished data) |
| | Grasshopper Entomopoxvirus | 14 |
| Fungi | *Beauveria bassiana* | 13 |
| | *Metarhizium anisopliae* | 13 |
| | *Colletotrichum truncatum* | 15 |
| | *Verticillium lecanii* | (Meyer, S., USDA-ARS, unpublished data) |
| | *Trichoderma spp.* | (Lewis, J., USDA-ARS, unpublished data) |
| | *Gliocladium spp.* | (Lewis, J., USDA-ARS, unpublished data) |
| Protozoa | *Nosema locustae* | (unpublished data) |
| Nematodes | *Steinernema carpocapsae* | (Nickle, W., USDA-ARS, unpublished data) |

## Other Uses

**Granular Formulations.** In addition to Bt, the starch and flour technology has been used with other microbial pesticides (Table III). Periera and Roberts (*13*) demonstrated the utility of starch for formulating *Beauveria bassiana* and *Metarhizium anisopliae* fungal mycelia. Formulations with starch stored better and enhanced sporulation compared with non-formulated mycelia or mycelia formulated in calcium alginate granules. Entomopoxvirus from grasshoppers was also formulated using the non-adherent granule technology (*14*). Additives to the final Miragel formulation included wheat germ oil, molasses, and charcoal. In the laboratory, grasshoppers preferred this formulation over others and, in the field, significantly higher infection levels were detected when the virus was encapsulated in starch than when coated on flaky wheat bran. Using the adherent granule technology, small amounts of chemical insecticide can be entrapped and will last in the feeding zone of the target pest for an extended period of time. In small plot studies, malathion entrapped in flour granules and applied once to field corn for control of a range of ear pests was as effective as five biweekly sprays of malathion (Dowd, P. F., USDA-ARS, unpublished data). The single granule application contained 500 times less malathion than the five sprays and laboratory evidence suggests that predators may not be as affected as much by the granules as by a spray application. Volatile feeding attractants were released in biologically effective amounts over a 12 day period as indicated by corn rootworm adult trap captures (*9*). In preliminary work, bioherbicides have also been tested with the granule technology. When *Colletotrichum truncatum* microsclerotia (*15*) were coated with pregelatinized flour, 100 times more conidia were produced compared with unformulated microsclerotia under laboratory conditions. These data suggest the flour contributed to the survival of propagules or actually stimulated increased conidia production. Greenhouse and field tests are underway to determine efficacy against *Hemp sesbania* seedlings.

**Sprayable Formulation.** The flour/sucrose sprays have been used for application of viruses and nematodes under greenhouse conditions (unpublished data). The nuclear polyhedrosis virus from *Autographa californica* was assayed against *Heliothis virescens* and found to survive well after application in a flour sucrose formulation. Virus polyhedral inclusion bodies (PIBS) were added to a suspension containing pregelatinized flour and sucrose and spread onto greenhouse grown cotton leaves. Newly emerged *H. virescens* larvae were then placed on excised leaf disks and allowed to feed for 4 days. Larvae then were transferred to artificial diet. At doses of $10^4$ and $10^6$ PIBS/ml, *H. virescens* mortality was 33% and 91%, respectively, after 8 days. Mortality from virus applied in water only was 0 and 76% for similar doses using the same procedures. *Steinernema carpocapsae*, a nematode, was applied in a spray containing flour and fructose and survived better than if applied in water only. Assays against Colorado potato beetle larvae demonstrated significant positive effects due to the formulation (Nickle, W., USDA-ARS, unpublished data).

## Summary

In the future, we will continue to see increased use of non-chemical tools to control pests that affect man, beast and agriculture. Many of these tools are still in the research stage of development and are not ready for wide scale use. Experience with microbial control agents such as Bt, however, have shown us that commercial success can be accomplished with environmentally friendly pest control products. Research with Bt has also led to knowledge about the biology and ecology of entomopathogenic bacteria that is directly transferrable to other insect pathogens. Environmental factors that cause loss of activity can be overcome. We have demonstrated that starch and flour can help to extend the activity of Bt, viruses, fungi, nematodes, and chemicals. Overall, the starch encapsulation technology has advanced to the point that commercial development is necessary to fully examine the technology under wide scale application. The process is extremely versatile and the applications are numerous. However, for the technology to truly become successful, an interested company must take the technology and adapt it to their specific needs. Partnerships between grain milling companies and pesticide companies will ensure that custom-made starches or flours with consistent properties can be made available for specific purposes. Cooperative interactions among private, state and federal scientists will only serve to hasten this process.

## Acknowledgments

In addition to those listed in the text as providing unpublished data, the authors wish to thank the following people for their help in contributing to this research: Dr. R. Gillespie, Dr. R. Behle, Dr. C. Hunter, Dr. L. Lewis, Mr. T. Fry, Ms. E. Bissett, Mr. J. Baumgardner, Ms. D. Black, Ms. G. Martin. Also thanks go to personnel at the Illinois Natural History Survey including Dr. C. Eastman, Dr. H. Oloumi, and Mr. D. Dazey for help with field work. We also thank Ms. B. Stanley for help in formatting the manuscript.

## Literature Cited

1. Morris, O. N. *Can. Entomol.* **1983**, *115*, pp. 1215-1217.
2. Sundaram, A; Leung, J. W.; Devisetty, B. N. In *Pesticide Formulations and Application Systems,* Berger, P. D.; Devisetty, B. N.; Hall, F. R. Eds; ASTM STP 1183; American Society for Testing and Materials: Philadelphia, PA, 1993; Vol. 13, pp 227-241.
3. Gillespie, R. L.; McGuire, M. R.; Shasha, B. S. *J. Econ. Entomol.* **1994**, *87*, pp. 452-457.
4. Dunkle, R. L.; Shasha, B. S. *Environ. Entomol.* **1988**, *17*, pp 120-128.
5. Bartelt, R. J.; McGuire, M. R.; Black, D. A. *Environ. Entomol.* **1990**, *19*, pp 182-189.
6. McGuire, M. R.; Shasha, B. S.; Lewis, L. C.; Bartelt, R. J.; and Kinney, K. *J. Econ. Entomol.* **1990**, *83*, pp 2207-2210.
7. Dunkle, R. L.; Shasha, B. S. *Environ. Entomol. 1989*, *18*, pp 1035-1041.

8. McGuire, M. R.; Shasha, B. S.;Lewis, L. C.; Nelsen, T. C. J. Econ. Entomol. **1994**, *87*, pp 631-637.
9. McGuire, M. R.; Shasha, B. S. *J. Econ. Entomol.* **1992**, 85, pp 1425-1433.
10. Shasha, B. S.; McGuire, M. R. *U. S. Patent Application* SN 07/913,565, 1992.
11. Guthrie, W. D. In *Toward Insect Resistant Maize for the Third World*; CIMMYT, Mexico; 1989, pp 46-59.
12. McGuire, M. R.; Shasha, B. S. *J. Econ. Entomol.* **1990**, *83*, pp 1813-1817.
13. Periera, R. M.; Roberts, D. W. *J. Econ. Entomol.* **1991**, *84*, pp 1657-1661.
14. McGuire, M. R.; Streett, D. A.; Shasha, B. S. *J. Econ. Entomol.* **1991**, *84*, pp 1652-1656.
15. Jackson, M. A.; Shasha, B. S.;Schisler, D. A.; Bothast, R. J. 5th International Mycology Congress, Vancouver, BC Abstract. 1994.

RECEIVED January 12, 1995

# Chapter 18

# Utilization Criteria for Mycoherbicides

**G. J. Weidemann[1], C. D. Boyette[2], and G. E. Templeton[1]**

**[1]Department of Plant Pathology, University of Arkansas, Fayetteville, AR 72701**
**[2]Southern Weed Science Laboratory, Agricultural Research Service, U.S. Department of Agriculture, P.O. Box 350, Stoneville, MS 38776–350**

As public pressures increase for reducing chemical inputs in agricultural production systems, industry is challenged to develop effective pest management strategies that result in reduced chemical inputs. The use of plant pathogenic fungi to control weeds provides an environmentally-friendly approach to weed management. Effective use of microbes as bioherbicides is dependent on developing formulation and delivery systems that ensure consistent performance. Advances in formulation technology can be used to overcome environmental restrictions such as limited free moisture and UV irradiation. Likewise, it also may be possible to use formulation adjuvants to overcome biological limitations, such as target weed susceptibility or pathogen infectivity. Examples of current research to overcome environmental or biological limitations and possible future approaches are presented.

Over the past four decades, chemical pesticides have become the primary component of many pest management systems and have accounted for many of the remarkable increases in crop productivity that have occurred during this period. Along with these enormous benefits has come the awareness that our heavy reliance on chemical pest control can lead to undesirable impacts on agricultural ecosystems and human and animal health.

0097–6156/95/0595–0238$12.00/0

concern about potential adverse effects has increased public pressures to reduce chemical inputs without sacrificing the high quality food and fiber they have come to expect. Current and future scientists in the public and private sector face the challenge of reducing chemical inputs in agricultural production systems while maintaining cost-effective pest management programs.

Herbicides now account for 85% of the pest control chemicals used in major field crops in the United States. Economic, alternative management strategies for weed control often are limited in most crop production systems. One management tool that has been underutilized is the use of biological control to supplement existing weed management practices. Biological control may never replace chemicals for weed control, but biologicals can help to reduce herbicide inputs in production systems and increase management flexibility as part of an integrated weed control program. Currently, an effective biological weed control approach is the use of endemic plant pathogenic fungi as, so called, mycoherbicides or bioherbicides. In this approach, naturally-occurring weed pathogens are produced and formulated on a commercial scale and applied to the target weed using conventional application technology to achieve levels of weed suppression comparable to that obtained using a chemical alone (*1,2*). Ideally, the biological is produced, formulated, marketed and used in a manner comparable to its chemical counterparts.

The use of plant pathogenic fungi as bioherbicides has a number of features that make them attractive for biological weed control (*3*). For example, many fungi are naturally-occurring on their respective weed hosts and as such are a natural part of the ecosystem and pose little environmental risk. As living agents, they have developed a parasitic relationship with certain host species and dramatic changes in host preference are unlikely. For many species, mass production on a commercial scale is possible using available technology (*4*) and costs often are competitive with chemical control (*5*).

To date, two fungi have been used as commercial bioherbicides in North America. Collego is a host-specialized strain of the fungus *Colletotrichum gloeosporioides* which was formulated as a wettable powder for control of northern jointvetch (*6*). From 1982 until 1992, Collego was sold as a two-part formulation consisting of dried fungal spores and a sugar solution to facilitate rehydration of the spores before application (*4,7*). DeVine is a strain of the fungus *Phytophthora palmivora* that was sold from 1981 until 1991 as a refrigerated liquid concentrate for control of stranglervine in citrus groves (*4*).

Several other fungi have been registered for use but not yet commercialized. BioMal is another host-specialized strain of *Colletotrichum gloeosporioides* registered in Canada and the United States for control of round-leaved mallow (*8*). The current formulation is composed of dried spores in silica gel added directly to the spray tank (*9*). The rust *Puccinia canaliculata* recently was registered for control of yellow nutsedge (*Cyperus esculentus*)(*10*).

Numerous other potential agents have been studied in detail but have not been commercialized (*1*). Biocontrol agents may not achieve commercialization for a variety of reasons. In many cases, the research and development process may reveal biological, technological or economic limitations to their use. Biological limitations include a host range that may be too broad or too narrow for commercial use, pathogen virulence that is too low to achieve the desired weed suppression, or environmental requirements that are too restrictive to ensure consistent performance (*3*). Technological limitations often include the inability to mass produce infective fungal propagules economically or failure to develop a high quality formulation that ensures consistent performance. Furthermore, the projected market size may be too limited to justify the research and development costs.

Significant increases in the use of biologicals for weed control are dependent on expanding our understanding of biological control organisms and developing methodologies for overcoming potential biological and technological constraints to use. Although molecular genetic approaches have been proposed for overcoming biological limitations (*11,12*), improvements in formulation and delivery systems also can be used to overcome many of the inherent limitations to effective use. This paper will address some of the critical research needs where formulation and delivery technology can improve performance, provide examples of current research underway to resolve some of these problems, and suggest future areas of research emphasis needed to improve biocontrol effectiveness.

## Overcoming Environmental Constraints

In contrast to many herbicides, biological control agents, as living organisms, are influenced greatly by environmental conditions following application to target weeds. For foliar pathogens, temperature, free moisture and protection from ultraviolet irradiation often are critical to plant infection. In particular, most foliar fungal pathogens require several hours of free moisture on plant surfaces for spore germination, infection structure formation and plant infection. Once the pathogen successfully penetrates the outer plant tissues and establishes a parasitic relationship with the host, water is supplied by the host cells and is no longer critical. In many geographic locations, free moisture in the form of dew or rainfall often is not sufficient and is a major constraint to consistent performance of biological agents (*13*). Formulation of a biological to delay evaporation of the application spray or reduce the free moisture requirement of the agent often is a key component of successful biological control.

Water-in-oil, or invert, emulsions have been used successfully to trap water around fungal spores and retard water loss by evaporation (*7,14-16*). Several invert formulations composed of paraffin, paraffin oil and an emulsifying agent have been used to eliminate or significantly reduce the free moisture requirement of several fungal weed control agents such as *Alternaria cassiae* (*9,17*), *A. alternata* (*18*), *A. angustiovoidea* (*18*) and *Colletotrichum truncatum* (*14*). Water-holding capacity of the formulation and pathogen efficacy were related to the paraffin content of the formulation and droplet size of the application (*15,19*). Unfortunately, the high viscosity of invert emulsions requires the use of specialized equipment, such as air-assist nozzles, making commercial acceptance of this technology unlikely. However, this research demonstrates that improvements in formulation technology can overcome or minimize the free moisture requirement of foliar pathogens used for weed control.

In addition to minimizing the free moisture requirement, invert emulsions have been shown to have other effects on biocontrol efficacy. Amsellem and co-workers (*20,21*) showed that the inoculum threshold of *A. cassiae* and *Alternaria crassa* could be reduced to one spore per droplet for plant infection greatly reducing the amount of inoculum required for successful weed control. In addition, host selectivity was altered to include several other weed hosts (*20*).

Oil-based formulations or oil-in-water emulsions also have been shown to improve biocontrol efficacy under moisture-limiting conditions. Vegetable oil emulsions consisting of an emulsifying agent and a vegetable oil reduced the dew period requirement of *Colletotrichum orbiculare* for cocklebur control (*22*). Application of *C. truncatum* in corn oil-water emulsions also reduced the dew period requirement (Table I), delayed the need for free moisture (Table II) and reduced the spray volume required for effective control of hemp sesbania (*Boyette, unpublished*). Several other humectants (*9*) or antidesiccants (*23*) also have been investigated to reduce the free-moisture requirement of biological control agents.

**Table I. Control of Hemp sesbania following application of *Colletotrichum truncatum* in corn oil or water as influenced by dew period**

| | Percent Control | |
|---|---|---|
| Dew Period (hr.)[a] | Corn oil | Water |
| 0 | 35 | 0 |
| 4 | 100 | 25 |
| 8 | 100 | 95 |
| 12 | 100 | 100 |
| 24 | 80 | 100 |

[a] Plants were sprayed to runoff with a spray mixture containing 1 X $10^7$ sp/ml and placed into a 25C dew chamber for the indicated time periods.

**Table II. Control of Hemp sesbania following application of *Colletotrichum truncatum* in corn oil or water as influenced by delay of free moisture**

| Dew | Percent Control | |
|---|---|---|
| Period Delay (hr.)[a] | Corn oil | Water |
| 0 | 100 | 100 |
| 4 | 95 | 20 |
| 8 | 93 | 5 |
| 12 | 85 | 0 |
| 24 | 80 | 0 |

[a] Plants were sprayed to runoff with a spray mixture containing 1 X $10^7$ sp/ml. Following the indicated delay period, the plants were placed into a dew chamber for 12 hr. at 25C.

Microencapsulation of biologicals in alginate, agarose, polyurethane and other materials has been used in commercial processes (*24,25*) and for formulation of insect biocontrol agents (*26,27*) and could be utilized for formulation of weed control agents as well.

Formulation also can be used to protect fungal inoculum from ultraviolet irradiation using UV blockers (*26,28*). UV irradiation often is fatal to fungal spores exposed to sunlight on plant surfaces, particularly those that are not melanized. The addition of UV blockers to the formulation could be used to prolong spore survival on plant surfaces.

For soilborne pathogens used as bioherbicides, short term fluctuations in environmental conditions generally are less critical because the soil environment is less subject to rapid changes in temperature or soil moisture. However, other factors, such as survival of the biocontrol agent and the influence of microbial antagonists, often become important considerations. Several fungal weed control agents have been applied to the soil as granular formulations (*7,29*). A common formulation consists of sodium alginate (a polysaccharide gum) and an inert filler, such as clay (*29,30*). The liquid mixture is combined with spores and hyphae of the fungus and added dropwise to a calcium chloride solution to form gel-like beads. The beads can be air-dried to form dry granules for storage and application.

Other amendments can be added to the gel mixture to increase efficacy or increase persistence. For example, nutritional amendments added to the alginate mixture of the weed control fungus *Fusarium solani* f.sp. *cucurbitae* influenced the efficacy and soil persistence of this fungus for control of Texas gourd (*Cucurbita texana*) (*31-33*). The fungus produces two types of spores in culture, macroconidia and microconidia, which differ in their ability to persist in the soil environment. By altering the nutritional amendment incorporated into alginate granules, the production of the more persistent macroconidia could be favored (*31*). Applications of granular formulations increased the control period for a single application over that obtained with single spray applications of macroconidia or microconidia (Table III) (*32*). Granules applied to the soil surface continued to produce spores on the granule surface until the granules were completely degraded, suggesting that modifications of the granular formulation could have increased persistence further. More recently, a granular formulation composed of wheat flour and clay has been used to provide effective control of several target weeds (*4,34*). In addition to nutritional amendments, chemical agents could be added to reduce colonization of the granule by microbial antagonists of the biocontrol agent, increasing efficacy and persistence.

**Table III. Control of Texas gourd with *Fusarium solani* f.sp. *cucurbitae***

| | | Percent Control | |
|---|---|---|---|
| | | Weeks After Application | |
| Formulation | Application | 6 | 12 |
| Microconidia | Spray | 91 | 52 |
| Macroconidia | Spray | 84 | 60 |
| Soyflour -alginate | Granules | 87 | 72 |
| Oatmeal -alginate | Granules | 92 | 71 |

SOURCE: Adapted from ref. 43.

**Virulence Enhancement**

In addition to minimizing environmental constraints, formulation can be used to increase the efficacy of a biocontrol agent by enhancing pathogen infectivity or by increasing the susceptibility of the target host. Formulation components can enhance pathogen infectivity by increasing spore viability, spore germination, infection structure development or metabolic activity associated with plant infection and disease development. Many aspects of pathogen viability and infectivity can be influenced during the fermentation and formulation process (*4,35*). Plant infection and weed control efficacy is influenced greatly by the number of viable infective units contacting the target weed and the capacity of each infective unit to cause plant infection (*21*). During production and formulation, care must be taken to maintain spore viability and ensure that the nutritional status of the spores are optimal for plant infection. Numerous studies on plant pathogenic fungi (*36*) have demonstrated that nutrition can have a profound effect on the infectivity of fungal spores, and this also has been demonstrated for several weed pathogens (*37-39*). For example, spores of *C. truncatum* produced in media with a 10:1 carbon:nitrogen ratio were more infective on the target weed hemp sesbania than spores produced in a medium with a 30:1 or 80:1 C:N ratio (*39*). Formulations of *C. truncatum* containing pregelatinized starch and casamino acids were more infective than treatments without the additives (*40*).

Spore germination and infection structure formation also can be influenced by nutrition and other formulation additives. Spore germination of *A. cassiae* was found to be influenced by the nutritional status of the fermentation broth, pH, and surfactant selected for use (*37*). Vegetable oil adjuvants enhanced the germination of two *Bipolaris* species used for germination of two *Bipolaris* species used for Johnsongrass control (*41*).

Formulation additives also can influence other aspects of infection or disease development resulting in increased or decreased weed control. Additions of several commercial surfactants containing $C_{16}$ and $C_{18}$ fatty acids to spore suspensions of the fungus *Microsphaeropsis amaranthi* resulted in increased control of redroot pigweed (*Amaranthus retroflexus*) (*Mintz, unpublished*) (Table IV). The compounds had little or no effect on spore germination or infection structure formation suggesting the role of other factors such as an influence on enzymes associated with plant penetration or an effect on the plant itself. Work with other fungal pathogens has shown that plant-derived fatty acid monomers can stimulate fungal cutinases required for enzymatic degradation of the plant cuticle needed for penetration (*42,43*). In a related study, formulations of *A. Crassa* containing pectin or water-soluble plant filtrates showed enhanced infectivity to jimsonweed and several additional weed species considered immune to the biocontrol agent (*Boyette, unpublished*) (Table V). The use of formulation additives to enhance infectivity and increase the effective host range presents intriguing prospects for expanding the host range of weed pathogens or increasing pathogen agressiveness to a target weed.

**Increasing host susceptibility**

Plant resistance to disease often is complex and may consist of structural barriers as well as preformed or induced biochemical defenses. Often, overall thriftiness of the host is important and plants exposed to biotic and abiotic stresses that reduce plant vigor may predispose the host to pathogen infection (*44*). Formulation additives or combination treatments with herbicides can be used as abiotic stress factors to increase weed susceptibility to a fungal pathogen. Most commonly, herbicides applied at recommended or reduced rates prior to application of a biocontrol agent, or in combination as a tank mix, may interact in an additive or synergistic manner and result in better weed control than either the biological or the chemical alone (*45*). For example, preemergence applications of *Fusarium solani* f.sp. *cucurbitae* tank mixed with trifluralin significantly reduced emergence and increased weed mortality of Texas gourd over that obtained with the fungus or trifluralin alone (*33*). Similarly, split applications with bromoxynil plus MCPA, imazethapyr, metribuzin, and sethoxydim increased round-leaved mallow control over that obtained with BioMal alone (*46*). Applications of the fungus *Drechslera monoceras* and

**Table IV. Influence of surfactants on control of redroot pigweed with *Microsphaeropsis amaranthi***

| Surfactant[a] | Percent Control |
|---|---|
| None | 33 |
| Activate Plus | 92 |
| Corn oil | 80 |
| Agri-Dex | 92 |
| Soydex | 93 |

[a]All surfactants were added to the spray mixture to give a final concentration of 0.5% (v/v). Plants were sprayed to runoff with a conidial suspension adjusted to 1 X $10^6$ sp/ml and placed into a dew chamber for 12 hr. at 28C.

**Table V. Influence of spray adjuvants on the host preference of *Alternaria crassa***

| Treatment[a] | Jimson weed | Hemp Sesbania | Eastern Black Nightshade | Showy Crotolaria |
|---|---|---|---|---|
| Fungus | 95 | 0 | 0 | 0 |
| Fungus+ pectin[b] | 100 | 100 | 100 | 100 |
| Fungus+ jimsonweed extract[c] | 100 | 85 | 75 | 80 |
| Fungus+ hemp sesbania extract[c] | 100 | 89 | 80 | 75 |

[a] Plants were sprayed to runoff with a conidial suspension adjusted to 1 X $10^6$ sp/ml and placed into a dew chamber for 16 hr. at 24C.
[b] Fruit pectin was added to give a final concentration of 1% (w/v).
[c] Plant extracts were prepared by macerating 50g of leaf and stem tissue in 1L of distilled water and strained through cheesecloth.

pyrazosulfuron-ethyl at reduced rates controlled barnyardgrass and other rice weeds at 20% of the rates required using chemicals alone (*Gohbara, unpublished*). Tank mix applications of *Colletotrichum coccodes* and the plant growth regulator thidiazuron significantly improved control of velvetleaf (*47*). Applications of the rust *Puccinia canaliculata* followed by paraquat (*10*) or imazaquin (*48*), or tank mix applications with bentazon (*49*) enhanced control of nutsedge.

In some cases, the chemical may have a direct effect on plant metabolism involved in plant defense that results in increased susceptibility to the biocontrol agent. For example, applications of glyphosate at sublethal doses increased the susceptibility of sicklepod to the fungus *Alternaria cassiae* by suppressing phytoalexin synthesis involved in plant defense against fungal infection (*50*). As plant pathologists continue to gain an understanding of the biochemical events involved in plant defense, the use of specific antimetabolites as formulation additives to increase weed susceptibility will become increasingly possible.

Plant defense against potential pathogens may be dependent on maintaining intact structural barriers to pathogen entry. Pathogens that enter host tissues directly must penetrate the waxy layer, cuticle and epidermis by a combination of enzymatic digestion and mechanical force (*51*). Disruption of these barriers often facilitates pathogen entry. Formulation additives can be used to remove or chemically alter barriers such as external waxes and the cuticle to promote pathogen infection. Considerable research has been done on herbicide formulations to promote penetration by the active compound by modifying external plant barriers (*52*). For example, herbicide uptake on johnsongrass was improved by the use of paraffinic oil as an adjuvant which partially dissolved the surface wax on leaves (*53*). The same or similar compounds can be used to enhance pathogen entry as long as they are not toxic to the pathogen.

**Future Research**

Current research suggests that much can be done to increase the efficacy of biocontrol agents using formulation technology to overcome biological or technological constraints to use. Research on many potential biologicals has been discontinued due to the discovery of a key limitation that could perhaps have been overcome using formulation technology. Development of a research base that demonstrates approaches to overcoming these constraints will provide a useful incentive for increased research on biologicals.

Future research should concentrate on several key areas of investigation. Research should increase on developing fermentation and formulation processes that produce inoculum with optimal infectivity in addition to high viability and good stability in storage. Although much has been done to produce biological products with a stable shelf life, little has been done to enhance infectivity once the agent is applied to the target weed. Also, further research is needed to develop formulations that provide optimal protection from disease-limiting environmental conditions. Previous work with invert emulsions and other oil-based formulations suggest that much can be done to ensure consistent performance using formulation technology. Finally, the limited studies demonstrating that pathogen agressiveness can be increased or host resistance can be decreased using formulation additives or low rates of weed control chemicals indicates that weed control effectiveness can be enhanced following application.

The commercial use of three fungi for weed control has demonstrated that fungi can be produced on a commercial scale and formulated to provide effective weed control. The challenge remains to build on these successes and to develop a research base that will facilitate the rapid development of additional biological agents.

**Literature Cited**

1. Charudattan, R. In *Microbial Control of Weeds*; TeBeest, D. O., Ed.; Chapman and Hall: New York, NY, 1991; pp. 24-57.
2. TeBeest, D. O.; Yang, X. B.; Cisar, C. R. *Ann. Rev. Phytopathol*. **1992**, *30*;637-657.
3. Templeton, G. E.; Heiny, D. K. In *New Directions in Biological Control: Alternatives for Suppressing Agricultural Pests and Diseases*; Baker, R. R.; Dunn, P. E., Eds.; Alan R. Liss, Inc.: New York, NY, 1990; pp. 279-286.
4. Boyette, C. D.; Quimby, P. C., Jr.; Connick, W. J.; Daigle, D. J.; Fulghum, F. E. In *Microbial Control of Weeds*; TeBeest, D. O., Ed.; Chapman and Hall: New York, NY, 1991; pp. 209-222.
5. Heiny, D. K.; Templeton, G. E. *Pesticide Interactions in Crop Production*; Altman, J., Ed.; CRC Press: Boca Raton, FL, 1993; pp. 395-408.
6. Daniel, J. T.; Templeton, G. E.; Smith, R. J., Jr.; Fox, W. T. *Weed Sci*. **1973**, *21*:303-307.
7. Connick, W. J., Jr.; Lewis, J. A.; Quimby, P. C., Jr. In *New Directions in Biological Control:*

*Alternatives for Suppressing Agricultural Pests and Diseases*; Baker, R. R.; Dunn, P. E., Eds.; Alan R. Liss, Inc.: New York, NY, 1990; pp. 345-372.
8. Mortensen, K. *Weed Sci*. **1988**, *36*:473-478.
9. Daigle, D. J.; Connick, W. J., Jr. In *Microbes and Microbial Products as Herbicides*; Hoagland, R. E., Ed.; ACS Symposium Series 439, American Chemical Society: Washington DC, 1990; pp. 288-304.
10. Phatak, S. C.; Callaway, M. B.; Vavrina, C. S. *Weed Technol*. **1987**, *1*:84-91.
11. Bailey, J. A. *Aspects Appl. Biol*. **1990**, *24*:33-38.
12. Charudattan, R. In *Biological Control in Agricultural IPM Systems*; Hoy, M. A.; Herzog, D. C., Eds.; Academic Press: Orlando, FL, 1985; pp. 347-372.
13. TeBeest, D. O. In *Microbial Control of Weeds*, TeBeest, D. O., Ed.; Chapman and Hall: New York, NY, 1991; pp. 97-114.
14. Boyette, C. D.; Quimby, P. C., Jr.; Bryson, C. T.; Egley, G. H.; Fulghum, F. E. *Weed Sci*. **1993**, *41*:497-500.
15. Daigle, D. J.; Connick, W. J., Jr.; Quimby, P. C., Jr.; Evans, J.; Trask-Morrell, B.; Fulghum, F. E. *Weed Technol*. **1990**, *4*:327-331.
16. Quimby, P. C., Jr.; Fulghum, F. E., Boyette, C. D.; Connick, W. J., Jr. In *Pesticide Formulation and Application Systems*, ASTM-STP 980; Hovde, D. A.; Beestman, G. B., Eds.; American Society for Testing and Materials: Philadelphia, PA, 1988, Vol. 8; pp. 264-270.
17. Connick, W. J., Jr.; Daigle, D. J.; Quimby, P. C., Jr. *Weed Technol*. **1991**, *5*:442-444.
18. Yang, S. Y.; Johnson, D. R.; Dowler, W. M.; Connick, W. J., Jr. *Phytopathology* **1993**, *83*:953-958.
19. Egley, G. H.; Hanks, J. E.; Boyette, C. D. *Weed Technol*. **1993**, 7:417-424.
20. Amsellem, Z.; Sharon, A.; and Gressel, J. *Phytopathology* **1991**, *81*:985-988.
21. Amsellem, Z.; Sharon, A.; Gressel, J.; and Quimby, P. C., Jr. *Phytopathology* **1990**, *80*:925-929.
22. Auld, B. A. *Crop Prot*. **1993**, *12*:477-479.
23. Shapiro, M.; McLane, W.; Bell, R. *J. Econ. Entomol*. **1985**, *78*:1437-1441.
24. Baker, C. A.; Henis, J. M. S. In *New Directions in Biological Control: Alternatives for Suppressing Agricultural Pests and Diseases*; Baker, R. R.; Dunn, P. E., Eds.; Alan R. Liss, Inc.: New York, NY, 1990; pp. 333-344.
25. Lawton, C. W.; Klei, H. E.; Sundstrom, D. W.; Voronko, P. J. In *Biotechnology and Bioengineering*, Symposium 17; Scott, C. D.,Ed.; John Wiley and Sons, Inc.: New York, NY, 1987; pp. 507-517.

26. Dunkle, R. L.; Shasha, B. S. *Envir. Entomol.* **1988**, *17*:120-126.
27. McGuire, M. R.; Shasha, B. S. *J. Econ. Entomol.* **1990**, *83*:1813-1817.
28. Morris, O. N. *Can. Entomol.* **1983**, *115*:1215-1227.
29. Connick, W. J., Jr. In *Pesticide Formulations: Innovations and Developments*; Cross, B.; Scher, H. B., Eds.; ACS Symposium Series 371, American Chemical Society: Washington DC, 1988; pp. 241-250.
30. Walker, H. L.; Connick, W. J., Jr. *Weed Sci.* **1983**, *31*:333-338.
31. Weidemann, G. J. *Plant Dis.* **1988**, *72*:757-759.
32. Weidemann, G. J. *Plant Dis.* **1988**, *72*:36-38.
33. Weidemann, G. J.; Templeton, G. E. *Weed Technol.* **1988**, *2*:271-274.
34. Connick, W. J., Jr.; Boyette, C. D.; McAlpine, J. R. *Biol. Control* **1991**, *1*:281-287.
35. Bannon, J. S.; White, J. C.; Long, D.; Riley, J. A.; Baragona, J.; Atkins, M.; Crowley, R. H. In *Microbes and Microbial Products as Herbicides*; Hoagland, R. E., Ed.; ACS Symposium Series 439, American Chemical Society: Washington DC, 1990; pp. 305-319.
36. Baker, R. R. In *Plant Disease*; Horsfall, J. G.; Cowling, E. B., Eds.; Academic Press: New York, NY, 1978, Vol II; pp. 137-157.
37. Daigle, D. J.; Cotty, P. J. *Weed Technol.* **1991**, *5*:82-86.
38. Jackson, M. A.; Schisler, D. A. *Appl. Envir. Microbiol.* **1992**, *58*:2260-2265.
39. Schisler, D. A.; Jackson, M. A.; Bothast, R. J. *Phytopathology* **1991**, *81*:587-590.
40. Bothast, R. J.; Schisler, D. A.; Jackson, M. A.; VanCauwenberge, J. E.; Slininger, P. E. In *Pesticide Formulations and Application Systems*; Berger, P. D.; Devissetty, B. N.; Hall, F. R.; Eds.; ASTM-STP, American Society for Testing and Materials: Philadelphia, PA, 1993, Vol.13; pp. 45-66.
41. Winder, R. S.; VanDyke, G. C. *Weed Sci.* **1990**, *38*:89-94.
42. Kolattukudy, P. E. *Ann. Rev. Phytopathol.* **1985**, *23*:223-250.
43. Trail, F.; Koller, W. *Physiol. Mol. Plant Pathol.* **1993**, *42*:205-220.
44. Ayres, P. G. *Ann. Rev. Phytopathol.* **1984**, *22*:53-75.
45. Altman, J.; Neate, S.; and Rovira, A. D. In *Microbes and Microbial Products as Herbicides*; Hoagland, R. E., Ed.; ACS Symposium Series 439, American Chemical Society: Washington DC, 1990; pp. 240-259.

46. Grant, N. T.; Prusinkiewicz, E.; Mortensen, K.; Makowski, R. M. D. *Weed Technol.* **1990**, *4*:716-723.
47. Wymore, L. A.; Watson, A. K.; Gottlieb, A. R. *Weed Sci.* **1987**, *35*:377-383.
48. Callaway, M. B.; Phatak, S. C.; Wells, H. D. *Trop. Pest Management* **1987**, *33*:22-26.
49. Bruckart, W. L.; Johnson, D. R.; Frank, J. R. *Weed Technol.* **1988**, *2*:299-303.
50. Sharon, A.; Amsellem, Z.; Gressel, J. *Plant Physiol.* **1992**, *98*:654-659.
51. Kolattukudy, P. E. In *Structure, Function and Biosynthesis of Plant Cell Walls*; Duggar, W. M.; Bartnicki-Garcia, S., Eds.; Waverly Press: Baltimore, MD, 1984; pp. 302-343.
52. Hull, H. M.; Morton, J. L.; Wharrie, J. R. *Bot. Rev.* **1975**, *41*:421-452.
53. McWhorter, C. G.; Barrentine, W. L. *Weed Sci.* **1988**, *36*:111-117.

RECEIVED January 12, 1995

# Chapter 19

# Biological Weed Control Technology

## An Overview

**N. E. Rees[1], P. C. Quimby[1], and J. R. Coulson[2]**

**[1]Range Weeds and Cereals Research Unit, Agricultural Research Service, U.S. Department of Agriculture, Montana State University, Bozeman, MT 59717–0056**
**[2]Insect Biocontrol Laboratory, Biological Control Documentation Center, Agricultural Research Service, U.S. Department of Agriculture, Building 004, Room 3, BARC-West, Beltsville, MD 20705**

An overview is presented describing the procedures used by USDA, Agriculture Research Service for locating, testing, and obtaining permission to introduce biological weed control agents into the United States. Information is provided for obtaining permits, selecting release sites, and monitoring the population increase of the weed's natural enemies. Precautions are rendered for avoiding problems in setting up and working a biological weed control program.

Most of the noxious weeds in the United States were introduced from foreign countries in a variety of ways. These included 1) exotic seeds contained in ballast soil which was dumped from old wooden ships in North American harbors; 2) exotic seeds mixed with farm seeds carried by early immigrants; 3) seeds that were purposely imported by nurseries for flower gardens, etc., and 4) plants which were deliberately introduced whose products were, at one time in our history, to be used in various businesses, e.g. the wool/sheep industry which used Dyers woad (*Isatis tinctoria* L.) to dye the wool, and teasel (*Dipsacus sylvestris* Huds.) to tease the wool.

Weeds are plants that are currently undesirable to humanity, or to the animals and/or land over which humankind has stewardship. Their undesirable properties may be physical, (e.g. sharp thorns or spines, tacky seeds, etc.) or chemical (e.g. toxic or irritating saps, oils or latex, etc). Generally, exotic plants arrive at a new location without their native enemies, and are therefore able to reproduce and spread, being limited only by their reproductive potential and anything unfavorable in the new environments.

There are many types of organisms that are natural enemies of plants. These include insects, mites, nematodes, pathogens, wild and domesticated herbivores, birds, humans, and sometimes other plants. Both plants and their natural enemies require similar amounts of the same critical components of the ecosystem, i.e. temperature, moisture, nutrients and a niche. Each factor must be available in the proper amount, the proper form, and at the proper time to satisfy each organism's requirements. Variation from the ideal configuration of these factors, while possibly not life threatening, may not allow the organism to attain its full potential.

Biological control is the utilization of living organisms (natural enemies) to depress the population of a specific organism which is a pest to society. The definition implies that society provides an effort to manipulate (alter) one or more existing element(s) of the environment to thus provide the advantage to the beneficial species. Generally, more than one species of natural enemy is needed to reduce the pest's population to levels tolerable to society, each one providing some level of detrimental influence to the pest organism. The effect may be obvious, i.e., the pest plant is defoliated, deformed or stunted because of the natural enemy, or subtle, such as is the case when the damage caused by secondary organisms is greater and at a later date than damage caused by the natural enemy. An example of the latter is where a plant is damaged by insect feeding, and the wound is later infected by a pathogen.

Because the majority of the noxious weeds in the United States are exotic, and because most of these exotic weeds are simply members of the plant community in their native lands, the best control methods are often to be found by studying the pest plant in its native surroundings to determine 1) what organisms are associated with it, 2) which of these are natural enemies, and 3) which of these natural enemies might be good candidates for introduction for biological control of the weed in its new environment. This method of control is a form of "Classical Biological Weed Control".

The eventual effect that a biological control agent will have on its host plant will be the result of 1) the density of the available host(s) compared to the density of the natural enemy, 2) the suitability of local biotic and abiotic conditions for the natural enemy in relationship to the suitability of those same conditions for the host, 3) the diversity in the way that the plant reproduces, i.e., seeds only, or seeds and vegetative reproduction, 4) the ability of the natural enemy to constantly stress the plant, 5) the ability of the plant to maintain and replace root reserves, and 6) the ability of the plant host to recover from the effects of its natural enemy or enemies. If other species of natural enemies are also utilizing the host plant, the impact of the one species may be enhanced or reduced, depending on how the different natural enemy species interact with each other.

## Procedures for Establishing Biocontrol of Target Weeds

There is a complex procedure in place by the United States Department of Agriculture, Agriculture Research Service for locating, screening, releasing and monitoring biocontrol agents that have been accidently or otherwise introduced into North America, and which have been targeted for study. These procedures are the result of many years of experience to perfect a program that is safe. Every effort is taken to ensure that the host range of each introduced biological weed control species is known, and that these agents are not parasitized or diseased. Because each of the host plants and their complement of natural enemies differ in the amount and diversity of testing they must endure, it is impossible to determine how long each set of tests will take. Nevertheless, the steps of this procedure are as follows:

**I- Determining the Suitability of the Target Plant for Biological Control Procedures.**
Quite often, pressure from the public drives the priority as to which target plants will be studied, and budget constraints limit the number of plant species that can be studied. Some pest plants may not appear to be good candidates for study, possibly because 1) the cost of study might far exceed that of the benefits, 2) because so little is known about a pest plant that it does not appear threatening enough to be of concern, or 3) because conflicts of interest exist.

For those that can be studied, the native land of the target species is identified and scientists begin to check the literature and to study the life cycle and associations of that plant. Those with few or no North American relative species are usually better candidates than are those with many close relatives. If the plant is difficult to locate, or does not attain the vigor, height or density that it does in North America, then it is considered to be a good candidate. Discovering potential biological weed controlling agents on the plant during the survey also assists in making this decision.

**II- Conducting Foreign Survey**. Once the target plant is approved for study, a survey in its native land is conducted, and natural enemies which are associated with the target plant are catalogued. The potential agents are reared, identified and tested to determine efficacy. For the United States, this is generally conducted through the USDA, ARS's European Biological Control Laboratory, and/or by the IIBC (International Institute of Biological Control). With the aid of published and unpublished literature, records, observations, etc., the various insect/pathogen species from the survey are evaluated as to being passive feeders, such as bees and butterflies, or as being destructive to the target plant. Those which are destructive are then tested to determine what other plant species they also damage. Those with limited host ranges are potential candidates for additional host specificity testing.

**III- Conducting Host Specificity Screening.** The purpose of conducting host specificity tests is to determine, without testing all plant species, the potential host range of candidate biocontrol agent species by exposing the candidate to a number of representative plant species from a number of plant groups. The plant species tested are selected from a centrifugal (concentric circle) plant matrix which recognizes the target plant species as the center, representatives of other species from the same subgenus located in the first ring surrounding the center, representatives of species from other subgenera but within the same genus located in the next ring, representatives from species of related genera of the same tribe in the next ring, etc., with each additional ring being less related to the target organism. In the next to last outer ring, families of economic or aesthetic value (but generally of no close relationship) are represented. In the

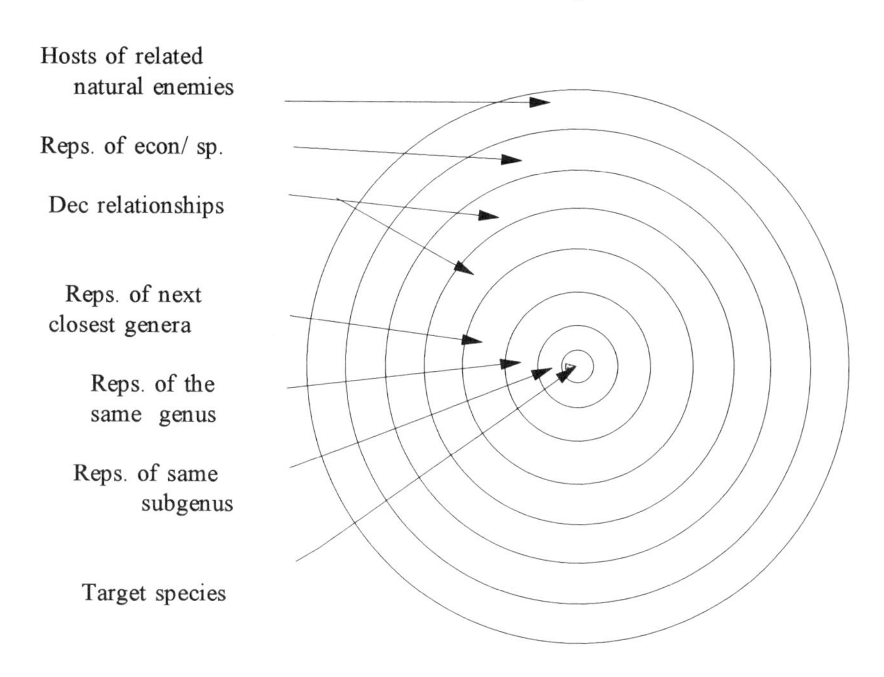


**Figure 1**. Centrifigal (concentric circle plant matrix for host specificity studies.

last ring, unrelated plants with biochemical or morphological characteristics in common with the target weed are also included, as are representatives of plants which are known to be attacked by other species related to the bioagent being tested. (Figure 1.)

The nature of screening tests depends on the target weed and control agent in question. The degree of specificity which needs to be demonstrated and the level of risk which is acceptable rely on the importance of the weed and the presence of non-target species closely related to it in the location where the weed is to be controlled.

In "no-choice" feeding and oviposition tests, (i.e. where the test subject can only eat or starve, and lay eggs on the plant provided to it) agents are isolated as male/female groups in containers (cages), each with a test plant, until the agents either die, feed and/or oviposit. Where the agent remains in the test and dies from apparent starvation without physically damaging the plant or ovipositing on the plant, the plant group is designated as outside the potential host range. Where feeding or oviposition occur, the test continues to determine if 1) the agents can survive in/on the test plant, 2) deposited eggs hatch, and/or 3) the agents can complete their life cycle in/on the test plant. The test then attempts to determine the degree of detrimental effect which is inflicted on the test plant. If the test species is of little concern, but other members may be, then these should be tested.

The artificial conditions of these tests can produce abnormal results and sometimes lead to the rejection of agents that are normally host specific under field conditions. Therefore, when possible, out-door cage and/or open-field testing of previously unlikely candidates may be conducted in the native land of the bioagent to yield more accurate results.

**IV- Petitioning.** Petitions are written at the beginning of each of the three investigative phases for introduction of a new weed biological control agents. The first petition requests permission to work on a plant project and the accompanying agents. At this point, it must be shown that the target weed would be a suitable candidate for a biological control program. The second petition requests permission to introduce biological control agents into quarantine for host specificity testing. Then when all testing to determine the host range of each candidate agent has been completed, a third petition is written which contains the results of the tests. Presently, these are written in the form of an Environmental Assessment (EA) which are, in reality, a measurement of risk, or a risk assessment.

Multiple copies of the petitions are sent to Plant Pest Quarantine (PPQ), a branch of USDA's Animal and Plant Inspection Service (APHIS). PPQ is that branch of the federal government responsible for issuing permits to transport and release insects into the United States. Associated with PPQ is an interagency group of professionals called the Technical Advisory Group (TAG) which is responsible for advising PPQ as to the accuracy and completeness of the testing. Specific members also insure that the results of the testing concern with the Endangered Species Act (U. S. Fish and Wildlife Service) and the Native Plant Act (National Parks Service) are addressed.

TAG may decide that 1) the candidate agent may be dangerous and recommend it not be introduced (generally the scientist is aware of bad results at the end of testing, and if the data indicates danger, would not write and submit the petition in the first place); 2) that the candidate needs more testing; or 3) that the agent appears safe and recommends that it be introduced. PPQ then considers the advice of TAG, but since TAG is only an advisory group, PPQ is not obligated to follow their recommendations, should they have additional concerns or information.

If more testing is required, the petition is returned and the additional data are collected before the petition is resubmitted. If, after careful study, PPQ does decide that all is in order, it then submits the petition for evaluation of the environmental assessment. Failure to pass this examination means that more testing must be completed and the petition resubmitted, but this time possibly in the form of an Environmental Impact Statement (EIS). Approval at this and the previous process will satisfy the remaining requirements and allow a permit to be issued. (2)

**V- Obtaining Permission to Make Field Releases.** Once a petition to introduce a new natural enemy of a plant into the United States has been approved, those wishing to be responsible for a release in their own state must complete a form PPQ-526, "Application and Permit to Move Live Plant Pests and Noxious Weeds." This application is also used for any organism that is shipped from state to state. The application is then sent to the Department of Agriculture in that state in which the release is to be made. The form must then be signed by someone in authority and sent to the USDA, APHIS office, PPQ.BATS (Biological Assessment and Taxonomic Support), Federal Building, Room 628, Hyattsville, MD 20782 for processing. When this is signed by PPQ, a copy will be returned to the applicant as an approval record. Generally these permits are valid for three years.

**VI- Checking All Incoming Shipments.** Once approval to introduce a natural enemy has been received by the researcher, collections are made and the agents shipped into a quarantine laboratory in the United States. Here a portion are killed, pinned and sent to a taxonomist (an authority for that group, generally associated with the USDA-ARS Systematic Entomological Laboratory (SEL)) to confirm that the species designation is accurate. At the same time, some insects are sent to an insect pathologist to determine whether or not they contain any parasites or pathogens. If a parasite is discovered, the colony can be reared through one generation to remove the parasites.

If a pathogen is detected, two courses are possible. Either the colony can be destroyed and, hopefully, another collection site located which is free from the pathogen, or the colony can be split up and reared in individual containers each containing one male and one female. Deposited eggs are kept under a "parent number" until after the females have ceased laying eggs. The adults are then sacrificed and examined for pathogens. Eggs from contaminated couples are destroyed while eggs from healthy couples are reared. This process continues until the colony is pathogen free.

**VII- Selecting a Release Site.** Several factors are important to the selection of sites for releasing and establishing a biocontrol agent. First, the site should be free from chemical treatments such as insecticides and herbicides, or at the least, free from the future use of these chemicals. The field should also be generally free from grazing animals during the initial establishment of the natural enemy, although grazing by wildlife is often insufficient to threaten the establishing colony. A release of 500 adults of the bioagent where one-half or more are female, is generally a logical number to ensure that if the location for release is suitable, the establishment will be successful.

The site should contain those elements of the environment which are necessary for the survival of the natural enemy, such as sufficient supply of the target plant, moisture (either dew, standing water, or sufficient moisture in the plant tissue), sufficient moisture so that the target plant does not suffer from moisture stress, sufficient flowering plants for adult nectar feeders, wind breaks for delicate insects, sunny areas for heat loving insects, shrubs and trees for insects that utilize them for protection and for resting during the day, and locations that are free from potential parasites and would-be predators such as ants, rodents, insect feeding birds, etc.

When temperature and moisture requirements are known or suspected, it is always advantageous to select climatic areas which meet these specifications, especially when making the initial releases. In many cases, however, more desirable areas may be located by determining which factors cause the more rapid increase of the natural enemy. This can be determined by measuring the progress of previous releases. At present, we know only where certain species can survive, and do not know most of their limitations or desirable essentials. It will take time for this to be discovered.

**VIII- Obtaining Cleared Agents from North American Sources.** Before planning any collections of natural enemies in another state for transport or shipment across state boarders, check with your State Department of Agriculture concerning regulations and restrictions on the importation and/or movement of biological weed control agents within or into your State, and whether or not a permit is required. Generally a PPQ-526 permit is required (see Sect. 5). It is the responsibility of the party receiving the natural enemies to obtain any permits, although sometime suppliers do provide this service. In some cases, your weed bureau or cooperative extension agent can provide assistance. These agencies also like to be kept informed and are often willing to work with you on this type of project.

Natural weed enemies can be obtained several ways. If one has seen a particular agent before, and is comfortable in recognizing it, the land manager can make the collections. However, make sure that permission is obtained from the land owner who owns the natural enemies before collections are made. Collectors who provide natural enemies for a profit are a source for those who do not wish to collect for themselves. In either circumstance, make sure that the natural enemies that you will release are the proper species, and are disease and parasite free (see Sect. VI).

**IX- Releasing Cleared Agents.** When the permit is obtained and the colony has been determined to be the proper species and free from parasites and diseases, it can be released on the target plant in an ecological condition which is believed best for the survival of the natural enemy. Generally 100 to 500 insects per release are sprinkled over an area of at least 150 square meters on the host plant of moderate density. Tall, dense stands of plants sometimes provide undesirable shading, whereas thinner stands allow more movement and better chance for establishment. Releases should be made in areas free from grazing and or general traffic, when possible. Herbicide and insecticide spraying should not be conducted in the immediate area or in areas close enough to allow drift of the spray. The initial release area should be protected for 5 to 10 years. Other sites may not need as much protection if the initial source is productive.

The judgement to use or not to use cages is generally made from a partial understanding of the life habits of the new biological control agent. If the agent is a strong flyer, has difficulty finding a mate, or if the release numbers are extremely low, cages are generally used. Flying insects, for example, require a large, tall cage in which they can conduct flights to attract a mate and visit multiple host plants, while small insects that do not readily disperse by themselves, can be confined in much smaller enclosures. Nevertheless, cages of any size produce what is known as a "cage effect", because there is limited space, there are "walls" in the environment on which to climb which reduces the time spent on or near the host plant, and unnatural and flexible "fissures" which may affect survival of the agent. Therefore, the larger the cage, the better.

Cage material may also be an important factor for those agents that spend much time on the cage fabric. For example, metal fabric holds the heat of the sun, and although the temperature within the cage may be acceptable, the metal may be many degrees warmer and thus slowly cook the resting agent. Fiberglass fabric cages can be damaged by chewing insects such as grasshoppers and crickets, although these insects may not be the confined subjects. Saran® fabric cages are excellent, but become discolored and weakened within a few years, depending upon use and care.

If you do not own the land on which you plan to conduct biological weed control, release of the natural enemy on any type of land (public or private) must be made with the consent of the land owner. It is often advisable to have an agreement stronger than just a hand shake because once the agents are released, they belong to the land owner. Therefore, a written cooperative agreement is advisable in which you maintain control of the agents, and which would protect the agent from possible chemical spraying, or other cultural practices such as grazing, mowing, plowing, etc., by the land owner. It is also advisable to write these agreements for protection of the land owner, should you leave the gate open, for example, and let his cow out. These agreements can be written for a limited time period, but five years is generally the shortest time needed for a biological control agent to build sufficiently to occupy ample area to survive most natural or chemical disasters. A period of ten years is more logical.

**X- Documentation of Importations and Releases.** Because classical biological control involves the introduction of exotic organisms into the United States, committed scientists keep detailed records of 1) all exotic materials being imported into quarantine facilities, 2) all shipments from quarantine facilities which are for field release or which will be used in laboratory cultures and studies, and 3) the transfers of established introduced bioagent species to new areas of the United States where these agents have not previously occurred. Specific forms are used in this documentation process, including forms for recording quarantine shipments, and for non-quarantine shipments and releases. Voucher specimens are sent for identification and are also retained by the quarantine facilities of agents received into quarantine, and when the agents are removed from quarantine and field released. Voucher specimens are also retained by various agencies and universities when transferring an agent from one location to another. These voucher specimens provide samples for identity verification, should the need arise, and for later taxonomic studies. Non-quarantine personnel involved in releases or recolonization of introduced biological control agents are asked to help document the dispersal of the agents by use of the forms or by providing pertinent data to the scientist evaluating the biological control program. These forms are the AD-943 which can be obtained from, and copies of which should, when completed, be sent to USDA-ARS-BA-IBL, Biological Control Documentation Center, Insect Biocontrol Laboratory, Bldg 004, Rm. 3, BARC-West, Beltsville, MD 20705. (1)

**XI- Monitoring Release Sites.** Once a biological control agent has been released, it becomes necessary to answer the question, "what's happening out there?" We begin to wonder "did the agent become established?"; "are its numbers increasing or declining?"; "is this species acting as expected in its effect on the pest?"; " is the population of the pest increasing or declining?"; " what are the biotic and abiotic forces acting on the bioagent?"; " what effect are these forces having on the newly released bioagent"?; and "how does the agent interrelate with other established organisms?" The most convenient method of answering these questions is to sample the locality to determine how the populations of the biocontrol agent and the target pest change, and in some cases, to observe the amount and type of damage actually inflicted on the target host. The best method of sampling depends on 1) how the agent utilizes its host, 2) the suspected density of the agent at the time of sampling, 3) the life cycle of both agent and host, 4) the desired accuracy of the data to be obtained, and 5) the amount of effort, labor and money which can be expended on taking the samples.

Monitoring can be conducted in three ways: The sampler 1) actively samples the study area, 2) uses traps with attractants to catch the desired species, or 3) employs traps which indiscriminately collects insects, mites, spiders, tics, nematodes, etc.

**I-Examples of active sampling by the sampler.**

**Observation.** Probably the "laziest" and often the least statistically reliable method of sampling, observation provides limited estimates. It reveals: 1) that both the target organism and the agent are present, 2) what the agent is doing during the time that it is being observed, 3) the type and amount of damage inflicted on the host, and 4) how this damage has physically affected the host. Since at least one stage of the insect's life is exposed on the plant or outside an insect host, one can easily verify the presence of the agent if the observation is made during this period of the insect's life cycle. As such, the densities cannot be calculate of the bioagent or host, the area occupied, or what the agent does in those hours when the observer is absent.

A better method for obtaining specific data combines observations within a defined boundary, such as rings of a known size, or staked out plots, and counts all of the specimens of a particular species in that given area. Of course the process must be repeated many times, then one can statistically utilize the data. Additional information can be obtained if one describes about what is happening on the stem, leaves, flowers, etc. of that plant, and the soil surface. This should be recorded along with time of day that counts were taken, approximate air and ground temperatures, absence of or approximate velocity of wind, direct sun or shade, moisture conditions, plant density and conditions, etc., presence or absence of the host, etc. If sufficient samples have been taken, one can usually determine if the target species is bunched or uniform throughout the area, and what type of conditions appeared to be favored within the ecosystem.

**Daubenmyer/Ring Samples.** When specific areas are to be sampled with exact results needed, squares, rectangles or rings of known area can be used. Daubenmyer plant sampling frames of specific size can be placed on the ground and all plant material within counted, measured, identified, clipped, sorted and/or weighed to determine the plant composition, canopy cover and biomass of the area. Multiple samples on a periodical basis over an extended area will provide data as to shifts in plant densities, plant composition, etc. Therefore, all samples where the exact sample area, time, conditions, etc. are known, can be statistically analyzed with a high degree of accuracy. The results will provide a reliable record of actual happenings.

Rings are often used to count the number of insects, such as grasshoppers, in a given area. If the insects are large enough to be obvious from a distance, then the number of insects is adjusted for the number that enter or leave during the period of observation. The resultant figure is the number of insects for that size sample, with numerous samples providing an average or mean for the area. If insects are too small to distinguish as one approaches, close counting of the insect or other organism within the ring, or enclosing the ring area in a screen and collecting all material therein, can provide the number of insects of the desired species per given area. Paired samples of treated areas versus untreated controls provide data showing the effect of particular treatments. Radial sampling with time at predetermined distances from a release point provides data as to the rate of population expansion for gregarious insects that move out gradually from a release point..

**Sweep Net Sampling.** Similar to observations, sweep net sampling is conducted when the agent is attracted to, or is in the vicinity of the host. With weed agents, this period is generally restricted to intervals when the agent is feeding on the flowers, leaves or stems, or when the adult agent is attempting to lay eggs. With insects, the agents are most frequent in the vicinity of the prey/hosts when they are feeding or attempting to parasitize the target organism. Therefore, one must be familiar with the natural agent-host association and phenology.

Sweep net sampling can be somewhat quantitative but should be done by the same individual at all times and in the same manner because each person has his own style of sweeping. Otherwise the results are generally a reflection of the collecting ability of each sweeper instead of actual variation in the population of the agent. Sweeping can be measured on the number of sweeps or the period of time sweeping. Amount and type of vegetation, excitability of the target organism, aspect of the terrain, air and ground temperature, and velocity of wind will all influence the number of organisms collected. Standardization of results can be made by comparing number of insects swept per given area with other numbers of insects observed or collected per given area by more precise means.

**Dissection.** When bioagents or parasites work within the host, both plant and insect host material can be collected and dissected. This provides information as to 1) the infestation or infection rate of the host population, 2) the number of agents per given host, 3) stage of development of both agent and host, 4) species composition of agents when numerous agents are involved, 5) location within the host where the agent resides, and 6) the amount of damage being inflicted by the agents. If the sample size is sufficient, data will provide very accurate indications of the true agent and host population in the surrounding territory.

**Suction or Vacuum.** Gasoline powered engines can provide power for vacuums in the field to collect live insects, spiders and mites from plants. However, unless all plant material is strongly agitated in the suction area, and the soil area also vacuumed, many specimens may be missed. Some species are very difficult to dislodge from debris and vegetation. Therefore, this is frequently not a reliable method to determine density unless the vegetation is checked afterwards to confirm that all organisms have been collected.

Vacuums are also used to catch flying insects, but the height of the vacuum mouth and time of day will determine which species may be obtained.

**Before and After Photographs.** This is one of the simplest methods for recording results when working with biocontrol agents of weeds, but is useless when attempting to measure numbers of insect bioagents. Photographic records from consecutive years, taken at approximately the same time each year from the same location and of the same horizon, may be compared to determine the gradual change in the plant community. This method does not provide information as to density of the target host, density of the biological control agent, type of damage inflicted on the target host, etc., but does display the conditions at the time of the photograph. Unlike memory, the photographic records are undeviating and display dramatic, visual differences which generally have a greater impact on an audience than do dry figures and tables.

**Digging.** Often, various stages of arthropods, nematodes, and fungi can be located in the soil or on the root system of the target organism. Depending on the size of the organism being sought, various mesh screens can be used to sift the soil and expose the desired material, water can sometimes be used to float the specimens. Berlese funnels can drive the organisms into collecting traps, and dissections of the root material of the target plant can reveal the bioagent in its natural surroundings.

**Berlese Funnel.** These are funnels with a heat source at the top which causes the soil inhabiting organisms to try to escape the heat. They are driven through the screen mesh or wire mesh material which is holding the soil, and falls through the funnel into a collecting container where it is generally killed. This is effective in collecting most soil inhabiting organisms including arthropods such as mites, spiders, insects, scorpions, etc.

**Water Sorting.** Sampling for mites, nematodes and other small worms is often done by dissolving small soil samples in water and checking the fluid under a microscope. By knowing the volume of soil washed and the number of organisms obtained, one can calculate the density of the population. Often samples are taken at continuous levels to determine at what depth in the soil profile the species is most concentrated.

**II- Examples of attractant traps are as follows.**

**Black Light.** Black light traps are commonly used for attracting flying insects. Not all types of insects will be attracted to the light, because not all species see the same wave lengths of light. Moths and butterflies, flies, some beetles, lacewings, etc. are most often collected. The black light is generally operated at night, and sometimes in the twilight periods.

For those species which are attracted to black light, material can be collected from long distances. Insects fly into the baffles, fall into the funnel and are collected in the lower trap which either kills them with chemicals or cold, or holds them until they can be collected. A black light against a bed sheet is sometimes used to attract specimens which are then hand collected. Either method is often used to see what is in the vicinity. However, unless much testing has determined the ratio of collected organisms to those available per given area in the field, this method is not quantitative.

**Pheromone.** These are attractants, or odors which are most often species specific. Usually, they are sex attractants and lure only males or females. If one can calculate the amount of area that the pheromone covers, one can sometimes determine the approximate density of a population. Only minute quantities of material are needed which is carried for great distances on the breeze. This method can attract insects into cages where they can be live-trapped without harm.

**Sound.** This method of attracting insects can also be used to determine the approximate density of a population if the organism can be attracted by sound and if one knows the area covered by the sound. Otherwise, like all other attractants, it only shows that there is something in the area at the time sampled. Sound is probably the least used sampling method and is generally very host specific. This method can be used for collecting live material without having to handle it.

**Entrance Traps.** These are screen traps with a passage which allows the organism to enter, but the portal is difficult to locate when it wishes to leave. These traps are often provided with some type of food attractant.

**III- Indiscriminate Sampling.**

**Pitfall Traps.** These are commonly cans, glass jars, or some other type of container which can be buried with the opening flush with the soil surface, and which can hold chemicals for killing the specimens. Because the preserving fluid can often evaporate and there is generally a long period of time between visits, regular antifreeze is often used. This method collects arthropods running or walking on the ground, some flying insects attracted to the sweet smell, and in some cases, small rodents. Moths, butterflies and some flies are often difficult to identify after being trapped in antifreeze, but most specimens are quite pliable even days after dying. Again, this is a general method of collecting organisms and does not provide an accurate indication of the density of most species.

**Sticky Traps.** These are similar to the old sticky fly traps. A strip or board is covered with a non-drying sticky material and left in the vicinity of the species one wishes to collect. Sometimes a food attractant, such as sugar, is added, or a bright yellow or orange color is used which attracts many flying species. The main problem here is that if proper species identification is to be made, one must clean the specimens which generally proves to be a very difficult and time consuming process.

**XII- Redistribution.** When a natural enemy of a target weed becomes established and increases in numbers so as to be easily collected, it is often advisable to redistribute the colony to new locations, especially when they do not appear to easily disperse themselves. This will provide new habitats in which to survive, and generally reduces the likelihood that the colony can be lost, should a local disaster strike the original release location.

Collection of material can be made several ways, depending upon the species involved. Sweep nets are generally the most effective, along with black lights, collection and rearing from galls, hand picking of individuals, etc. In any event, collections should be made so as to cause the agent a minimum of discomfort and stress, and captivity should be kept to a period as short as possible. The longer the period of time that an insect is kept, the more eggs that the female will deposit in the container, leaving fewer eggs for the new establishment. Therefore, paraphrasing an old phrase, "time are eggs, and eggs are money".

Collections can be made by the party desiring the agents, or by collectors wishing to make a profit. The Federal Agency which is responsible for redistribution of bioagents is APHIS (Animal & Plant Health Inspection Service). APHIS generally creates nurseries for increasing insect numbers, and eventually turns the nurseries over to state agencies after a period of time.

As noted above, such redistribution, or recolonization, of introduced biological control agents should be documented, if such distribution is a significant distance from the area from which the collections are made, e.g., more than 25 miles, and particularly, if distribution is made to another County or State. The USDA Biological Shipment Record - Non-Quarantine (Form AD-943) is available for use for such shipments or recolonization, if desired, or otherwise indicates the kind of information that should be recorded and provided to the scientist(s) responsible for evaluating the particular biological control program. Supplies of the form are available from the responsible research scientist(s), APHIS, or USDA-ARS-BA-IBL, Biological Control Documentation Center, Insect Biocontrol Laboratory, Bldg 004, Rm. 3, BARC-West, Beltsville, MD 20705. (Coulson, 1992)

***Caution:*** Make sure that all insects to be transferred to the new location are the proper species. If any question exists, have all insects checked by an experienced entomologist or the SEL (Systematic Entomology Laboratory, USDA, ARS, Center-West, 10300 Baltimore Ave., Room 101A, Bldg 046, Beltsville, MD 20705-2350: Ph. (301) 504-7041) before they are released. It is also advisable to have insects being transported over long distances checked by an insect pathologist to insure that the colony that is being transferred is clean, and will not be hindered by its own controlling agents, such as parasites or pathogens.

After making collections, insects should be sorted from other species, counted and stored in containers or cages which will allow them to move about and feed on their host plant until time of transport or shipment. In cool storage, many natural enemies can be stored for several weeks if they are provided with intermittent periods of warm temperatures to allowed feeding and move about. However, it is best to ship or transfer quickly and release insects as soon as possible after collection to enable them to deposit most of their eggs at the release site rather than in the carton or cage.

The best containers are tight cardboard or paper cans, such as the old style ice cream cartons. Plastic or glass cartons are not recommended because they allow moisture to build up, which increases insect mortality. Shipment should last no longer than several days, because eggs are laid in the container when the natural enemies warm up, and ice packs generally only last 12 to 24 hours.

For travel, the insects should be transferred to containers which include part of the host plant for food, and as something on which to hold. Lids should be taped around the edges to prevent escape. Do not ship parts of the plant which might reproduce at the new location, such as root material or seed. The cartons containing insects and fresh plant material can then be stored in refrigerators at temperatures safe to store lettuce until shipment/transfer or in cool ice chests during shipment/transfer. Make sure that cartons and ice packs are not loose but firm within the shipping containers.

**XII- What You Should and Should Not Do.** The successes of many biological weed control programs have made the use of natural enemies a desired procedure for many land managers. Through talking with their neighbors, observing before and after photographs of the effects of the bioagent on a patch of the host plant, observing first hand these effects on field trips, reading and listening to news releases and articles, and then considering their weed control costs, they have been sold on the effectiveness biological weed control in solving some of their problems. Therefore, ***They want some of these natural enemies, and THEY WANT THEM NOW*!**

The forces which sometimes cause a frenzy to obtain biological weed control agents are many but are interrelated. These include 1) weed laws and the pressures for their compliance, 2) an awareness that there is a major weed problem, 3) the need for proper land management, 4) the costs of chemical and other types of control, 5) reduction of grazing, farming and recreational areas due to the presence of the weed, 6) reduced productivity, and 7) now and then the loss or injury of livestock. There are many other types of control methods, ie., chemical, mechanical, cultural and integrated. However, biological control is by far the least expensive overall because once established, the natural enemies are self perpetuating.

It is at this point that problems can arise if proper precautions are not taken, especially if the land manager rushes blindly into a bioweed control program without understanding what is happening, and without the proper education. The most apparent problems are obvious. By blindly ordering agents without knowing what is needed, or under what conditions each agent species does best, one can waste money on the wrong agents, or release the correct agents in areas where they may not survive. By not checking the weeds to see if the natural enemies are already in the area, one may purchase and release a minute number on top of a large number of agents that are already established, and money is wasted. By purchasing agents that have not been tested for pathogens or parasites, one can release biocontrol agents of the biocontrol agents that they need. Again, money is wasted. If the bioagent is not properly identified, one might receive an organism that appears very similar to the natural enemy, but one that has a completely different food range. Again, money is wasted. In this case, not only is the money lost that purchased the erroneous agents, but more money will be needed to control the introduced pest. More important, the land manager may not realize the mix up and will blame the bioagent. This hurts the concept of biocontrol for that land manager.

**The best way to avoid problems is to take the following precautions.**

1). **Become educated** on what biological weed control is, how it works, which natural enemies will work best on your weed, which of these agents will work best under your land management and environmental conditions, how to integrate biocontrol into your land management practices, how to monitor the progression of the natural enemy build up and its impact on the weed, and how to insure that you obtain the proper, healthy species of agent. Much of this information can be obtained through and from local weed control departments, state and federal departments of Agriculture, some areas of the Bureau of Land Management, Forest Service, BIA, USDA-APHIS and the USDA-ARS Rangeland Weeds & Cereal Research Unit, in Bozeman, MT.

2). **Work with personnel from your local weed control department.** They need to be informed of where bioagents are located to protect them from potential spray programs, and in some cases, be able to report to the State department of agriculture that the program is operating in your area. They can also be of great assistance in supplying you with update information, and helping to make your integrated program work.

3). **Obtain a statement** through the supplier (when purchasing weed natural enemies) from a reliable taxonomist, on the authenticity of the agent, and a report from a reliable insect pathologist that the colony that you receive is disease and parasite free. If you collect the agents yourself, send a sample of your collection to a reliable taxonomist and another sample to a reliable insect pathologist before making a release.

4). **Select the best natural enemy or enemies for your environmental area**, and the best areas on your property which meet the requirements of the natural enemy, and exclude that area from grazing, spray programs, and general disturbance for a period of at least 5 to 10 years. To avoid duplication, make sure before you obtain the new agents that the natural enemy has not already found your patch of weeds. From this core, the agents can increase in number and area occupied, and eventually will migrate to those weed infested areas outside the exclosure. Also, from this nursery, and when numbers warrant (generally after 3 or 4 years), collections can be made for releases at areas distant from the establishment. Removal of these "starter colonies" at this time should not noticeably diminish the effect of the established colony.

5). **Treat your weed's natural enemies like livestock**, because they are. Each species has its own living requirements as to the temperature range within which it can endure, its moisture requirements, the ecological nitch within which it can survive and flourish, and the type of food that it needs. Insects, for example, are animals which differ from cows and sheep only by three conditions; 1) size, 2) three pair of legs rather than two, and 3) the fact that they wear their skeleton on the outside rather than on the inside. Mites differ from insects in the number of legs (4 pair), and nematodes are missing both legs and an exoskeleton. Pathogens are a different type of organism, but provide the same end result with the same level of safety.

6). **Do not halt normal weed control programs** (including chemical control) outside the areas occupied by the weed's natural enemies while you are waiting for biocontrol to work. Rather, determine each year how far out from the center the colony has progressed, and after calculating a buffer zone, continue the programs. Although the numbers of the bioagent should greatly increase each year, it is more difficult for the population of the agent to "catch up" to that of the weed if the weed is allowed to continue its yearly unrestricted increase.

7). **Take photographs** of the colony site at the same time each year from the same location and with the same horizon in the photograph. Although all of us have good minds, we sometimes forget how bad the weed problems were in years past if we do not have some type of record to remind us. These photographs will display how extensive the weed infestation was, and how prominent is the impact each year by the bioagents. It also provides justification for your expenditures in time, effort and money.

The use of natural enemies to take care of your weed problem can be a very rewarding experience if it is done correctly. It is a tool to be incorporated with other weed control and land management practices. It requires much knowledge, as does other facets of ranching and farming, and sometimes as much effort to make it successful..

**XIV- The Potential of Biological Weed Control.** We now have sufficient knowledge to know that the science of biological weed control is still in the very early stages. However, we are now beginning to understand the basics of how the interactions and interrelationships influence the survival of various organisms, and how these processes can be manipulated to benefit one organism or another.

Concepts of biological weed control are constantly being revised as new knowledge becomes available. Recent concepts which provided possible solutions actually were in conflict with each other. For example, we thought that as natural enemies lived at the expense of their hosts, sufficient hosts would be allowed to survive so as to continue the livelihood of the natural enemies. If one examines this concept closely, one can see that it suggests that there is some thinking on the part of the natural enemies that would purposely allow some plants to survive, or restrict their own population.

Now we perceive a different circumstance. When a new species of organism has evolved through mutation, and that organism is able to reproduce and survive, it carries with it a full compliment of genetic parameters that determine within what temperature ranges, moisture ranges, host or food ranges, niche ranges, etc. it can survive. Plants are producers and thus utilize the soils, being restricted in the area they occupy by temperature, moisture and soil extremes. The environment in relation to the genetic parameters of that species determines how far north and south this species will survive (in conjunction with the elevation), while oceans provide the east-west boundaries. and can provide north and south boundaries if the genetic potential exceeds the environmental limits and available land area.

New and existing natural enemies also have their own genetic constraints which determine where or how they can survive. In some areas, the plant and natural enemy relationships are overlapping, in others they nearly duplicate. Therefore, the plant's natural enemy can live in an environment which is genetically acceptable, and feed, grow and reproduce, being limited only by its own potential and any unfavorable forces in that environment. If the environment favors the natural enemy, it is then possible that the natural enemy may completely remove all of its host from that part of the environment and be forced to live with its host in those areas where the environment is more favorable for the host's survival.

This is what has happened in the native lands of those exotic plants that have since become our weeds With classical biological control, our aim in not to control weeds through black magic as some think, but to recreate that balance of nature that is normal in the introduced weed's homeland. With the proper balance of natural enemies, these weeds may someday become as difficult to find in the United States as they are in their native land(s).

**Literature Cited**

(1) Coulson, J. R.; *Crop Prot.* **1992,** 11:195-205.

(2) Coulson, J. R.; Soper, R. S.; Williams, D. W. (ed); Appendix 3: 151-194. *In Biological Control Quarantine: Needs and Procedures. Proceedings of a Workshop. USDA, ARS*; **1991**, ARS-99:336 pp.

RECEIVED January 12, 1995

# FOREST BIORATIONALS

# Chapter 20

# Use of Insect Pheromones To Manage Forest Insects

## Present and Future

**Patrick J. Shea**

**Pacific Southwest Research Station, Forest Service, U.S. Department of Agriculture, Davis, CA 95616**

Research and development of insect pheromones for forest pest management has been underway for 15 years. Pheromones offer a particularly unique approach to pest management because they often affect critical behaviors associated with the reproductive process i.e., mating, oviposition, host feeding, and host finding. There are several general tactics or strategies used when utilizing insect pheromones for forest pest management purposes: (1) monitoring; (2) mass trapping; (3) mating disruption; (4) antiaggregation. Development of population management strategies for lepidopterous pests has concentrated on using the mating disruption strategy. Whereas with Scolytidae (bark beetles) research has been focused primarily on development of antiaggregation pheromones. A major hurdle in deploying pheromones in forested environments has been the lack of efficient and effective formulation and delivery systems. Lastly, as with other pesticides, pheromones used in mitigating the effects of pest are regulated by the US-EPA's registration process.

The western United States contain forests of extraordinary beauty and commercial value. Competition for these resources is becoming increasingly intense and controversial. These increased and divergent demands often create conflicts with the traditional uses of forest, especially those under public management, and are compounded by high per capita consumptions of wood fiber and other natural resources. The public is concerned about timber harvesting and the methods used to protect forests from the devastating effects of insects, yet they have difficulty connecting these management scenarios with every day use of forest products.

For purposes of perspective the following data will illustrate the amount of damage some forest insects are currectly causing. The following statistics can be found in the Forest Service's annual report on insect and disease conditions in the United States *(1)*.

In 1990 gypsy moth (Lymantria dispar [L.]) defoliated 2.9 million hectares of hardwood forests in the northeastern United States and Michigan. Over half of the defoliation took place in Pennsylvania. In the south, southern pine beetle (Dendroctonus frontalis Zimmermann) was declared in outbreak status on more than 2.4 million hectares of southern pines, primarily in east Texas and South Carolina. In California bark beetles have accounted for an estimated 1.4 million cubic meters of merchantable sized trees being killed during the recent drought. Finally, in Alaska spruce beetle (Dendroctonus rufipennis [Kirby]) has been in outbreak status for the last 7-8 years and in 1993 alone accounted for losses of several hundred thousand cubic meters of white (Picea glauca [Moench] Voss) and Lutz (P. Lutzii Little) spruce scattered over 121,000 hectares of new outbreaks.

When management of tree mortality is necessary, we should strive to take action that is as environmentally sensitive as possible. The public expects the scientific community to produce alternatives to broad spectrum chemical insecticides for pest control, but there are few accepted, efficacious, and environmentally sound tactics available today.

For decades, entomologists have had a basic understanding of the roles of some important semiochemicals in forest insect biology and their potential for pest management applications *(2), (3), (4)*. Semiochemical-based forest pest management has only recently become an economically feasible reality because of changes in public perception, advances in pheromone synthesis and analytic chemistry procedures, new neurophysiological assays, and most importantly technological improvements in pheromone formulation and release devices.

Before pheromone-based management strategies can be employed several important conditions must be met. First, a thorough understanding of the structure and function of the chemical(s) used by the target insect is required. Second, synthetic methods of production are required to supply sufficient quantities for research and development purposes. Third, suitable, consistent, and efficient deployment or release devices must be available. Once these major preconditions are satisfied pheromones can be used in several different scenarios to meet pest management objectives.

## Use of Pheromones

Biomonitoring refers to the use of pheromones for the purposes of survey, detection/distribution, or the occurrence of a critical biological event. The latter uses refer to such events as emergence from diapause, flight periodicity, or arrival on a target crop. The precise occurrence of these events may be important to application of an insecticide to increase performance. It may also trigger the action of cultural or mechanical treatments such as the harvesting and removal of bark beetle infested trees *(5), (6)*.

Mating disruption refers to the use of pheromones (primarily sex attractants) for the purpose of interfering with effective communication between males and females. The objective of this strategy is to permeate an insect's environment with synthetic pheromones so that mating communication is disrupted. This disruption is probably a result of two factors; lack of orientation to the point source and habituation to the pheromone. In addition, there is an implicit assumption that the effectiveness of disruption should increase as the density of the population decreases i.e., as density decreases, distance between individuals increases and pheromone mediated communication becomes progressively important for locating suitable mates *(7)*.

Mass trapping involves the use of synthetic pheromones in combination with some type of trap or other device to capture a significant portion of the populations so that damage or mortality to the resources is reduced or prevented. This approach is particularly suitable for isolated coniferous seed orchards or other "island"-like situations because: (1) the entire pest population is in close proximity to the baited traps; (2) the probability of immediate reinvasion of the target insects from the surrounding area is much reduced. This tactic has been operationally employed against ambrosia beetles infesting logs in sorting areas throughout British Columbia *(8)*, *(9)*.

Antiaggregation is a strategy gaining increasing attention and employs the use of compounds referred to as antiaggregation pheromones *(10)*, *(11)*, *(12)*. These compounds are limited to bark beetles and play an important ecological role in bark beetle tree colonization. Antiaggregation pheromones mediate the negative effects of intraspecific competition by regulating attack density. In effect these compounds signal other responding beetles that the substrate tree is completely utilized and another tree or log must be found.

For the remainder of this discussion I will describe the different management scenarios where pheromone based pest management programs are being developed or applied.

## Seed Orchards

Coniferous seed orchards are an important source of genetically improved seed needed for the large reforestation programs taking place throughout the United States. These managed seed orchards most closely resemble agricultural environments where pheromones are also being developed for pest management in annual crop situations *(13)*. Insects often cause considerable damage to cones and seeds resulting in significant losses. These losses have stimulated research into control measures that use semiochemicals. Currently, pheromones are used primarily to survey the presence or absence of pests and to monitor for population changes and insect phenology. Pheromones for many of the principal seed and cone insects in the southeast have been identified and active research is underway to develop pheromone-based insect management programs for several species of coneworms (disruption) and the white pine cone beetle (trap-out). In the western United States the research program is in its infancy and is concentrating on the identification, synthesis, and field bioassay of several insects' importance.

## Defoliators

The effect of insect defoliators on forest and shade tree resources can be very devastating and unsightly. Infestations can be spread over large areas and occur in a very short period of time. Stand and tree mortality, changes in stand composition, loss of growth, increased susceptibility of trees to other pests, and the general unsightliness are impacts that can be attributed to defoliating insects.

Gypsy moth pheromone-baited traps have been used for approximately 18 years for detecting and delineating established populations in the eastern United States. Rapid detection of new introductions into the western and southern United States has been made possible by a nationwide system of pheromone-baited traps. In recent years mass trapping, used in conjunction with pesticide applications, has become a standard technique in eradication efforts. Recently, tests of mating disruption in areas with low populations have shown great promise in some areas and in some years.

At the present time there are no operational programs that utilize pheromones for control of eastern or western spruce budworms. Current research is investigating the use of pheromone-baited traps to monitor populations through time and relate trap catches to the next year's defoliation. Population control through mating diruption has been extensively researched in eastern Canada with mixed results.

Monitoring of Douglas-fir tussock moth populations in the western United States has utilized pheromone-baited traps for many years. The technique was developed as an "early warning system" to forecast pending outbreaks and has been quite accurate in some areas. Mating disruption has also been extensively tested and found to have substantial impact on tussock moth populations. Field tests conducted in Oregon, Idaho, and California over a period of several years have consistently resulted in reduced mating success of female Douglas-fir tussock moths. This in turn led to lower larval populations in the next generation and subsequently reduced levels of defoliation. Interestingly, the aerial application of the Douglas-fir tussock moth pheromone had no detectable effect on non-target arthropods especially parasitoid and invertebrate predators.

## Plantations

Shoot and tip insects can be serious problems in coniferous plantations where high monetary investments are at risk because of the initial costs involved in plantation establishment. Western pine shoot borer, Eucosma sonomana (Kearfoot) larvae mine the pith of terminal shoots of ponderosa pine, Pinus ponderosa (Lawson), causing an estimated 25% loss of height growth *(14)*. In addition, terminals can also be killed outright which promotes multiple leaders and further economic degradation. Trees in the 1.5 to 15 meter height range appear to be the most susceptible to damage. Mating and oviposition occur in early spring and hatching larvae immediately bore into the shoots and begin feeding. In midsummer larvae exit the mined terminal shoot, fall to the ground, and pupate in the forest litter. The pupae overwinter in the soil and adults emerge early next spring to mate and search for oviposition sites. Females attract males for mating by release of exact mixtures of (Z)-9- and (E)-9 dodecenyl acetates in a 4:1 ratio. The pine shoot borer is an ideal subject for deployment of the mating disruption technique because: (1) it is univoltine; (2) infestations and damage can be accurately assessed; (3) its generally low population density suggests vulnerability to control by disruption of the mating communication.

Field studies to test the mating disruption strategy for reducing the damage caused by western pine shoot borer were conducted over several years in Oregon and California *(7)*. Controlled release formulations using polyvinyl chloride pellets and ties applied manually and Conrel fibers and Hercon flakes applied from aircraft have been field tested with surprisingly consistent results. Damage was reduced by about 70 to 90 percent compared to untreated controls. These results were obtained regardless of whether the material was applied by ground application (4 and 8 hectares) or aerial application (20, 100, 600 hectares) (Table I). A somewhat unexpected result was the apparent lack of a dose response based on the dose (grams/hectare) applied (Table I).

**Table I. Summary Results of Several Field Tests to Demonstrate Mating Disruption of Western Pine Shoot Borer**

| *Area Treated* | *Dose (g/ha)* | *Release Device* | *Damage Reduction %* |
|---|---|---|---|
| 8 | 3.5 | PVC | 83 |
| 4 | 14.0 | PVC | 84 |
| 20 | 15.0 | Conrel | 67 |
| 100 | 20.0 | Hercon | 88 |
| 600 | 10.0 | Conrel | 76 |

SOURCE: Reproduced from reference 7.

In addition to the above results tests have also been conducted in both Idaho and Montana with similar results. Clearly, the mating disruption strategy can be used to reduce damage caused by the western pine shoot borer. Until recently, when the registrant allowed the EPA registration label to lapse, this pheromone pest management strategy was available throughout the western United States.

**Bark Beetles**

In the past two decades, outbreaks of bark beetles in western North America have focused renewed attention toward these devastating forest insects. The impact of present outbreaks, and the anticipated impact of future outbreaks have been strong stimuli for the continuation of basic research on semiochemical-based communication systems, and for the development of semiochemicals for monitoring and management of bark beetles.

The following two examples will illustrate the present direction of research aimed at developing these materials for management of bark beetles in the western United States.

Torrey pine (Pinus torreyana Parry ex. Carr.) is a beautiful and comparatively delicate species with a life span of only about 200 years. It is also the rarest pine on the North American continent existing as an intact ecosystem in an approximately 40 hectare stand in Torrey Pine State Reserve, La Jolla, CA. In the late 1800s this unique resource only numbered several hundred trees, whereas in 1991 there were approximately 7000 trees in the stand. This island population of coniferous trees lies within a sprawling urban setting which is bounded on the west by the Pacific Ocean and the north, east and south by urban settlements. Between the winter of 1989 and spring of 1991 Reserve officials and patrons observed hundreds of Torrey pines dying within the Reserve.

Subsequently the mortality was determined to have been caused by a native bark beetle, Ips paraconfusus Lanier, the California five-spinned ips (Shea and Neustein, in press). Discussions with Reserve personnel concluded that the infestation was the result of population build-up by ips in trees that were blown down by a severe storm during the winter of 1989.

By the winter of 1991 in excess of 840 trees, or 12% of the last remaining natural stand of Torrey pine, had been killed by ips beetles and there was no reason to believe that the infestation would subside on its own. The Reserve managers considered several unattractive intervention options to reduce or eliminate the escalating tree mortality. Application of insecticides to treat currently infested trees and kill developing larvae and to protect uninfested trees from new attacks was rejected as too toxic and indiscriminate and logistically impossible. Watering the Torrey pine in hopes of strengthening their defense against ips attack was considered not feasible for so large a stand. Thinning the stand with a limited logging operation was unacceptable not only from an esthetic standpoint but also because of the probable damage that would occur to the many species of "sensitive" plants in the Reserve.

The strategy selected to abate the potentially devastating effects of this bark beetle is commonly referred to as the "trap-out" strategy. The strategy has many variations that include employment of trap trees, toxic trap trees, baited traps of various configurations all of which have the basic objective of concentrating large numbers of beetles at a few or many specific spots and removing or killing them. It is assumed, sometimes incorrectly, that by removing a large number of beetles from the surrounding population that tree mortality will be reduced.

Beginning in May of 1991, 10 sets of three black Lindgren funnel traps were strategically placed 25-40 meters apart in a line approximately 400 meters long. Traps were hung on dead Torrey pines approximately 30 meters inside the dead portion of the stand and parallel to the green uninfested stand (Figure 1). Traps were baited with the Ips paraconfusus pheromones formulated in controlled release bubble caps. The specific pheromones were: (+)50%/(-)50% ipsenol, (+) cis-verbenol, and (+)97%/(-)3% ipsdienol. In addition to the aggregation pheromones, 10 sets of two antiaggregation pheromones formulated as bubble caps were placed approximately 30 meters into the green uninfested stand and parallel to the funnel trap line. The antiaggregation pheromones were: (-)86%/(+)14% verbenone and (-)50%/(+)50% ipsdienol. All pheromones were formulated to last 70 days.

During the first year of trapping approximately 156,000 ips beetles were removed from the Reserve. The highest number of beetles caught in any one week was about 29,000 during week number 4 (Figure 2). The numbers of beetles caught in subsequent weeks declined steadily except for a small surge of beetles during week 16. Weekly trap catches of beetles approached zero during the middle of October 1991 and trapping was discontinued (Figure 2). At the beginning of the 1991 trapping period 38 trees were currently infested. At the end of this trapping period only two additional trees were killed (August 1991).

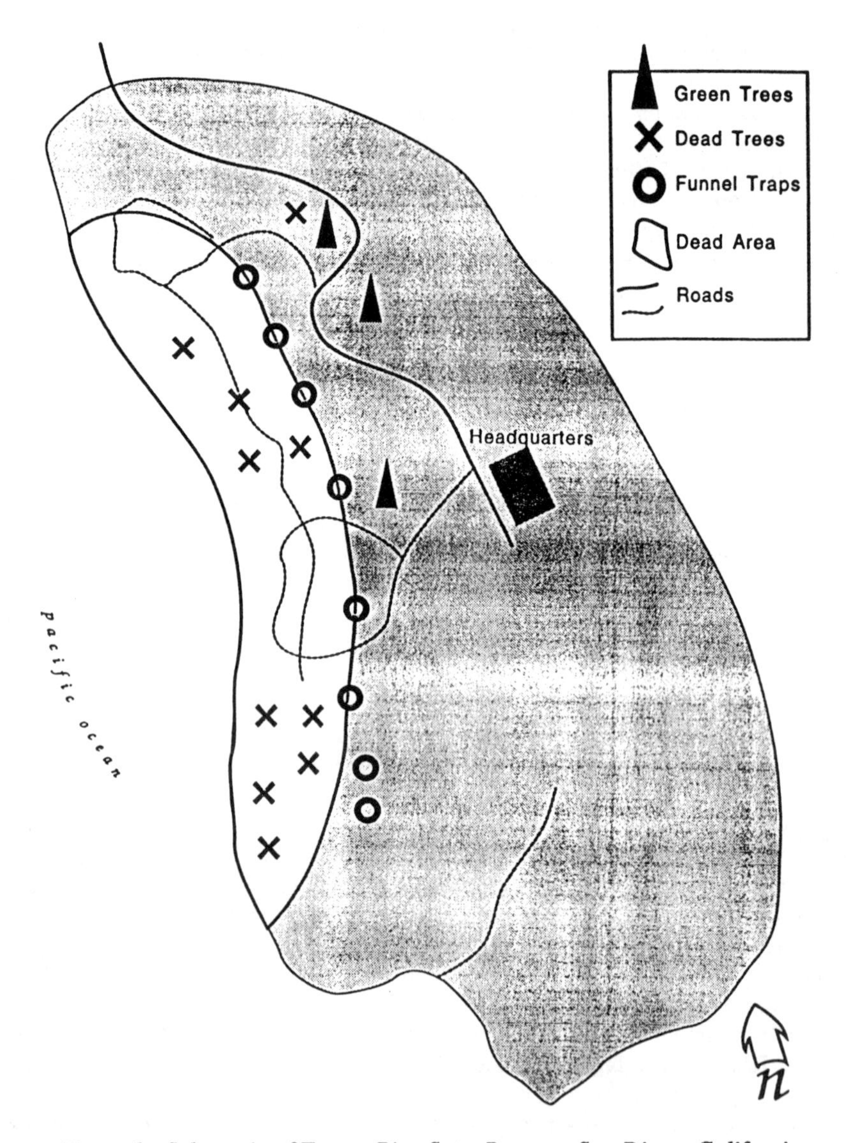


*Figure 1. Schematic of Torrey Pine State Reserve, San Diego, California.*

Upon visiting the Reserve in February of 1992 I noticed a number (14) of Torrey pine turning red, indicating an increase in ips beetles. The trapping program was immediately resumed. In 1992 trapping was conducted over a 32 week period during which time 158,000 beetles were trapped and removed from the Reserve. Weekly trap catches never reached the levels found in 1991 but during week number 6 in excess of 18,000 ips beetles were caught (Figure 3). Again, only two additional trees were killed by ips beetle during the 1992 trapping program.

Except for a brief period between December 1992 and March of 1993 trapping continued through September of 1993. To date no additional trees have been killed by ips beetles since July of 1992.

Before the semiochemical based management program was instituted in the Reserve approximately 840 trees (13% of the total number of Torrey pine in the Reserve) were killed by ips beetles. After the program was started 42 additional Torrey pines were killed, but only 4 of these trees were killed during periods of active trapping.

In summary, the problem presented by the situation at Torrey Pine State Reserve and the environmental setting were ideal for deployment of the trap-out strategy. The size of the forest and insect populations were relatively small, i.e. neither were part of a more extensive coniferous landscape that could harbor additional populations of ips that could reinvade the Reserve. This condition made for an ideally isolated "island" that was exceedingly conducive to manipulation by a pheromone-base management strategy. Because this was not a controlled experiment it cannot be claimed with any statistical assurance that the trapping program caused the decline in Torrey pine mortality. However, other explanations seem lacking.

The last example involves the use of an aerially applied antiaggregation compound in a strategy that attempts to discourage beetles from infesting trees within a specific area *(12)*. The mountain pine beetle, Dendroctonus ponderosae Hopkins, is a native bark beetle that has been at epidemic levels in western North America for more than 25 years. Its primary host is lodgepole pine, Pinus contorta var. latifolia Englem. These losses severely affect timber production and disrupt management of wildlife populations both of which adversely affect local, regional, and national economies.

As described before in addition to aggregation compounds, bark beetles produce antiaggregation compounds to reduce the negative impacts of intraspecific competition. Mountain pine beetles utilize a specific enantiomeric blend of verbenone as their antiaggregation compound. It is naturally derived from three sources: (1) female beetles; (2) auto-oxidation of trans-verbenol and cis-verbenol and (3) oxidation of these same compounds by microorganisms associated with mountain pine beetle. Simply put, the message contained in verbenone is that all the available space for colonization is occupied and suitable hosts must be found elsewhere.

We attempted to manipulate mountain pine beetle populations on replicated 10 hectare plots with an aerial application of (-)86%/(+)14% verbenone impregnated in translucent polyethylene beads. The verbenone was loaded in the beads at a rate of 1.2% active ingredient by weight resulting in 5.35 grams of verbenone per pound of beads. Each treatment plot received 720 grams loaded beads per hectare. Applications were made using an underslung bucket system attached to a Bell 47B Soloy conversion. Beads were applied before emergence of beetles and sampling for infestation rates was done after all beetle flight had ceased.

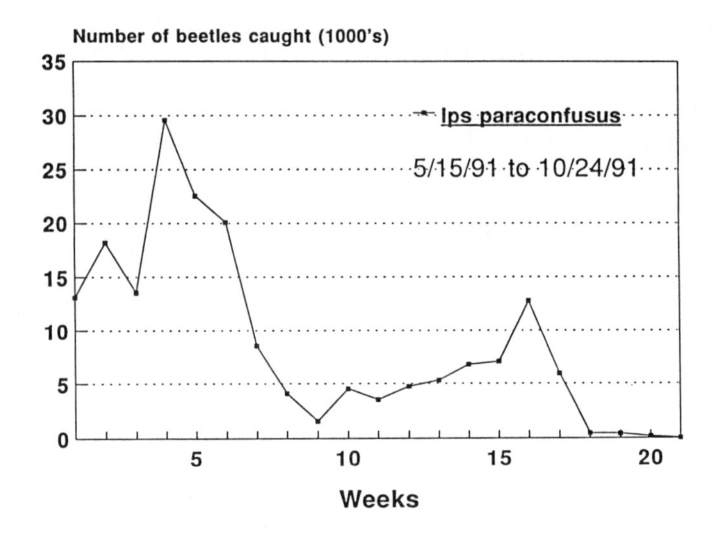


*Figure 2. Trap catch of beetles at Torrey Pine State Reserve, 1991.*

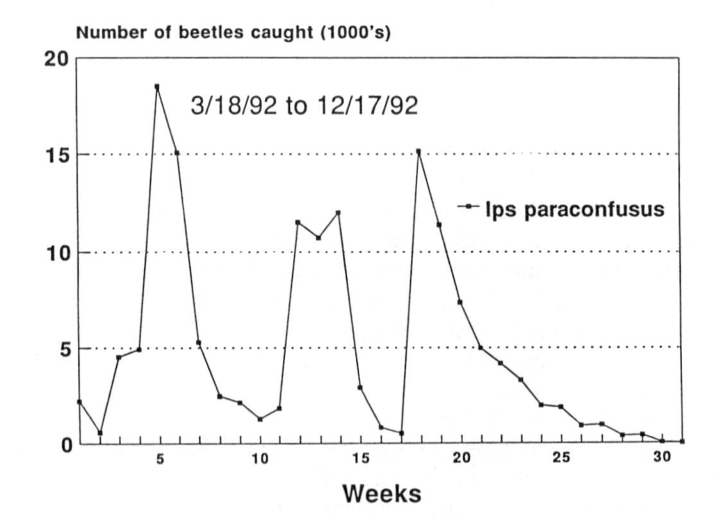


*Figure 3. Trap catch of beetles at Torrey Pine State Reserve, 1992.*

The results of this experiment indicate that the ratio of new to old attacks of mountain pine beetle in currently infested stands can be significantly reduced (Table II).

**Table II. Summary Results of the Effect of Verbenone Beads on Mountain Pine Beetle Attacks**

| | *Verbenone Treatments* | *Control* | *t* |
|---|---|---|---|
| Mean # '87 attacks/ha | 52.1 | 52.3 | 0.01 |
| Mean # '88 attacks/ha | 9.91 | 40.4 | 1.95 |
| Mean ratio '88:'87 attacks | 0.20 | 0.85 | 4.40 |
| Mean # unsucc. attacks/ha | 18.8 | 6.20 | 3.15 |

SOURCE: Reproduced from reference 12.

The ratio of 1988 to 1987 attacks in the treated plots was significantly lower than that found in the untreated plots. In addition the number of attacks per hectare in the treated vs. untreated, while not statistically different, were greatly lowered (Table II). Lastly, the mean number of unsuccessful attacks per hectare was significantly greater on the treated than on the untreated plots, indicating that the trees had a greater ability to deter colonization successfully (Table II). This phenomenon is quite interesting and helps explain the mode of action of verbenone. Because verbenone was permeating the air space the population of beetles in the area could not mount a mass attack via the aggregation pheromone and therefore could not overcome the natural defense systems of the trees under attack.

This study represents the first successful demonstration of an aerially applied antiaggregation pheromone to reduce tree mortality in a large area infested with bark beetles. I surmise that the reduction in tree mortality is the result of verbenone affecting mountain pine beetle in two different modes. As mountain pine beetles emerge from trees within the plot they immediately receive a signal that the stand is sufficiently infested and they should search elsewhere for susceptible trees. This conclusion is supported by the high level of unsuccessful attacks experienced in the treated plots vs. the control plots. Additionally beetles entering from outside the verbenone treated plots also receive the signal that the stand is sufficiently infested and are diverted or if they enter they continue to fly through the plots without attempting to attack.

Aerial application of verbenone, or perhaps any antiaggregation pheromone, offers the forest land manager expanded opportunities for managing the damaging effects of bark beetles. Other possibilities include combining pheromone treatments with partial cutting or in combination with aggregation compounds to influence or direct the beetles' final destination.

## Future

In concluding I want to provide a few thoughts about the future of semiochemical-based management systems for management of forest insects. Semiochemicals are known for a number of economically important forect insects and few species of minor importance. However, it is uncertain for any of these species whether the complete blend of semiochemicals used to mediate host selection and reproduction in nature is known. For example, for several species there are recent discoveries of new pheromones or kairomones that enhance the effects of other semiochemicals that have been known for some time. Of note is the relatively recent discovery of lanierone and its significant enhancement of the attractiveness of the aggregation pheromones of Ips pini.

Recent technologies have revolutionized the isolation, identification, and synthesis of semiochemicals. These include: the development of techniques for micro-capture and analysis of volatiles emanating from individual bark beetle galleries; the analysis of candidate semiochemicals by coupled gas chromatography-single cell receptor detection; techniques for derivatization and gas chromatographic analysis of minute amounts of enantiomers of terpene alcohols; the separation of pheromone enantiomers by chiral gas chromatographic columns; and the use of enzymes and yeast to synthesize chiral pheromones of absolute purity. One might even guess that there will be major advances in the use of biotechnology, e.g. in the identification of trace semiochemicals biomagnified through tissue culture of relevant insect or plant cells, and the use of engineered microorganisms in the synthesis of semiochemicals that are beyond practical or financial limitations of conventional organic chemistry.

I suspect there will be a major effort to develop reliable, consistent and efficient delivery systems. By delivery systems I mean both release devices and application equipment. In forestry this is a major impediment to the development of practical semiochemical based management systems. We must have delivery systems with consistent release characteristics under highly variable field conditions. We need practical application systems that can deliver pheromones by air over uneven terrain and place them in selected micro-sites i.e. canopy vs. ground. Semiochemicals will never be the "magic bullet" that some simplistic thinking forest managers seek. However, it remains the challenge of the scientific community to develop semiochemical based methods for forest pest management purposes and to demonstrate their efficacy to forest managers and to the public at large.

## Literature Cited

1. *Forest Insect and Disease Conditions in the United States - 1991.* USDA Forest Service, Forest Pest Management, Washington D.C. pp. 131.

2. Birch, M.C.; Svihra, P.; Paine, T.D.; Miller, J.C. *J. Chem. Ecol.* 1980, *6*, 395-414.
3. Borden, J.H. In *Comprehensive Insect Physiology Biochemistry and Pharmacology*; Kerkut, G.A.; Gilbert, L.I., Eds.; Pergamon Press: Oxford, 1985, *Vol 12*; pp. 257-285.
4. Borden, J.H. *Holartic Ecology.* 1989, *12*, pp. 501-510.
5. McLean, J.A. *J. Entomol. Soc. Brit. Colum.* 1980, *77*, pp. 20-24.
6. Shea, P.J.; Haverty, M.I.; Daterman, G.E. In *Proceedings 2nd Conference of the Cone and Seed Insect Working Party S2.07-01.* Roques, A., Ed.; Recherche Agronomique (France) Publ. Station de Zoologie Forestiere INRA-CRF: Ardon 45160, Olivet, France, 1986; pp. 341-347.
7. Sower, L.L.; Daterman, G.E.; Sartwell, G. In *Management of Insect Pests with Semiochemicals Concepts and Practices*; Mitchell, E., Ed.; Plenum Press: New York/London, 1981; pp. 351-364.
8. McLean, J.A.; Borden, J.H. *J. Econ. Entomol.*, *72*, pp. 165-172.
9. McLean, J.A.; Stokkink, E. In *Integrated Control of Scolytidae Bark Beetles*; VPI and State University: Blacksburg, VA, 1988; pp. 179-187.
10. McGregor, M.D.; Furniss, M.M.; Oaks, R.D.; Gibson, K.E.; Meyer, H.E. *J. of Forestry*, *82*, pp. 613-616.
11. Amman, G.A.; Thier, R.W.; McGregor, M.D.; Schmitz, R.F. *Can. J. For. Res.*, *19*, pp. 60-64.
12. Shea, P.J.; McGregor, M.D.; Daterman, G.E. *Can. J. For. Res.*, *22*, pp. 436-441.
13. Rothchild, G.H.L. In *Management of Insect Pests with Semiochemicals Concepts and Practices*; Mitchell, E., Ed.; Plenum Press: New York/London, 1981; pp. 207-228.
14. Stoszek, K.J. *J. Forestry*, *71*, pp. 701-706.

RECEIVED January 12, 1995

Chapter 21

# Impact of *Bacillus thuringiensis* on Nontarget Lepidopteran Species in Broad-Leaved Forests

**R. C. Reardon[1] and D. L. Wagner[2]**

**[1]National Center of Forest Health Management, Forest Service, U.S. Department of Agriculture, Morgantown, WV 26505**
**[2]Department of Ecology and Evolutionary Biology, University of Connecticut, Storrs, CT 06269**

Bacillus thuringiensis variety kurstaki (Btk) is the only commercially produced biological insecticide available for use in suppression and eradication programs against the gypsy moth, Lymantria dispar (L.). There have been few multi-year laboratory and field studies designed specifically to evaluate the impacts of Btk on non-target native lepidopteran species. The susceptibility of lepidopteran larvae to Btk must often be evaluated on a species-by-species basis.

Since 1980, approximately 1.7 million hectares (ha) have been treated with Bacillus thuringiensis variety kurstaki (Btk) in the eastern United States as part of the Federal and State Gypsy Moth, Lymantria dispar (L.), Cooperative Suppression Program. During this interval, one application of Btk was applied per year to suppress populations of the European strain of the gypsy moth that was introduced and established in the United States since the 1860's, however, two or three applications per year were used in eradication efforts in Oregon (1985-87) and Utah (1988-1993) against the European strain and, in Washington and Oregon (1992), on approximately 200,000 ha against the Asian strain. Also, two applications of Btk were used in 1994 to eradicate an infestation of European, Asian and hybrid strains of the gypsy moth (introduced via military cargo shipped from Germany) on approximately 50,000 ha in eastern North Carolina. In Ontario, Canada, between 1985 and 1994, approximately 250,000 ha were treated with Btk to control the European Strain. No doubt, Btk usage will continue to increase because gypsy moth is already established in approximately 30 million ha of forest land in North America and about 240 million ha are believed to be susceptible to gypsy moth infestation.

Here we provide an overview of : (1) Btk characteristics that are likely to influence non-target species, (2) documented impacts of Btk on non-target lepidopteran species in association with gypsy moth populations; and (3) recommendations for improving Btk efficacy for target pest species while minimizing impact to non-target species.

0097–6156/95/0595–0284$12.00/0

## Bacillus thuringiensis var. kurstaki characteristics

**Commercial formulations.** In 1970, Dulmage (1) isolated the HD-1 strain of Btk and it became commercially available shortly thereafter. It is used today for production of most Btk formulations used to control defoliating forest Lepidoptera in North America. The HD-1 strain is a serotype 3a3b, and the crystal has a fairly broad spectrum of activity against a large number of Lepidoptera. Four companies produce various types of formulations (e.g., aqueous flowable suspension, nonaqueous emulsifiable suspension, oil flowable) of the HD-1 strain of Btk for use against gypsy moth. Each formulation contains inert ingredients which are unique and various additives (e.g., stickers) can be included to produce the final tank mix (2).

**Mode of action.** The mode of action of Btk is complex and poorly understood. Commercial formulations of Btk contain both the spore (or endospore) and crystal (or parasporal body). The crystal is a protein matrix of large molecules of inactive protoxins that are not toxic to insects until solubilized in the gut. In many lepidopteran pests, the toxin subunits, when ingested separately, are the major cause of mortality; the spore effect is believed to be minimal. In susceptible insects, the alkaline midgut environment (pH>8.0) and proteolytic (protein-splitting) enzymes, dissolve ingested crystals and release smaller delta-endotoxins. These proteins, also known as the insecticidal crystal proteins (ICPs), bind to and force through specific receptor sites on the midgut membrane forming an ion-selective channel. This results in a perforation of the gut and leakage of gut contents, including spores, into the hemolymph. At this point, gut paralysis occurs, the larva stops feeding, and death follows in a few hours to a few days. In less susceptible insects, the spore penetrates into the hemolymph where conditions permit spore germination and bacterial (vegetative cell) multiplication to take place, resulting in a septicemia, that contributes to or causes larval death (2). If a sublethal dose is ingested, the larva may stop feeding; weight gain and development may also be slowed. In some cases, damaged cells in the midgut are replaced and the larva eventually recovers and resumes feeding (3).

As part of its mode of action, Btk can germinate, multiply and resporulate in the infected insect's hemolymph; however, vegetative cells, spores and crystals are not abundantly produced under such conditions. Since the insect integument does not rupture, spores and crystals are not released to contaminate foliage that might be consumed by other susceptible species. Usually the diseased caterpillars fall to the ground, and the Btk toxins are degraded in the soil. Under favorable conditions, Btk spores can germinate and grow in moist soil, deriving essential nutrients from decaying plants. The spore can persist in soil (and other protected sites) longer than the crystal toxins (4) staying viable for several months and, under ideal conditions, for years. Natural Btk epizootics have never been observed in a forest insect population. Consequently, to control forest insect pests, Btk must be applied annually in the manner of a conventional stomach-poison type of insecticide. Btk cannot be expected to infect subsequent generations of the gypsy moth (2).

The insecticidal crystal proteins (ICP's) of Btk were first classified according to the genes that encode them (5). The cryI (A to E) groups with their subgroups, e.g., cryIA (a, b or c) are toxic to lepidopteran larvae. The commercial formulations produced with the HD-1 strain generally contain one or up to three of the cryIA ICPs. Recent studies of these purified ICPs against the gypsy moth showed that cryIA(a) and cryIA(b) are significantly more toxic than cryIA(c). This is not necessarily the case with all lepidopteran larvae. Spores alone have no effect on gypsy moth larvae. The addition of a very small amount of spores to a low concentration of the ICPs, significantly increases mortality to 100 percent as a result of lethal septicemia. This interaction between Btk and the ICPs does not appear to be specific. Other bacteria that are part of the forest microflora also show significant synergism with the cryIA(a) and cryIA(c) ICPs. These observations suggest that once the midgut is perforated, these insects become susceptible to nonspecific infections by bacterial opportunists and that other forest microflora can act synergistically with Btk.

**Potency.** The potency of Btk preparations is determined by parallel bioassays with the HD-1-S-1980 standard on artificial diet with 4-day-old cabbage loopers (Trichoplusia ni). Since insecticidal activity varies greatly among insect species this method often results in a misrepresentation of the actual efficacy against a given pest species. Nevertheless, all bioassays should include this international standard, primarily because differences in larval batches and variation in fermentation dramatically effect formulation potency.

Comparisons of preparations for efficacy need to take into consideration not only the $LC_{50}$ (that dose needed to kill at least 50 percent of the larvae) but the slope of the regression. The slope shows the dose-response relationship over a range of doses, i.e., the regression coefficient. This information is important because many preparations will have similar $LC_{50}$ but differ significantly at the 95 percent level of effectiveness ($LC_{95}$) - the level of insecticidal activity often required to significantly reduce a pest population to acceptable levels. Radcliffe and Yendol (6) documented that for third instar gypsy moth larvae the $LC_{50}$ was 2.7 (range 1.9-3.4) international units (IU)/larva and the $LC_{95}$ was 21.1 (13.6-48.5) IU/larva.

**Efficacy.** The timing of Btk application for gypsy moth is generally dictated by (1) the degree of foliar expansion, (2) larval stage of development, and (3) the size (biomass) of the larvae within an instar (there is an inverse relationship between susceptibility to Bt and larval size). In general, the application timing requires a subjective judgment and varies at a particular site between years due to differences in weather and population density (7).

In the laboratory, Yendol et al (8) showed that when given a choice, gypsy moth larvae consumed more untreated leaf disks or those sprayed with the lowest Btk concentrations than those receiving the highest concentrations. Bryant and Yendol (9) showed that a given dose of Btk per unit oak leaf surface area ($cm^2$) was more effective against gypsy moth when applied at a higher density of small drops (50 to 150 μm) than at a lower density of larger drops (>150 μm).

The optimal drop size and drop density of Btk on foliage needed to control gypsy moth have not yet been determined. During application, however, a wide range of drop sizes from 50-500 μm can usually be generated and deposited with different types of nozzles and atomizers. Typically, drop sizes between 75 and 250 μm volume median diameter (VMD) are used in gypsy moth suppression. Also, data is insufficient to support the exclusive use of a particular atomizer or nozzle (e.g., flat fan, hollow cone, Micronair, Beecomist) over another. Presently, a wide range of drop sizes and types of nozzles are used for both rotary and fixed-wing aircraft (7).

The current trend is towards increasing the dose and decreasing the total volume of Bt applied. Typically, doses of 50-90 BIU/ha are applied undiluted in volumes of 1.8 to 4.7 L/ha for one application. Since there exists only minimal replicated results supporting the effectiveness of one dose and volume combination over others, there is a broad range of doses and diluted and undiluted volumes presently applied in suppression programs.

Yendol et al (10) showed that the distribution of Btk deposit within a broadleaved forest canopy following aerial application was highly variable; however, deposit differences between upper and lower canopy levels or directionally within canopy level, were not significant. Deposition tends to be log normal, where many leaves contain less than the average dose, balanced by relatively few highly dosed leaves. For a typical application of undiluted Bt at a dose of 60 BIU/ha and rate of 4.7 L/ha, droplet densities can range from 1.3 to 6.0 droplets/$cm^2$ of foliage. Droplet sizes during these applications commonly ranged from 80 to 226 μm VMD.

**Technology.** The aerial application technology used to apply Btk to broadleaved forests for suppressing gypsy moth populations was developed during the 1960's for application of chemical insecticides. In general, that technology was not very efficient at maximizing deposit on target foliage; therefore, with present technology approximately 50% of the applied Btk does not reach the target. The results of spray tower and laboratory bioassays using Btk indicate that small droplets (<150 μm) are more effective (e.g., deposition and efficacy) than larger droplets although small droplets are more prone to drift.

Once the most efficacious range in droplet sizes and densities are determined then the appropriate commercially available delivery system will need to be specified or equipment manufactured. The characterization of droplet spectra for various formulations of Btk using an array of settings for different nozzles and atomizers operated at various pressures, development of spread factors, quantification of evaporation rates, and characterization of the foliage profiles for broadleaved forests have provided critical data for the continued development of the Forest Service Cramer-Barry-Grimm (FSCBG) canopy deposition and penetration model. Anderson et al (11) compared the deposition of aerially applied Btk in an oak forest with predicted deposit using the FSCBG model. In this study, the deposit concentration and spatial distribution of Btk were extremely variable among individual spray runs, primarily due to rapidly changing and somewhat unpredictable local atmospheric conditions. Nevertheless, the FSCBG model predicted the average Btk distribution accurately enough to demonstrate that it can be a reliable tool for estimating deposition in broadleaved canopies.

**Persistence.** Loss of residual toxicity of Btk on foliage can result from degradation by sunlight, leaf temperatures, drying, being washed off by rain, microbial degradation, and leaf chemistry (12-13). Solar radiation appears to be the key factor affecting survival of Btk spores and crystals deposited on foliage (14). In a series of Btk bioassays, the half-life of Btk insecticidal activity for early stage gypsy moth larvae in the field has been estimated at 12-32 hours (23). In spite of this short half-life a deposition of 75 IU/cm$^2$ from a 90 BIU/ha application will give, on the average, insecticidal activity against early stage gypsy moth larvae of at least an $LC_{50}$ for 4 to 6 days.

## Bacillus thuringiensis var. kurstaki non-target impacts

**General.** Many safety tests have been performed with Btk (15). None of the vertebrates tested showed any abnormal reaction to Btk in terms of external symptoms or internal pathologies (16-17). Nevertheless, vertebrate species that rely on lepidopterans as a food source (e.g., Virginia Big-eared bat, insectivorous birds) have the potential of being indirectly affected by Btk suppression programs. For example, Rodenhouse and Holmes (18) showed that a reduction in biomass of lepidopteran larvae following Btk application led to significantly fewer nesting attempts of certain birds. In another study, Bellocq et al (19) showed that the application of Btk increased immigration rates and led to dietary shifts in shrews.

Many Lepidoptera that co-occur with a pest species are also susceptible to Btk. Of particular concern would be non-target impacts on lepidopterans that are important as pollinators, in the suppression of weedy plants, and other ecosystem functions. For example, James et al (20) showed that Btk is toxic to late, but not early, instar larvae of the cinnabar moth (Tyria jacobaeae), which is an important species in the control of the noxious weed, tansy ragwort. Impacts on rare and endangered species are also of special concern.

There have been numerous field efforts to determine the potential impacts of Btk on non-target arthropods. Unfortunately, most of these evaluations were conducted as a minor component of an operational suppression or eradication program; in general, for studies up-to-this date, data suffer from lack of adequate replication and appropriate controls, pre- and multi-year post-treatment monitoring, and from inadequate sampling techniques. There have been few multi-year laboratory and field studies designed specifically to evaluate the impacts of

Btk on non-target native lepidopteran species in association with gypsy moth populations.

**Native Lepidoptera - laboratory studies.** Schweitzer et al (21) evaluated the susceptibility of 41 species of Lepidoptera (representing seven families) to two Btk formulations, Foray 48B and Dipel 8AF. The Btk was applied neat (undiluted) to host seedlings or foliage bouquets at a dose that approximated the field application of 90 and 100 BIU's/ha, respectively, using a Mini-Beecomist atomizer in a cylindrical spray tower. Larvae of each species were tested at an instar in which they are likely to occur when Btk is applied for gypsy moth suppression. Larval mortality was monitored daily for 5 to 7 days; pupal survivorship was also recorded. Significant mortality was recorded for 22 of the 41 species of Lepidoptera evaluated in the Foray assays. Eleven of the 25 species of Noctuidae assayed against Foray showed significant mortality. Early instar larvae of native species were more susceptible to Btk -- 17 of the 19 species evaluated in the first or second instar with Foray died following treatment, however 8 of the 24 species assayed in later (third to last) instars showed significant mortality. Significant intrageneric differences in response to Foray were recorded for three genera: Lithophane, Catocala and Dasychira. For example, only six of eight Catocala species were found to be susceptible to the Foray formulation. In one species, the red spotted purple (Limenitis arthemis), larvae survived exposure to Btk, fed to maturity, then died in the pupal stage. Fourteen species were exposed to Dipel 8AF treated foliage, seven of these showed significant mortality. The authors concluded that the susceptibility of lepidopteran larvae must often be evaluated on a species-by-species basis. Their data also emphasized the importance of screening the appropriate instar when conducting non-target studies.

**Native Lepidoptera - field studies.** One of the initial field studies to document the impacts of Btk on non-target Lepidoptera was conducted by Miller (22) in Lane County, Oregon. The project involved three applications of the Foray 48B formulation of Btk applied at 40 BIU/ha to eradicate a gypsy moth infestation. Miller reported reductions in both species richness and abundance of lepidopteran larvae during the treatment year and in the first post-treatment year. Larval abundance recovered after two years while species richness remained reduced into the third year.

In another study, Sample et al (23) evaluated the impact of Btk and forest defoliation by gypsy moth on native non-target Lepidoptera. The evaluation was conducted in Grant and Pendleton Counties in West Virginia from 1990 through 1992. Adult and larval lepidopterans were collected weekly at 24 plots, each 20 ha, representing six replicates of four treatments: unsprayed without gypsy moth, unsprayed with gypsy moth, sprayed without gypsy moth, and sprayed with gypsy moth. In May 1991, Btk as Foray 48B was applied at 90 BIU/ha for one application in a total volume of 7.0 L/ha. Collecting methods included black light trapping for adult moths and foliage pruning to collect lepidopteran larvae. A single application was found to reduce both species richness and abundance of larval and adult non-target Lepidoptera. Species with early-season larvae experienced the greatest impact. While effects of Btk application were evident among lepidopteran larvae in 1991 (the treatment year), effects on adults were not apparent until the year following treatment (1992). Total number of microlepidoptera appeared not to be impacted perhaps because most feed within some kind of leaf shelter or insufficient sampling.

Richness and abundance of some larval and adult Lepidoptera also were reduced in the unsprayed with gypsy moth (i.e. gypsy moth defoliation) plots. Sample and his colleaques reported that impacts were most apparent in oak-feeding species and in the Notodontidae and Lasiocampidae, two families with species that feed primarily on trees. They also stated that while Btk application and defoliation reduced the abundance of native Lepidoptera, environmental conditions such as weather may have an equal or greater influence on population fluctuation.

Btk residue samples were collected on artificial collectors and foliage and analyzed at five of the 12 sprayed plots. Mean (± SE) drop density was 21.84 ± 3.87 drops/$cm^2$ foliage. Mean (± SE) drop size was 90.5 ± 40.75 µm. Btk residues declined rapidly following application; the half life of residues ranged from 12.7 *h* to 18.5 *h* post-treatment. No residues were detected beyond 96 *h* post-treatment. Initial (within 2 *h* of application) activity of Btk, as determined by foliage bioassays on early instar gypsy moth, ranged from 41.3 to 73.3% mortality. The decline in Btk activity mirrored residue levels; no mortality was observed beyond 96 *h* post-treatment.

Wagner et al (24) conducted a study over 3 years (1991-1993) to determine the impacts of Btk on non-target Lepidoptera in ten 20-ha plots in Rockbridge County, Virginia. Five of the plots were sprayed with Foray 48B at 90 BIU/ha for 9.4 L/ha in one application in 1992. Lepidopteran larvae were collected from foliage samples (15-20 cm branch tips) using a bucket truck from two strata (oak canopy, oak subcanopy) and blueberry understory and from burlap bands on oak boles; and adult moths using light traps.

Approximately 12,000 larvae representing 14 families and over 130 species were collected in 1992. The relative abundance of 16 of the 19 most common taxa collected in foliage samples decreased in the treatment plots; 12 of these were microlepidopterans. Eleven of the 12 most common macrolepidoptera taxa under burlap decreased in relative abundance following treatment. There were no apparent differences between treated and untreated plots in the total numbers of either macro- or microlepidopterans one year following treatment in the 1993 foliage samples, although a few species remained much less common in the treatment plots. Light trapping data are still being analyzed.

## Recommendations

The demand for commercial development of environmentally friendly microbial insecticides to manage the gypsy moth will intensify as the gypsy moth continues to spread throughout North America. Presently available commercially produced

Btk formulations are specific to lepidopteran larvae which is preferable to previously used broader spectrum chemical insecticides. Nevertheless, the general public is becoming more sensitive to impacts of Btk on non-target Lepidoptera.

Efforts to identify the minimally effective droplet sizes and densities, and dose need to be intensified such that these data can be input into the FSCBG model. Thereby, the appropriate aerial application technology and Btk formulation can be selected to deliver an effective deposit on the foliage. Concurrently, aerial application technology needs to be improved in an effort to maximize deposit of short-lived microbial insecticides such as Btk.

Technical developments in genetic engineering and molecular biology are providing opportunities for development of genetically manipulated strains of Btk and transfer of toxin-coding genes into other bacteria (cloning) or plant specie (transgenic plants). In the near future, it may be possible to engineer more taxon specific strains of Btk that can be developed and commercially produced.

There are approximately 11,000+ species of named Lepidoptera in North America and many more are not named. Biologies, life histories and ecosystem functions are not known for most species which greatly complicates attempts to evaluate the impacts of the aerial application of Btk. Venables (25) used a rating system for macrolepidoptera based on the presence of larvae at the time of gypsy moth treatment, larval hosts and larval habitat. She estimated that 92-94% of the 223 baseline species identified from broadleaved forests in parks in western Maryland and northern Virginia would show some measure of vulnerability to Btk.

Nevertheless, based on the laboratory studies conducted by Schweitzer et al (21), it is clear that it would be difficult to accurately predict the impact of Btk on any one lepidopteran species. Differences in susceptibility are likely to result from differences in feeding habits, larval stage present at the time of aerial application, and habitat preferences.

Laboratory assays will continue to be important in the documentation of nontarget effects, especially if data is needed on specific taxa. Laboratory studies will be essential for documenting impacts on those species that routinely occur at low densities in the wild, including rare and endangered species. Rigorous field evaluations with replicated control and treatment plots are very much needed, not only to document the impacts of Btk on forest Lepidoptera, but also other components of forest ecosystems such as the insectivorous mammal and bird communities. Because of the inherent difficulties of finding appropriate controls for forests that are slated to be treated with Btk, greater emphasis needs to be placed on baseline monitoring of lepidopteran communities prior to the initiation of suppression activities. More data on species diversity and function in natural forest ecosystems would enable the selection of indicator taxa that would signal trends in the condition of specific habitats and ecosystem dynamics.

## Literature Cited

1. Dulmage, H.T. *J. Invertebr. Pathol.* **1970**, *15*:232-239.

2. Reardon, R.C.; Dubois, N.R.; McLane, W. *USDA Forest Service FHM-NC-01-94.* **1994**, 32pp.
3. Fast, P.G.; Regniere, J. *Can. Ent.* **1984**, *116*:123-130.
4. West, A.W.; Crook, N.F.; Burgess, H.D. *J. Invertebr. Pathol.* **1984**, *43*:150-155.
5. Hofte, H.; Whiteley, H.R. *Microbial Review* **1989**, *53*:242-255.
6. Radcliffe, S.L.; Yendol, W.G. *J. Environ. Sci. Health B* **1993**, *28*:91-104.
7. Reardon, R.C.; Mierzejewski, K.; Bryant, J.; Twardus, D.; Yendol, W. *USDA Forest Service NA-TP-20* **1990**, 167pp.
8. Yendol, W.G.; Hamlen, R.A.; Rosario, S.B. *J. Econ. Entomol.* **1975**, *68*:25-27.
9. Bryant, J.E.; Yendol, W.G. *J. Econ. Entomol.* **1988**, *81*:130-134.
10. Yendol, W.G.; Bryant, J.E.; McManus, M.L. *J. Econ. Entomol.* **1990**, *83*:1732-1739.
11. Anderson, D.E.; Miller, D.R.; Wang, Y.; Yendol, W.G.; Mierzejewski, K.; McManus, M.L. *J. Appl. Meteor.* **1992**, *31*:1457-1466.
12. Kushner, D.J.; Harvey, G.T. *J. Insect Pathol.* **1962**, *4*:155-184.
13. vanFrankenhuyzen, K.; Mystrom, C.W. *J. Econ. Entomol.* **1989**, *82*:868-872.
14. Pozsgay, M.; Fast, P.G.; Kaplan, H.; Carey, P. *J. Invertebr. Pathol.* **1987**, *50*:246-253.
15. Otvos, I.S.; Vanderveen, S. *Forestry Canada* **1993**, 81pp.
16. Harper, J.D. *Auburn Univ.* **1974**, 64pp.
17. Boberschmidt, L.; Saari, S.; Sasseman, J.; Skinner, L. *USDA Ag. Handbook No. 685* **1989**, 578pp.
18. Rodenhouse, N.; Holmes, R. *Ecology* **1992**, *73*:357-372.
19. Bellocq, M.I.; Bendell, J.F.; Cadogan, B.L. *Can. J. Zoology* **1992**, *70*:505-510.
20. James, R.R.; Miller, J.C.; Lighthart, B. *J. Econ. Entomol.* **1992**, *86*:334-339.
21. Schweitzer, D.F.; Peacock, J.W.; Carter, J.L. *J. Econ. Entomol.* **1994**, In prep.
22. Miller, J.C. *Amer. Entomol.* **1990**, *36*:135-139.
23. Sample, B.E.; Butler, L.; Zivkovich, C.; Whitmore, R. *Can. Ent.* **1994**, In press.
24. Wagner, D.L.; Carter, J.L.; Peacock, J.W.; Talley, S.E. *J. Econ. Entomol.* **1995**, In prep.
25. Venables, B.A. *USDI National Park Service Report.* **1990**, 51pp.

RECEIVED February 17, 1995

# Author Index



# Affiliation Index

# Subject Index

*Production: Meg Marshall*
*Indexing: Deborah H. Steiner*
*Acquisition: Michelle D. Althuis*
*Cover design: Cornithia Allen Harris*

*Printed and bound by Maple Press, York, PA*

668.65
/Biorational pest control
8980